Oncogenic Signaling of Growth Factor Receptors in Cancer: Mechanisms and Therapeutic Opportunities

Oncogenic Signaling of Growth Factor Receptors in Cancer: Mechanisms and Therapeutic Opportunities

Editor

Anica Dricu

MDPI • Basel • Beijing • Wuhan • Barcelona • Belgrade • Manchester • Tokyo • Cluj • Tianjin

MDPI

Editor
Anica Dricu
Department of Biochemistry
University of Medicine and Pharmacy
Craiova
Romania

Editorial Office
MDPI
St. Alban-Anlage 66
4052 Basel, Switzerland

This is a reprint of articles from the Special Issue published online in the open access journal *International Journal of Molecular Sciences* (ISSN 1422-0067) (available at: www.mdpi.com/journal/ijms/special_issues/Oncogenic_Signaling).

For citation purposes, cite each article independently as indicated on the article page online and as indicated below:

LastName, A.A.; LastName, B.B.; LastName, C.C. Article Title. *Journal Name* **Year**, *Volume Number*, Page Range.

ISBN 978-3-0365-7337-3 (Hbk)
ISBN 978-3-0365-7336-6 (PDF)

Contents

About the Editor

Anica Dricu

In 1997, Anica Dricu obtained a PhD in Experimental Pathology at the Karolinska Institutet in Stockholm. She then joined its Oncology Pathology Department, becoming Docent in Experimental Oncology in 2006, at the same Medical University, Karolinska Institutet. From 2009, she has been a professor at the University of Medicine and Pharmacy of Craiova.

Her current research addresses the role of growth factor receptors (GFRs) in the diagnosis and therapy of solid tumours. The dysregulation of GFR signaling in malignant diseases makes these membrane receptors important diagnostic, predictive and prognostic biomarkers. Targeted therapy, also called precision medicine or personalized medicine, also involves GFRs. She focuses particularly on small molecule cancer drugs, used to target GFR activation and downstream signaling. These drugs have a significant impact on the modern modality in treating cancers.

She has published several articles, and she serves as a member of several editorial and review boards for peer-reviewed journals.

Preface to "Oncogenic Signaling of Growth Factor Receptors in Cancer: Mechanisms and Therapeutic Opportunities"

At the molecular level, the activation of growth factor receptors (GFRs) induces a mitogenic response and maintains cancer cell growth. The majority of malignant diseases are related to aberrant intra- and intercellular communication, associated with the GFR-mediated pathways. Moreover, the evasion of apoptotic signals and the requirement of angiogenesis were also found to be of fundamental importance for tumor progression and metastasis. In this context, a high expression of GFRs aids blood vessel formation, cell migration, and the inhibition of apoptosis. GFR-directed therapy that would theoretically selectively kill malignant cells and reduce the toxicity associated with nonselective conventional chemotherapy may be a promising treatment for cancer.

Many intracellular proteins involved in GFR signal transduction can also function as oncogenes. Mutations affecting key proteins in the RAS/MAPK and PI3K/AKT pathways are known to be crucial in maintaining the malignancy of different types of cancers. This information has guided the development of compounds designed to target one or more of these pathways in cancer cells.

Even though there have been important advances in our understanding of GFRs and their signaling, certain essential information is still lacking, and these membrane receptors are still being laboriously studied by several research groups, to find therapeutic solutions to unmet medical needs.

This Special Issue will cover the latest preclinical and clinical progress made in the areas associated with GFRs' oncogenic signaling.

Anica Dricu

Editor

International Journal of
Molecular Sciences

MDPI

Editorial

Oncogenic Signalling of Growth Factor Receptors in Cancer: Mechanisms and Therapeutic Opportunities

Anica Dricu

Department of Biochemistry, Faculty of Medicine, University of Medicine and Pharmacy, 200349 Craiova, Romania; anica.dricu@live.co.uk

Cancer is a common name for several distinct diseases caused by uncontrolled cell growth and proliferation. More than 200 types of cancer are described in the literature, each of them with its own identity given by specific gene, protein or hormone signatures. However, concerted and redundant dysregulations of mitogenic pathways arising from growth factor receptors (GFRs) are common events in all cancer types [1,2].

These sophisticated membrane-spanning proteins harmonize the information flow from several sources, controlling the mitogenic network in the normal cell. The complexity of GFRs function is supported by their multiple regulatory mechanisms, including feedback loops, multidirectional cross-communication and redundancy in downstream signalling. Recent large-scale studies identified alterations in genes and proteins of several GFRs such as epidermal growth factor receptor (EGFR), platelet-derived growth factor receptor α/β (PDGFR α/β), vascular endothelial growth factor receptors (VEGFRs), IGF-1R, fibroblast growth factor receptor (FGFR), etc. [3].

The majority of malignant diseases are related to aberrant intra- and intercellular communication, associated with subverted GFRs pathways. At the molecular level, the overactivation of GFRs induces a mitogenic response and maintains cancer cell growth. Four main mechanisms are known to generate aberrant activation of GFRs in malignant diseases: autocrine/paracrine activation, genomic amplification, chromosomal rearrangements and gain-of-function mutations [4,5].

GFs mediate their mitogenic function by binding to and activating GFRs with intrinsic tyrosine kinase (TKs) activity. Cancer cells produce GFs or reprogram and force other cells to produce GFs according to their own needs, becoming independent of endocrine signalling and finally leading to constitutive receptors activation in tumours [6–8].

GFRs gene amplification, also known as genomic DNA copy number amplification, has been found in a wide variety of tumours, causing receptor protein upregulation and overactivation, inducing oncogenic behaviour and resistance to therapy [9,10].

Chromosome rearrangements mechanism is a usual condition of malignant cells, in which a fragment of chromosomes is deleted or inverted, giving rise to fusion proteins that are responsible for the formation of several types of malignancies. The BCR-ABL fusion oncoprotein, which fuses the ABL1 tyrosine kinase gene on chromosome 9 to the BCR gene on chromosome 22, was the first tyrosine kinase fusion identified [11]. Chromosome rearrangements leading to fusion proteins are also found in many solid cancers, such as breast cancer, brain tumours, lung cancer, colorectal cancer, etc. [12–15].

Gain-of-function mutations can exercise mitogenic functions by stimulation of growth factors or by inducing constitutive activation of GFRs, driving uncontrolled cell proliferation and tumour progression [16].

Once activated, GFRs trigger a wave of intracellular signalling events, mediated by two major pathways: mitogen-activated protein kinase (MAPK) and phosphoinositide 3-kinases (PI3K) cascades [17].

Many intracellular proteins involved in rat sarcoma virus (RAS)/MAPK or PI3K/AKT pathways can also function as oncogenes. Mutations affecting key proteins in RAS/MAPK

Citation: Dricu, A. Oncogenic Signalling of Growth Factor Receptors in Cancer: Mechanisms and Therapeutic Opportunities. *Int. J. Mol. Sci.* **2022**, *23*, 7376. https://doi.org/10.3390/ijms23137376

Received: 20 May 2022
Accepted: 29 June 2022
Published: 2 July 2022

or PI3K/AKT pathways are known to be crucial in maintaining the malignancy of different types of cancers [18–20].

Many effector proteins in GFRs signal transduction, such as PI3K, extracellular signal-regulated kinase 1/2 (ERK1/2) or MAPK can act as junction for multiple signalling pathways [21]. It is also well demonstrated that mutations in mammalian target of rapamycin (mTOR), RAS or rapidly accelerated fibrosarcoma (RAF) are very common in malignant diseases [22].

Crosstalk and collaboration between GFRs and other protein families are constantly being discovered, making the receptor signalling system far more complex. For example, G protein-coupled receptors (GPCRs) can engage GFRs to mediate cell proliferation, differentiation, and vice versa, several GFs use GPCRs proteins to exert their mitogenic signal signalling [23].

Moreover, the evasion of apoptotic signals and the requirement of angiogenesis were also found to be of fundamental importance for tumour progression and metastasis. In this context, high expression of GFRs aids blood vessel formation, cell migration and the inhibition of apoptosis [24,25].

All this information has guided the development of compounds, designed to target one or more of these pathways in cancer cells. A vast variety of GFR signalling inhibitors have been developed, many of which have been approved by the Food and Drug Administration (FDA). While some FDA-approved inhibitors are selective for individual GFRs (e.g., Alectinib, Afatinib, Dacomitinib, Erlotinib, Gefitinib, Lapatinib, etc.), others demonstrate efficiency by inhibiting several receptors (e.g., Dasatinib, Lestaurtinib, Imatinib, Ponatinib, Vandetanib, etc.). However, the development of novel therapeutic strategies for cancer treatment is tightly restricted by the similarities between the normal and malignant cells. GFR-directed therapy that would theoretically selectively kill malignant cells and reduce the toxicity associated with nonselective conventional chemotherapy may be a promising treatment for cancer. Based on this rationale, different strategies have been developed to inhibit the oncogenic effects of GFRs (e.g., small-molecule inhibitors, monoclonal antibodies, siRNA, antisense oligodeoxynucleotides, triple helix, dominant-negative mutants, etc.).

This Special Issue will cover the latest preclinical and clinical progress made in the areas associated with GFRs' oncogenic signalling.

Funding: Grant PN-III-P4-ID-PCE2020-1649, by the UEFISCDI Authority, Romania.

Conflicts of Interest: The author declares no conflict of interest.

References

1. Mongre, R.K.; Mishra, C.B.; Shukla, A.K.; Prakash, A.; Jung, S.; Ashraf-Uz-Zaman, M.; Lee, M.S. Emerging Importance of Tyrosine Kinase Inhibitors against Cancer: Quo Vadis to Cure? *Int. J. Mol. Sci.* **2021**, *22*, 11659. [CrossRef] [PubMed]
2. Carapancea, M.; Alexandru, O.; Fetea, A.S.; Dragutescu, L.; Castro, J.; Georgescu, A.; Popa-Wagner, A.; Bäcklund, M.L.; Lewensohn, R.; Dricu, A. Growth factor receptors signaling in glioblastoma cells: Therapeutic implications. *J. Neurooncol.* **2009**, *92*, 137–147. [CrossRef] [PubMed]
3. Wang, Z.; Zhang, L.; Xu, W.; Li, J.; Liu, Y.; Zeng, X.; Zhong, M.; Zhu, Y. The Multi-Omics Analysis of Key Genes Regulating EGFR-TKI Resistance, Immune Infiltration, SCLC Transformation in EGFR-Mutant NSCLC. *J. Inflamm. Res.* **2022**, *15*, 649–667. [CrossRef]
4. Chioni, A.M.; Grose, R.P. Biological Significance and Targeting of the FGFR Axis in Cancer. *Cancers* **2021**, *13*, 5681. [CrossRef]
5. Dricu, A.; Kanter, L.; Wang, M.; Nilsson, G.; Hjertman, M.; Wejde, J.; Larsson, O. Expression of the insulin-like growth factor 1 receptor (IGF-1R) in breast cancer cells: Evidence for a regulatory role of dolichyl phosphate in the transition from an intracellular to an extracellular IGF-1 pathway. *Glycobiology* **1999**, *9*, 571–579. [CrossRef] [PubMed]
6. Singh, A.B.; Harris, R.C. Autocrine, paracrine and juxtacrine signaling by EGFR ligands. *Cell Signal.* **2005**, *17*, 1183–1193. [CrossRef] [PubMed]
7. Walsh, J.H.; Karnes, W.E.; Cuttitta, F.; Walker, A. Autocrine growth factors and solid tumor malignancy. *West J. Med.* **1991**, *155*, 152–163.
8. Kentsis, A.; Reed, C.; Rice, K.L.; Sanda, T.; Rodig, S.J.; Tholouli, E.; Christie, A.; Valk, P.J.; Delwel, R.; Ngo, V.; et al. Autocrine activation of the MET receptor tyrosine kinase in acute myeloid leukemia. *Nat. Med.* **2012**, *18*, 1118–1122. [CrossRef]

9. Hechtman, J.F.; Polydorides, A.D. HER2/neu gene amplification and protein overexpression in gastric and gastroesophageal junction adenocarcinoma: A review of histopathology, diagnostic testing, and clinical implications. *Arch. Pathol. Lab. Med.* **2012**, *136*, 691–697. [CrossRef]
10. Yu, H.A.; Arcila, M.E.; Rekhtman, N.; Sima, C.S.; Zakowski, M.F.; Pao, W.; Kris, M.G.; Miller, V.A.; Ladanyi, M.; Riely, G.J. Analysis of tumor specimens at the time of acquired resistance to EGFR-TKI therapy in 155 patients with EGFR-mutant lung cancers. *Clin. Cancer Res.* **2013**, *19*, 2240–2247. [CrossRef]
11. Spiers, A.S. The clinical features of chronic granulocytic leukaemia. *Clin. Haematol.* **1977**, *6*, 77–95. [CrossRef]
12. Han, T.; Schatoff, E.M.; Murphy, C.; Zafra, M.P.; Wilkinson, J.E.; Elemento, O.; Dow, L.E. R-Spondin chromosome rearrangements drive Wnt-dependent tumour initiation and maintenance in the intestine. *Nat. Commun.* **2017**, *8*, 15945. [CrossRef] [PubMed]
13. Stransky, N.; Cerami, E.; Schalm, S.; Kim, J.L.; Lengauer, C. The landscape of kinase fusions in cancer. *Nat. Commun.* **2014**, *5*, 4846. [CrossRef] [PubMed]
14. Wang, M.; Nilsson, G.; Carlberg, M.; Dricu, A.; Wejde, J.; Kreicbergs, A.; Larsson, O. Specific and sensitive detection of the EWS/FLI1 fusion protein in Ewing's sarcoma by Western blotting. *Virchows Arch.* **1998**, *432*, 131–134. [CrossRef]
15. Mitelman, F.; Johansson, B.; Mertens, F. The impact of translocations and gene fusions on cancer causation. *Nat. Rev. Cancer* **2007**, *7*, 233–245. [CrossRef]
16. Isozaki, K.; Hirota, S. Gain-of-Function Mutations of Receptor Tyrosine Kinases in Gastrointestinal Stromal Tumors. *Curr. Genom.* **2006**, *7*, 469–475. [CrossRef]
17. Schlessinger, J. Common and distinct elements in cellular signaling via EGF and FGF receptors. *Science* **2004**, *306*, 1506–1507. [CrossRef]
18. Downward, J. Targeting RAS signalling pathways in cancer therapy. *Nat. Rev. Cancer* **2003**, *3*, 11–22. [CrossRef]
19. Zhang, J.; Grubor, V.; Love, C.L.; Banerjee, A.; Richards, K.L.; Mieczkowski, P.A.; Dunphy, C.; Choi, W.; Au, W.Y.; Srivastava, G.; et al. Genetic heterogeneity of diffuse large B-cell lymphoma. *Proc. Natl. Acad. Sci. USA* **2013**, *110*, 1398–1403. [CrossRef]
20. Stern, D.F. Keeping Tumors Out of the MAPK Fitness Zone. *Cancer Discov.* **2018**, *8*, 20–23. [CrossRef]
21. Dhillon, A.S.; Hagan, S.; Rath, O.; Kolch, W. MAP kinase signalling pathways in cancer. *Oncogene* **2007**, *26*, 3279–3290. [CrossRef] [PubMed]
22. Ambrogio, C.; Köhler, J.; Zhou, Z.W.; Wang, H.; Paranal, R.; Li, J.; Capelletti, M.; Caffarra, C.; Li, S.; Lv, Q.; et al. KRAS Dimerization Impacts MEK Inhibitor Sensitivity and Oncogenic Activity of Mutant KRAS. *Cell* **2018**, *172*, 857–868.e15. [CrossRef]
23. Cattaneo, F.; Guerra, G.; Parisi, M.; De Marinis, M.; Tafuri, D.; Cinelli, M.; Ammendola, R. Cell-surface receptors transactivation mediated by g protein-coupled receptors. *Int. J. Mol. Sci.* **2014**, *15*, 19700–19728. [CrossRef] [PubMed]
24. Neophytou, C.M.; Trougakos, I.P.; Erin, N.; Papageorgis, P. Apoptosis Deregulation and the Development of Cancer Multi-Drug Resistance. *Cancers* **2021**, *13*, 4363. [CrossRef] [PubMed]
25. Serban, F.; Artene, S.A.; Georgescu, A.M.; Purcaru, S.O.; Tache, D.E.; Alexandru, O.; Dricu, A. Epidermal growth factor, latrophilin, and seven transmembrane domain-containing protein 1 marker, a novel angiogenesis marker. *Onco Targets Ther.* **2015**, *8*, 3767–3774. [CrossRef] [PubMed]

International Journal of *Molecular Sciences*

MDPI

Review

Transcriptomic Crosstalk between Gliomas and Telencephalic Neural Stem and Progenitor Cells for Defining Heterogeneity and Targeted Signaling Pathways

Roxana Deleanu [1], Laura Cristina Ceafalan [2,3] and Anica Dricu [4,*]

1 Institute for Neuroscience, Medical University of Innsbruck, 6020 Innsbruck, Austria; Irina-Roxana.Deleanu@i-med.ac.at
2 Department of Cell Biology and Histology, Carol Davila University of Medicine and Pharmacy, 050474 Bucharest, Romania; lauraceafalan@yahoo.com
3 Victor Babeș National Institute of Pathology, 050096 Bucharest, Romania
4 Department of Biochmistry, University of Medicine and Pharmacy, 710204 Craiova, Romania
* Correspondence: anica.dricu@live.co.uk

Abstract: Recent studies have begun to reveal surprising levels of cell diversity in the human brain, both in adults and during development. Distinctive cellular phenotypes point to complex molecular profiles, cellular hierarchies and signaling pathways in neural stem cells, progenitor cells, neuronal and glial cells. Several recent reports have suggested that neural stem and progenitor cell types found in the developing and adult brain share several properties and phenotypes with cells from brain primary tumors, such as gliomas. This transcriptomic crosstalk may help us to better understand the cell hierarchies and signaling pathways in both gliomas and the normal brain, and, by clarifying the phenotypes of cells at the origin of the tumor, to therapeutically address their most relevant signaling pathways.

Keywords: single-cell RNA-seq; primary cerebral tumors; glioma; cancer stem-like cells; human neural stem cells; human neural progenitors; neurogenesis; gliogenesis

Citation: Deleanu, R.; Ceafalan, L.C.; Dricu, A. Transcriptomic Crosstalk between Gliomas and Telencephalic Neural Stem and Progenitor Cells for Defining Heterogeneity and Targeted Signaling Pathways. *Int. J. Mol. Sci.* **2021**, *22*, 13211. https://doi.org/10.3390/ijms222413211

Academic Editors: Giuseppe Lombardi

Received: 10 November 2021
Accepted: 30 November 2021
Published: 8 December 2021

1. Introduction

Deciphering the composition of the human brain and its primary tumors continues to be one of the central concerns in neuroscience and neuro-oncology. Over recent years, several international large-scale efforts have been devoted to analyzing and understanding normal and diseased brain composition, development and functions—including its oncological aspects. However, the classification and characterization of the cells of the brain and brain tumors is very challenging, and no consensus has yet been achieved.

The adult human brain is composed of two main cell compartments: a larger differentiated cell compartment which contains several hundred billion neuronal, glial and non-neural cells [1,2], and a much smaller compartment of potentially proliferative cells—including neural stem and progenitor cells—responsible for the modest cell turnover of the adult brain. However, new neurons and glia are continuously produced from various types of stem and progenitor cells throughout life in restricted brain areas. Unfortunately, some of these cells can suffer genetic alterations, producing different types of primary brain tumors—mainly gliomas—with an incidence of around 5 new cases per 100,000 people every year [3].

Malignant gliomas, of which glioblastomas are the most common [3], are associated with different genomic alterations and are some of the most lethal cancers. Several glioma subtypes relate to mutations in the enzyme isocitrate dehydrogenase (IDH). Mutations in *IDH1*, or less frequently *IDH2*, define two major classes of malignant gliomas: astrocytoma (IDH-A) and oligodendroglioma (IDH-O). Their distinct morphology and oligodendroglial or astrocytic marker expression suggest different glial lineages. However, a mixture of cells with histological features of neoplastic astrocytic and oligodendroglial cells are frequently observed within individual tumors, making the existence of distinct glial lineages in different IDH-mutated gliomas questionable [4].

IDH wild-type (IDHwt) glioblastoma is the most prevalent form of adult primary brain cancer. Analysis of whole-tumor transcriptomic data has shown that IDHwt glioblastoma includes three main subtypes: proneural (MGH26); classical (MGH30); and mesenchymal (MGH28, MGH29) [4,5]. However, all IDHwt glioblastoma individual tumors are highly heterogeneous, each containing different percentages of neoplastic cell types also present in all the other subtypes [4].

Moreover, all these malignant tumors are composed of two cellular compartments: a larger differentiated cell compartment, and a smaller compartment of cells with stem and progenitor features generically named "cancer stem-like cells" (CSCs), which means that they can self-renew and differentiate into multiple cell types, continually contributing to the tumor maintenance. CSCs isolated from different glioma, named glioma stem-like cells (GSCs), show variability with respect to marker expression, proliferation and differentiation, pointing to interpatient and intratumoral heterogeneity within the proliferative cell compartment as well [6]. GSCs were first identified through isolation of CD133-positive cells from primary glioblastoma, demonstrating that these cells were necessary and sufficient to give rise to an ectopic tumor [7]. A number of other neural stem cell markers, such as SOX2 and nestin, were used to validate the stemness of the GSC populations [6]. However, the current understanding of the stem and progenitor like-cell phenotypes within tumors and of their contribution to tumor heterogeneity and maintenance is still limited [8]. Although the GSC compartment is small in comparison to the differentiated compartment, it is still clinically relevant. Several studies have shown that GSCs resist radiotherapy and chemotherapy and are the main contributors to cancer recurrence. Presently, there are no clinically approved treatments specifically targeting GSCs—this is considered the main cause of the poor response to treatment in malignant gliomas [3,8].

In order to solve this problem, several recent studies have aimed to clarify the heterogeneity of GSCs and how their phenotypes link to the neural stem cells or the glial progenitor cells in the developing and adult brain. As most of the malignant glioma localize at the cerebral level, and many in the vicinity of the adult neural stem cell niches [4], the neural stem and progenitor cells in the developing and adult human telencephalon are the most relevant populations to be addressed. The updated knowledge about the neural development stages, cell hierarchies and phenotypes in humans is presented briefly, with the aim of better understanding the links they have with the neural stem, progenitor and differentiated cells in adult telencephalon, but also to address the hypothesis that different progenitors are specifically affected in different types of gliomas.

Recent single-cell high-throughput approaches have allowed the taking of huge steps ahead in the definition of cellular identities and have provided unprecedented details on cellular diversity. The deep sequences of single cells or nuclei combined with bioinformatics tools provide the scale for an unbiased survey of molecular expression [9]. These tools can now overcome some of the previous difficulties associated with the scarcity of human brain tissue and can be applied to relatively small neurosurgical or postmortem samples [10,11]. Several reported transcriptomic profiles coming from different human telencephalon stages and regions have increased our current understanding of the developmental dynamics of the three major neural cell types in the human brain: neurons, astrocytes and oligodendrocytes.

At the same time, many databases nowadays include single cell sequencing of primary tumor specimens from human brains. These molecular profiles have captured the great extent of intra-tumor heterogeneity and have identified different GSC populations in different classically defined gliomas. Some recent studies, which are further briefly overviewed, have reported that several proliferating cells inside adult glioma samples show a remarkable similarity to different neural stem and progenitor cells found during normal telencephalic development. This supports the connection between neural development, neural stem cell niches and cancer biology, requiring a deeper consideration and opening new perspectives in cancer therapy.

2. Neural Cell Types in the Human Developing Telencephalon

The development of the nervous system in general implies a precise temporal and spatial generation for each cell type, following the stages of neural induction, patterning/proliferation, neurogenesis, gliogenesis and functional maturation [12–18].

Neural induction starts in the early human embryo at the middle of gestation week (GW) 3 in the midline anterior ectoderm, which transforms into the neuroectoderm. The neuroectoderm first organizes as a neural plate, which further extends and forms neural folds; they gradually fuse to form the neural tube, which is entirely closed at the end of the GW 4. In parallel with the neural tube forming and closing, the cells in the neuroectoderm transform. The initial neuroepithelial (NE) cells express the transcription factors (TF) PAX6 and SOX2, intermediate filaments such as nestin (NES), and adherent junction proteins such as N-cadherin (NCAD), zonula occludens 1 (ZO1) and prominin 1 (PROM1 or CD133) [19–21]. NE cells begin a transition into more elongated, radial-oriented cells called radial glia (RG), with their somas located in the ventricular zone (VZ) of the neural tube wall, their apical processes in contact with the internal surface (lumen or ventricle), and their basal processes contacting the external surface (pia matter) of the neural tube. In addition, by *pattering*, these RG acquire different identities in the anterior–posterior (A–P) and dorsal–ventral (D–V) axes due to gradients of morphogens produced by different organizer centers. A–P patterning starts in the head region by defining the forebrain (prosencephalon) and continues with the midbrain and the hindbrain. The forebrain further divides into the telencephalon and diencephalon. Parallel D–V patterning in the telencephalon leads to the definition of two main regions: the dorsal telencephalon or pallium and the ventral telencephalon or subpallium (Figure 1).

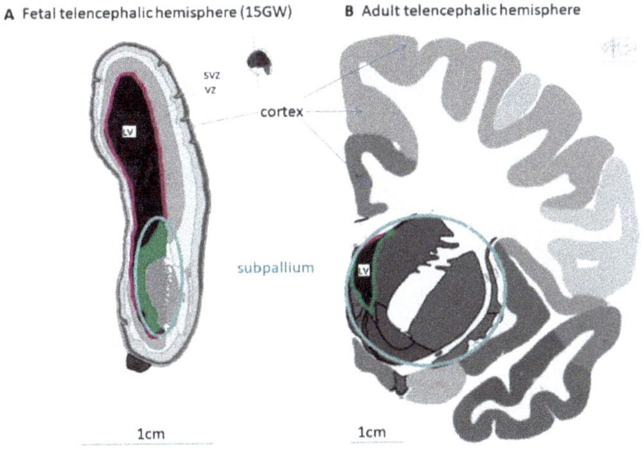

Figure 1. Morphological structures of the fetal and adult human telencephalon. Coronal sections from fetal (**A**) and adult (**B**) telencephalic hemispheres showing the pallium with the cortex, the subpallium and the lateral ventricle (LV). Images modified from the BrainSpan Reference Atlases for 15 gestational week and 34-year-old human brains sectioned at the rostral level (https://atlas.brain-map.org/, accessed on 20 August 2021).

Concomitantly, RG situated at different positions in the neural tube *proliferate* at different rates in response to local mitogens [16,22], forming morphologically defined domains and subdomains (Figure 1A). Several subtypes of RG with distinct behavioral, morphological, and transcriptional signatures have recently been described in prenatal human development [23,24]. Some RG, called ventricular radial glia (vRG, also called apical RG), remain situated in the VZ and continue to express NE cell markers, while others gradually move their somas into the newly formed subventricular zone (SVZ). After GW 9, many of them lose contact with the lumen of the neural tube and become the outer radial glia (oRG), locating and proliferating into the outer SVZ (oSVZ) and specifically expressing *HOPX* and the cell surface marker PTPRZ1, while losing the expression of some vRG markers [24]. By GW 34, some vRG detach basally and form truncated radial glia (tRG), which will be present during all of the following fetal stages [23]. The features of RG division are complex and tightly controlled in time and space. Symmetrical or asymmetrical mitosis occurs in response to different signals, parts of the FGF-MAPK cascade—including platelet-derived growth factor (PDGF) signaling [25]—the PI3K/PTEN/AKT pathway [26], hedgehog signaling [27], N-cadherin/β-catenin signaling [28], and Notch signaling [29].

Neurogenesis, the developmental process by which the neurons are generated, includes the production of neuronal progenitor cells and their differentiation into mature neurons. Between GW 5 and 28, RG dominate in all VZ, SVZ and oSVZ regions and produce most of the neurons in the human telencephalon, mainly indirectly, by generating transit-amplifying progenitors such as the intermediate (or basal) progenitor cells (IPCs) [30]. By asymmetric division, both vRG and oRG generate IPCs, which may undergo few symmetric divisions in response to different local mitogens. IPCs further differentiate into neurons, which use the RG processes as scaffolds to migrate superficially toward the basal (or marginal) zone of the neural tube [31]. The location of the neurons links tightly to the timing of their generation. This temporal patterning results in the sequential generation of specific types of neurons and is a fundamental process of neuronal diversification [18,32]. Generally, long projection neurons (excitatory and inhibitory) are produced first, followed by interneuron production. Excitatory neurons (EN), including glutamatergic neurons, are produced from pallial domains, while inhibitory neurons (IN), including GABAergic and cholinergic neurons, are produced from subpallial domains [16]. The neurons born in the subpallium can remain in ventral regions or migrate to the dorsal regions, as is the case with cortical inhibitory neurons [14,22,33]. During cerebral cortex formation, the first neurons migrating from the dorsal SVZ into the telencephalon mature into cortico-thalamic neurons. The most recently born neurons migrate past the earlier-born neurons and mature into intra-telencephalic neurons [18,34], while the migrating interneurons reach the cortex with a specific temporal and spatial distribution [22].

Gliogenesis, the developmental process by which glial cells are generated, includes the production of glial progenitor cells and their differentiation into mature glia. In the brain, these are macroglia (astrocytes, ependymal cells and oligodendrocytes) and microglia. Gliogenesis in the human telencephalon mainly occurs after the completion of neurogenesis, when the remaining RG in the VZ and SVZ proceed to produce macroglia directly or indirectly, while microglial cells have mesodermal origins [35]. The mode of transition from RG to mature astrocytes, ependymal cells and oligodendrocytes is still controversial and several models are proposed.

Most of the studies on astrocytogenesis have relied on the detection of intermediate filament proteins such as glial fibrillary acidic protein (GFAP), which are expressed by mature astrocytes but also by fetal and postnatal RG [36], making the interpretation of their expression difficult. Different types of RG could be direct or indirect sources of protoplasmic astrocytes in the gray matter and fibrous astrocytes in the white matter, as well as of the ependymal cells lining the ventricular system, which in the telencephalon form the lateral ventricles (LV; Figure 1). By GW 34, tRG appear to transform mainly into astrocytes, both fibrous and protoplasmic, that populate widespread regions of the brain parenchyma and express mature astrocytic markers such as AQP4 and APOE [35]. In the postnatal period, a few astrocytes can also undergo symmetric division and generate daughter astrocytes—a process that can also be detected in adult life [35]. Immature ependymal cells also arise from RG in parallel with astrocytes, and, by GW 34, they start to differentiate into mature ependymal cells, responsible for producing cerebrospinal fluid. From postnatal day 10 to adulthood, all ependymal cells lining the LV acquire a mature multi-ciliated morphology [37].

Oligodendrocytes are the myelin-forming glia in the central nervous system. As shown by several lineage-tracing studies in animal models, oligodendrocyte origins are both spatially and temporary diverse [38–40]. Multiple progenitor domains generate oligodendrocyte progenitor cells (OPCs) at distinct embryonic stages in the developing mouse telencephalon and several similar phenotypes have been identified in the human embryonic and fetal brain. Many studies have identified a variety of molecular markers for OPCs: platelet-derived growth factor receptor α (PDGFRα), chondroitin sulfate proteoglycan 4 (CSPG4) (also known as NG2), basic helix-loop-helix TFs OLIG1 and OLIG2, as well as the Sox family of high mobility group-containing TFs. OPCs begin to emerge as early as GW 5 and are continuously produced throughout the rest of the prenatal period, while migrating toward various brain regions, thereby becoming abundant and widespread already at birth [39]. Early born OPCs emerge from the most ventral progenitor domains and spread to more dorsal domains at later prenatal stages, with progenitors in the pallium becoming the major source of OPCs at early postnatal stages. OPCs derived from ventral progenitor domains mainly populate the subpallium, while those from both ventral and dorsal domains differentiate into oligodendrocytes which later myelinate the axons of the neurons in the dorsal telencephalon, such as the neocortex and corpus callosum. OPCs can stay undifferentiated for a long period of time during fetal and early postnatal stages, eventually becoming myelin-expressing mature oligodendrocytes at later stages [38,40]. Starting at early postnatal stages, OPCs begin to differentiate first into pre-myelinating immature oligodendrocytes (Pre-OL), defined by the expression of TFs NKX2.2 and NKX6.2, which subsequently undergo maturation into myelin-expressing oligodendrocytes [38]. While there are few oligodendrocytes and little myelination before birth, their number and the myelin produced by them in the white matter expand rapidly after birth until approximately 5 years of age [41].

The precise cell hierarchies and mechanisms that control the transition from RG to macroglia in different human telencephalic regions, as well as the ontogenic relationship between oligodendrocytes and astrocytes are still subjects of debate. The occurrence of transit-amplifying glial IPCs and migrating glioblasts, such as the neurogenic IPCs and migrating neuroblasts in neurogenesis, was also speculated for the astrocyte and oligodendrocyte lineages, but their exact identity and timing are currently unknown [35]. Cell culture experiments have demonstrated that OPC-like cells behave as bi-potent progenitors that can differentiate into both oligodendrocytes and astrocytes. Several in vivo studies, however, have demonstrated that OPCs become oligodendrocytes almost exclusively. Yet, some recent studies using genetic lineage-tracing methods have provided evidence that a fraction of PDGFRα/NG2-expressing cells differentiate not only into oligodendrocytes, but also into astrocytes and/or neurons in certain regions of the brain, although such cells seem to be relatively rare. A stepwise differentiation of RG via a bipotent glial progenitor cell (GPC), which may share markers and differentiate into both OPCs and astrocytes, has been proposed [42,43]. Recent data coming from single cell transcriptomics have started to clarify this issue (Section 4).

3. Neural Cell Types in the Human Adult Telencephalon

Differentiated neural cells such as neurons, oligodendrocytes, astrocytes and ependymal cells compose the large compartment of the adult telencephalon. Less than 1% of the neurons in the human brain, which are situated at pallial and subpallial levels in hippocampus and striatum, respectively, are replaced during adult life [44–46]. Unlike neurogenesis, adult gliogenesis remains active in the whole brain. Most OPCs generated during embryonic and early postnatal periods remain undifferentiated in the mature brain parenchyma throughout life, gradually becoming mature oligodendrocytes and replacing existing oligodendrocytes that are lost physiologically or after injury. The production of oligodendrocytes and myelin is very active postnatally and in early childhood—especially in the white matter—gradually decreasing toward the adult stages, when only 1 in 300 oligodendrocytes is replaced every year [41]. However, it is likely that postnatal gliogenesis is dynamically modulated in humans by learning new skills, with neuronal activity in general being associated with the generation of new oligodendrocytes and increased myelination in working brain regions. The production of new astrocytes in the adult brain parenchyma in normal conditions is even lower than the production of oligodendrocytes [47]. Somewhat unexpectedly, in certain pathological situations such as experimental stroke, astrocytes in the brain parenchyma can acquire an activated stem cell behavior, enter a neurogenic program, and give rise to new neurons, as well as new astrocytes. This supports the hypothesis of a widespread distribution of quiescent or "dormant" adult neural stem cells (NSCs) with an astrocytic-like phenotype [48–50].

However, it is recognized that new telencephalic neurons can normally be produced by adult NSCs, which are present only in two distinct niches: a pallial one situated in the subgranular zone (SGZ) of the hippocampal dentate gyrus [45] and a subpallial one in the SVZ lining the LV [51]. While hippocampal neurogenesis in the SGZ is mostly similar in mice and humans, the neurogenesis in SVZ is different. Large numbers of neurons generated from SVZ progenitors are continuously added to the olfactory bulb in adult rodents, but adult olfactory bulb neurogenesis could not be detected in adult humans [52]. Instead, human SVZ progenitors produce a subpopulation of interneurons detected in the striatum, which mainly express the marker calretinin [53]. However, adult neurogenesis at very low levels in other areas cannot be disregarded and continues to be an active area of exploration [49].

Adult NSCs in both the SVZ and SGZ derive from prenatal RG and share astrocyte morphology and markers, such as GFAP, but also stem cell markers such as CD133, SOX2 and nestin. The SVZ lining the LV in the human brain is composed of four distinct layers: a monolayer of ependymal cells alongside the ventricular cavity; a hypocellular space containing mainly GFAP-positive cellular processes; a dense layer containing mainly cells expressing GFAP (both adult NSCs and astrocytes), a smaller population of proliferating and migrating cells; and a transition zone adjacent to the parenchyma, mainly composed of oligodendrocytes and microglia. The proliferative marker KI67 (or the gene *MKI67*) is expressed in a limited number of cells in the SVZ, reflecting a very small number of cycling cells, which may co-express GFAP and SOX2 with OLIG2 or ASCL1, and the number of which decreases with age [37]. Although the adult NSCs, derived from various locations of the LV wall, can self-renew and behave as multipotent progenitors in both human and rodents—meaning that they produce all three neural lineages in vitro—whether each individual cell indeed produces both neurons and glia in vivo remains uncertain. Adult NSCs from the rodent SVZ, named B cells, undergo asymmetric cell divisions to give rise to a new B cell (one of the hallmarks of a stem cell), as well as to a transit-amplifying progenitor cell—also known as a type C cell. The progenitors further differentiate to only one distinct subset of neurons that migrate toward the olfactory bulb [54,55]. It is also currently unknown whether the same neurogenic adult NSCs can produce glial cell types in vivo. Again, a stepwise differentiation of the adult NSCs via intermediate precursor cells such as the bipotent GPCs has been proposed in rodents [42,43] and partially clarified by the scRNA-seq profiling of the adult SVZ in rodents and the transcriptomic crosstalk with human prenatal telencephalic progenitors (Section 4).

4. Single-Cell Transcriptomics of the Human Telencephalon

Transcriptomic sequencing is a technique that uses high-throughput, next-generation sequencing approaches to reveal the presence and quantity of RNA in a biological sample at a given moment. Recent advances in RNA-seq include single cell and single nucleus RNA-seq (scRNA-seq and snRNA-seq, respectively) [10,56,57]. Unlike bulk RNA sequencing, which interrogates average gene expression in cell populations that are in most cases heterogeneous [58], scRNA-seq can elucidate heterogeneity and allow cell-type specific transcriptomic profiling to be performed. The recent advent of high-throughput microfluidic systems with droplet-based profiling techniques has further advanced the precision of sc/sn RNA-seq profiling. Initially limited to only a few hundred cells per experiment, due to advances in experimental technologies, more than 1 million single cell transcriptomes can be profiled nowadays [59].

Several studies have performed single-cell sequencing of human embryonic fetal and adult samples of the human telencephalon [60–69]. In order to analyse the complex sets of single-cell data, robust computational methodologies need to be applied [56]. Unsupervised approaches use clustering followed by cluster annotation of cell types based on differentially expressed marker genes [70], while supervised approaches use a reference panel of labelled transcriptomes to guide both clustering and cell type identification [71]. Clustering of human telencephalic cell types has been obtained using unsupervised and supervised methods, or by using a combination of both. Each cluster was attributed with "unique markers", which are the genes expressed only in that type of cell among all the cells sampled, as well as with "combinatorial markers", which are differentially expressed genes that are not restricted to a single cell type. A consensus approach was proposed for both the clustering paradigms in order to increase the accuracy of the clustering and the precision of cell type annotation [72], which is expected to be applied in future studies.

Nowakowski et al. performed scRNA-seq in human pallial and subpallial samples across prenatal stages from GW 6 to 37 [73]. Using unbiased clustering followed by a supervised approach using a reference panel of labelled transcriptomes, they identified transcriptionally distinct cell clusters and subclusters corresponding to RG, dorsal IPCs or excitatory neuron progenitors (ENP), excitatory neurons (EN), inhibitory neuron progenitors (INP), inhibitory neurons (IN), astrocytes (Astros), and OPCs (Figure 2A), as well as non-neural cell types: microglia, choroid plexus cells, mural cells, and endothelial cells. As expected, the proliferation gene *MKI67* was expressed in the known transit-amplifying populations of INPs and IPCs, as well as in a subpopulation of RG; *PDGFRA* and *OLIG1* were expressed mainly in the OPC cluster; the oRG gene *HOPX* was widely expressed in the RG cluster, and the mature astrocytic gene *AQP4* was restricted to the astrocyte cluster (Figure 2B). In addition, the lineage reconstruction method enabled the inference of gene expression trajectories from heterogeneous developmental tissue. Correlation of sample age with gene co-expression networks defined a maturation score, or "pseudoage". Analysis of gene expression trajectories across "pseudoage" confirmed that early human cortical RG showed enriched expression of proneural transcription factors, whereas genes involved in gliogenesis are upregulated later in development. In addition, microdissected VZ and OSVZ samples were used to define a lamina score, or "pseudolamina". Gene co-expression networks correlated with "pseudoage" and "pseudolamina" supported the classification of RG subtypes as vRG, oRG and tRG. These transcriptomic results indicate multiple signalling pathways which act during RG diversification and show the temporal and topographical hierarchy in dorsal and ventral telencephalic lineages of developing cortical neurons; within the dorsal telencephalon, these temporal and typological differences define progenitors across cortical areas, while topographical distinctions predominate across maturing neurons [73].

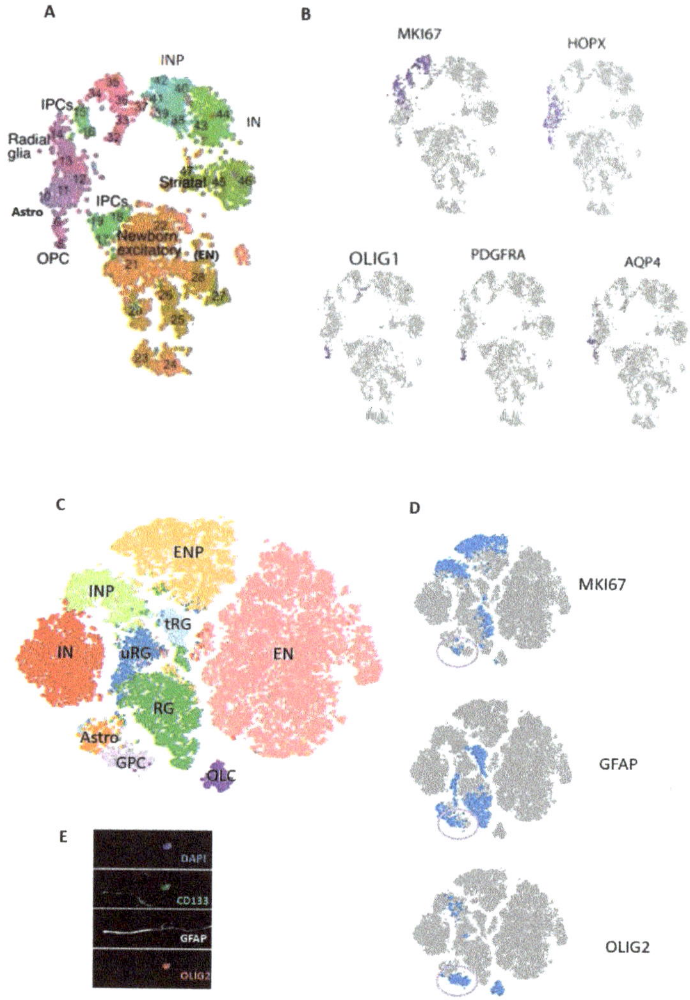

Figure 2. Single-cell RNA sequencing in the developing human telencephalon. (**A**). Plot of neural cells from pallial and subpallial human samples across prenatal stages (gestational weeks 6–37), colored by cluster and subcluster cell assignments, corresponding to different types of radial glia, intermediate progenitor cells (IPCs), excitatory neurons (ENs), inhibitory neuron progenitors (INPs), inhibitory neurons (INs), astrocytes (Astros), and oligodendrocyte progenitor cells (OPCs). (**B**). The same cluster representation as in (**A**), showing the cell/cluster-related expression of cycling and cell0specific genes (RG, OPCs and Astros). (**C**). Plot of neural cells isolated from the telencephalon of four human fetuses (gestation weeks 13–21), in which data sets from total neural cells and CD133+-selected cells were combined, colored by clusters representing three types of radial glia (marked as RG, truncated RG (tRG) and unknown RG (uRG), excitatory neuronal progenitors (ENPs), ENs, INPs, INs, Astros, glial progenitor cells (GPCs) and oligo-lineage cells (OLCs). (**D**). The same-cluster representation as in (**C**), showing the cell/cluster-related expression of the genes of the cycling and cell-specific genes (RG, OPCs and Astros); some cells in the GPC cluster co-express these markers. (**E**). Immunofluorescence image of human fetal telencephalic cells in primary culture co-expressing the proteins CD133, OLIG2, and GFAP. DAPI nuclear staining (blue). (**A**,**B**) adapted from [73], and (**C**–**E**) adapted from [74].

Couturier et al. performed scRNA-seq on freshly isolated cells from the telencephalon of four human fetuses ranging from GW 13 to 21 [74]. Fluorescence-activated cell sorting (FACS) was used to remove most of the non-neural cells such as microglia and endothelial cells, and CD133-positive selection improved the resolution of the neural stem and progenitor cell populations. Data sets of total and CD133-positive selected cells from all fetal brains were combined in silico after excluding ependymal cells and were used for unbiased grouping of cells into 10 clusters (Figure 2C). Differential gene expression analysis of these clusters identified important genes in each cluster. Most of these clusters correspond to previously defined cell populations in the developing human telencephalon [73], such as excitatory neuron progenitors (ENPs, corresponding to dorsal IPCs), ENs, INPs, INs, astrocytes (Astros) and oligodendrocyte lineage cells (OLCs, which may include OPCs and Pre-OLs), as well as three different RG clusters. However, the RG clusters were found to only partially correspond to the expression found in the pallial vRG, oRG, tRG or subpallial RG, and they were labelled as RG, tRG and uRG (undefined RG; Figure 2C). The proliferation gene *MKI67* was expressed, as expected, in the transit-amplifying neuronal progenitor clusters, but also in RG subpopulations—especially in the RG cluster—while *GFAP* was expressed in subpopulations in both the tRG and RG clusters. This suggests that the RG and uRG clusters correspond to mixed pallial and subpallial subpopulations of previously defined oRG and vRG, respectively.

Interestingly, a glial progenitor cell (GPC) cluster was detected at all gestational ages and strongly expressed oligodendrocyte lineage genes (e.g., *OLIG1*, *OLIG2*, and *PDGFRA*), glial/astrocytic lineage genes (e.g., *GFAP*, *SOX9*, *HOPX*, *HEPACAM*, and *VIM*), and progenitor genes (e.g., *ASCL1*, *MKI67*, and *HES6*). However, it did not express several differentiation markers found in astrocytes or oligodendrocyte-lineage cell clusters (Figure 2D). This mixed gene signature partially differs from the signature of previously defined OPCs [73], but may be compatible with that of the proposed bidirectional GPC [42,43]. Notably, this GPC signature was almost exclusively identified in CD133-positive sorted cells, in a fairly small cluster, which likely explains why it was not previously detected in sequenced unsorted brain cell populations in both the prenatal and adult brain [60–69]. Importantly, the confirmation of cells co-expressing astro-like and OPC-like markers was done in a primary culture at first passage obtained from one of the cell-sequenced fetal brains (Figure 2E), as well as in the SVZ of the adult human brain (Figure 3C) [74]. This indicates a special subtype of progenitor population that is present from the fetal to the adult stages in the human brain. Additional bioinformatics helped to define a roadmap of the developmental-related trajectories in the human prenatal telencephalon, where these GPCs are linked to three differentiated neural lineages—represented here by interneurons, astrocytes and immature oligodendrocytes—while tRG link solely with the astrocyte lineage [74].

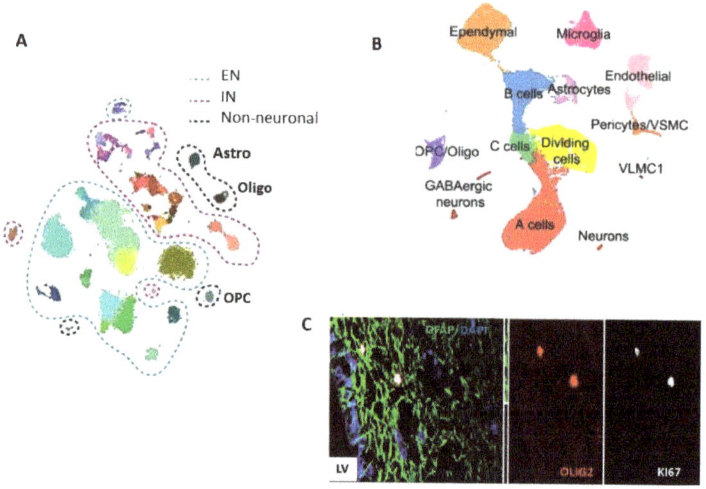

Figure 3. Single-cell RNA sequencing in the adult telencephalon. (**A**). Plot of cells from human adult cortex samples showing clusters and subclusters of neuronal populations (excitatory neurons—ENs and inhibitory neurons—INs), as well as small non-neuronal clusters of astrocytes (Astros), oligodendrocytes (Oligos) and oligodendrocyte progenitor cells (OPCs; adapted from Hodge et al., 2019). (**B**). Plot of cells from adult mouse subventricular zone (SVZ) samples (adapted from Dulken et al., 2017). (**C**). Immunofluorescence image of an adult human SVZ sample. Distribution of the cells lining the lateral ventricle (LV) and co-expressing KI67, OLIG2, and GFAP markers. DAPI nuclear staining (blue; adapted from [74]).

Velmeshev et al. performed snRNA-seq of human adult cortex samples and unbiased clustering, followed by annotation according to expression of known cell type markers, which identified 17 cell types, including subtypes of EN, IN and astrocytes [66]. Hodge et al. provided a more detailed transcriptomic map of the cells in the human adult cortex by following the same computational analysis as used in the mouse cortical cells previously profiled by Tasic et al. [64,75]. The transcriptional analysis of nuclei isolated from samples of human cortex revealed 69 neuronal and 6 non-neuronal clusters. From the neuronal clusters, 24 represented different types of EN, and 45 represented different types of IN. The major clusters of non-neuronal cells expressed SCL1A3 and included two astrocyte (Astro) types with different laminar distributions, OPCs, oligodendrocytes (Oligos) (Figure 3A), microglia, and endothelial cells. Astrocytes in the first cluster expressed higher *GFAP* and *AQP4* levels than the astrocytes of the second cluster; the first group may represent the interlaminar and fibrous types (from the connected white matter), while the second group represent the protoplasmic type. Cells in the OPC cluster expressed a high level of *PDGFRα*, which was also expressed at lower levels in the IN subpopulation, suggesting a common developmental-related pathway [64]. Addressing the composition and cell hierarchies of the known proliferative and neurogenic niches in the adult telencephalon, a single-cell RNA sequencing experiment investigated the rodent SVZ [76] (Figure 3B). Interestingly, the clusters representing the SVZ astrocytes, B cells and ependymal cells were closely related, while C, A, and the dividing cells form a quasi-continuum. OPCs formed a small separate cluster, but their hierarchic relationship with a subcluster of C cells, which may be the GPCs, should be further explored. To complement the studies in rodents, similar single-cell transcriptomic studies in human SVZ are expected to confirm these neurogenesis and gliogenesis pathways, but also to clarify the dilemma of the "dormant" NSCs in the adult human telencephalon [49].

5. Single-Cell Transcriptomics in Gliomas

Paralleling the extensive development of single cell transcriptomics of normal brain tissue, many databases and reports include single cell sequencing of brain tumor specimens. Several strategies have been used to exclude non-malignant cells, which are critical components of the brain tumor microenvironment, and to properly group malignant cells. FACS with negative selection for non-neural cells and computational filtering by using copy number variation (CNV) are the most used approaches to classify cells as belonging to malignant or normal tissues [74,77,78].

In addition, different known genetic alterations and expression of cell cycling, stemness and class-specific genes were investigated at the single cell level, and in some cases compared with their expression in normal human brain cells. The generated high-throughput molecular profiles captured to a great extent the intra-tumor heterogeneity and identified several populations of GSCs in different gliomas.

5.1. IDH Mutant Gliomas

With the aim of understanding the differences between the two major types of IDH-mutated diffuse gliomas—including the cells of origin—samples from oligodendroglioma (IDH-O) and astrocytoma (IDH-A) were first sequenced at the single-cell level by Tirosh et al. [79]. Each tumor included in the study contained a large population of cells with confirmed *IDH1* or *IDH2* mutations and co-deletion of chromosome 1p and 19q arms, as well as tumor-specific CNVs. Highly consistent across all IDH-O and IDH-A tumors, two prominent cell clusters expressed distinct lineage markers of oligodendrocytes and astrocytes, respectively. One cluster was strongly associated with the high expression of oligodendrocyte markers (such as *OLIG1/2*) and the low expression of astrocytic markers (such as *GFAP* and *APOE*), while the other cluster had the opposite expression patterns. A smaller cluster highly expressed genes related to neurodevelopment and neural stem cells, such as *SOX2*, *SOX4* and *ASCL1*. Remarkably, cells with OPC gene expression, which were suggested to represent the origin of oligodendrogliomas in a mouse model [80], did not form a separate cluster. All proliferating cells found in each tumor (1.5–8%), consistent with Ki67 expression, grouped in the small cluster. Together, this analysis revealed three main expression patterns recapitulating a stem/progenitor program of early neural development and subsequent differentiation into oligodendrocytes and astrocytes. The single-cell profiles suggested that the tumor-initiating cells in IDH-mutated gliomas more closely resembled the NSC type than a more committed glial progenitor type such as OPCs.

Following the study of Tirosh et al. [79], Venteicher et al. combined scRNA-seq results from ten IDH-A and six IDH-O tumor samples with bulk data from large cohorts from The Cancer Genome Atlas (TCGA) [81]. They found that differences in bulk expression profiles between IDH-A and IDH-O were explained primarily by genetic alterations and the composition of the tumor microenvironment (TME), but not by distinct glial expression programs in the malignant cells. Again, in both IDH-A and IDH-O tumors, only a small proportion of cells (~4% on average) were in a proliferative stage, co-expressing cycling and putative stem cell markers. Single-cell approaches showed again that undifferentiated cells from both tumor types exhibited increased similarity in gene expression programs, further suggesting a shared cell of origin for both IDH-A and IDH-O. Thus, IDH-mutant gliomas as defined by genetics and histopathology as differing in terms of genetics and TME but, examined at single-cell resolution, all contain three subpopulations of malignant cells: two non-cycling differentiated glial lineages—astrocyte-like and oligodendrocyte-like cells—as well as one cycling undifferentiated subpopulation that resembles NSCs.

Together, the studies on IDH mutant gliomas represent a shift in understanding the histogenesis of glial tumors and support a model where IDH mutant glioma subclasses share developmental programs and putative lineages of glial differentiation, but differ primarily by the genetic mutations and the number of macrophages and microglia in the TME [81].

5.2. Glioblastoma

IDHwt glioblastoma was the first brain tumor investigated at the a single-cell transcriptome level [78]. As for IDH mutant gliomas, it was shown clearly that bulk transcriptomics did not capture the true diversity of transcriptional subtypes within a tumor but detected only the dominant transcriptional program. While the classification of IDHwt glioblastoma via bulk transcriptomics includes the proneural, classical and mesenchymal subtypes [4,5], the scRNA-seq showed that all tumor samples consisted of heterogeneous mixtures with individual cells corresponding to different glioblastoma subtypes. Panoramic analysis of the chromosomal landscape identified chromosomal aberrations in each tumor cell, such as the gain of chromosome 7 and the loss of chromosome 10, the two most common genetic alterations in glioblastomas [82]. Single-cell transcriptomic analysis have revealed that IDHwt glioblastoma samples contain multiple cell states with distinct transcriptional programs and have provided inferential evidence for dynamic transitions; cell cycle-related genes were active in 1.4% to 21.9% of malignant cells. Application of the stemness signature revealed stemness gradients in all tumors, modestly anti-correlated to the cell cycle signature and consistent with the notion that NSCs divide at lower overall rates, as compared with IPCs. The stemness signature was stronger in individual cells from samples of proneural and classical subtypes. In contrast, cells of the neural subtype were more like oligodendrocytes. These findings suggested parallels between intratumoral cellular heterogeneity in glioblastomas and cellular diversity in the developing brain, with respective subsets of tumor cells resembling a stem cell and progenitor compartment, an astrocytic lineage, or an oligodendrocytic lineage. The analysis also revealed "hybrid" states in which a single cell scored highly for two subtypes, most commonly classical and proneural (progenitor states) or mesenchymal (differentiated states).

To further understand glioblastoma transcriptional and genetic heterogeneity, Neftel et al. addressed an integrative approach, combining scRNA-seq, analysis of bulk specimens and lineage tracing in glioblastoma models [83]. They found that malignant cells in glioblastomas may be grouped into four categories: neural progenitor-like (NPC-like), OPC-like, astrocyte-like (AC-like) and mesenchymal like (MES-like) states. While each glioblastoma sample contained cells in multiple states, the relative frequency of each state varied between tumors. Furthermore, by coupling scRNA-seq to uniquely barcoded single cells in vivo, Neftel et al. demonstrated the plasticity between states and the potential of a single mutated cell to generate all four states. This work provided a roadmap of the cellular programs of malignant cells in glioblastomas, as well as their plasticity and modulation by genetic drivers, but did not address the origin of the malignant cell types [83].

Bhaduri et al. analyzed the glioblastoma samples by using scRNA-seq and the previously described transcriptional signatures of the developing human brain [73] and adult cortex [66]. Despite their heterogeneous composition, each tumour contained a distinctive combination of transcriptionally defined cancer cell types (Figure 4A) and a mitotic index of 20% or higher. By exploring the cell types associated with CNV, the authors found dividing RG-like cells, IP-like cells and OP-like cells, which were expected, but also neuronal-like cells which expressed *MKI67* (Figure 4B) [84].

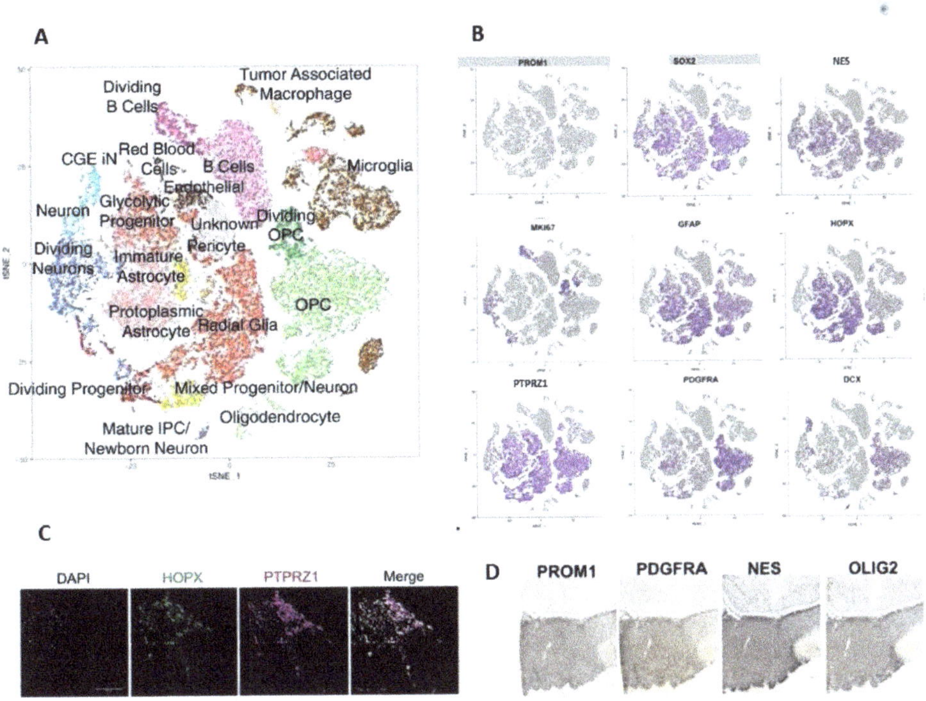

Figure 4. Single-cell RNA sequencing in glioblastoma. (**A**). Plot of cells from glioblastoma samples from 16 patients showing clusters colored by attributed cell types. (**B**). The same cluster representation as in (**A**), showing the cell-type/cluster-related expression of several genes expressed during neural development in stem and progenitor populations. (**C**). Immunofluorescence image of a glioblastoma sample showing co-expression of the oRG markers HOPX and PTPRZ1; DAPI nuclear staining (blue); scale bar: 10 µm. (**D**). Glioblastoma samples from a glioma repository, showing the expression of the proteins PROM1, PDGFRA, nestin (NES) and OLIG2. (Adapted from [84]).

Almost all identified cell types expressed at least one marker associated with stemness and previously known GSC markers. A number of previous studies have shown the potential for a variety of cell types to become GSCs, such as OPCs, astrocytes, and neuronal cell types [85–88]. However, the gene combinations previously associated with GSC stemness were expressed uniquely for each individual tumour. While several progenitor genes such as *SOX2* and *NES* were expressed broadly, *PROM1* (expressed as CD133), a marker that has been shown to be sufficient to give rise to ectopic tumours [7], was very sparsely expressed (Figure 4B). Thus, the cell types that make up glioblastomas can be found in various combinations across tumours, but the cocktail of stemness markers co-expressed within a GSC cell type is largely specific to every individual tumour. By further exploring a glioblastoma repository [89], Bhaduri et al. observed that individual tumours co-expressed a variety of GSC marker genes such as *PROM1* (CD133), *PDGFRA*, *NES* and *OLIG2* (Figure 4D), which also confirmed the expression found at the single cell transcriptomic level (Figure 4B). These results from orthogonal datasets support the hypothesis that diverse sets of GSCs can be found within a single tumour, characterized by heterogeneous marker gene combinations. The transcriptomic profiles of glioblastomas suggest that programs associated with stemness are broadly expressed, and that the activation of stemness programs indicated by these GSC marker genes can occur in almost any cell type within the tumour. Additionally, a distinct cell type within the glioblastoma atlas expressed oRG marker genes [84]. The oRG network was strongly recapitulated in glioblastoma, with the same hub genes such as *PTPRZ1* and *HOPX* [23,24] (Figure 4B), and with confirmed expression at the protein level (Figure 4C). This suggests that re-expression of the developmental oRG signature in GSCs is associated with a dynamic cell behaviour characteristic of prenatal oRG cells [84].

The oRG-like population in glioblastoma cells was further enriched using FACS PTPRZ1-positive selection and re-analyzed by scRNA-seq. Even though the PTPRZ1-positive sorted population was not homogenous, it was significantly enriched for RG-like cells compared to the PTPRZ1-negative population [90,91]. To functionally investigate the PTPRZ1-positive, negative and unsorted cell populations, they were labelled with a GFP-expressing adenovirus and transplanted into an in vitro model of human brain organoids. Two weeks after transplantation, the tumour cell populations were composed primarily of either neuronal or astrocytic cells. Both PTPRZ1-positive and negative cells expressed canonical GSC markers in each of these populations, consistent with the earlier observation that different glioblastoma cell types express stemness markers. The expression of GSC markers uniformly decreased after transplantation, while differentiated cell types within each population increased. Together, these results supported the PTPRZ1-positive oRG-like glioblastoma cells as being one of several GSC cell types and showed their invasive nature and involvement in tumor propagation [84].

Couturier et al. made a step forward in defining the cells of origin and heterogeneity in glioblastoma by comparing the previously established lineage hierarchy of the developing human brain (Figure 5A) to the transcriptome at the single-cell level of both whole-tumor samples and CD133-positive selected samples (with the aim of increasing the proportion of GSCs), and after CNV selection. Plotted on the roadmap for human prenatal telencephalic cells (Figure 5A), the whole tumor samples mainly mapped onto a GPC cluster, an oligo-lineage cluster, an astrocyte cluster, a tRG cluster, and an interneuron cluster—but also onto a non-defined intermediate population (Figure 5B). In the population enriched by CD133-positive selection (named GSC in Figure 5C), most of the plotted cells expressed GPC genes, but some expressed neuronal and astrocytic genes. These data suggest that the GSC-enriched population is also heterogeneous but organizes into subpopulations resembling a developing brain [74].

Almost all cycling cancer cells had high glial progenitor scores (Figure 5D). These data also show that GPC-like cancer cells are the cell types with the highest rates of proliferation—more than the cancer cells undergoing lineage differentiation. This model reveals a GPC-centered organization in both the whole-tumor and the GSC-enriched population. Remarkably, the GPC signature was the only one robustly expressed in all patients. The identification of highly proliferative GP-like cells was in contrast to previous works [78,83], where such a population was not found. Protein marker panels, representative of each cancer cell type, and single-cell proteomic analysis, were additionally used to validate this result. Together, these analyses suggest that astrocytic, mesenchymal, oligodendrocytic, and neuronal-like glioblastoma cells are more differentiated than GP-like cells, and that they are one of the originators of the trilineage hierarchy in glioblastomas [74].

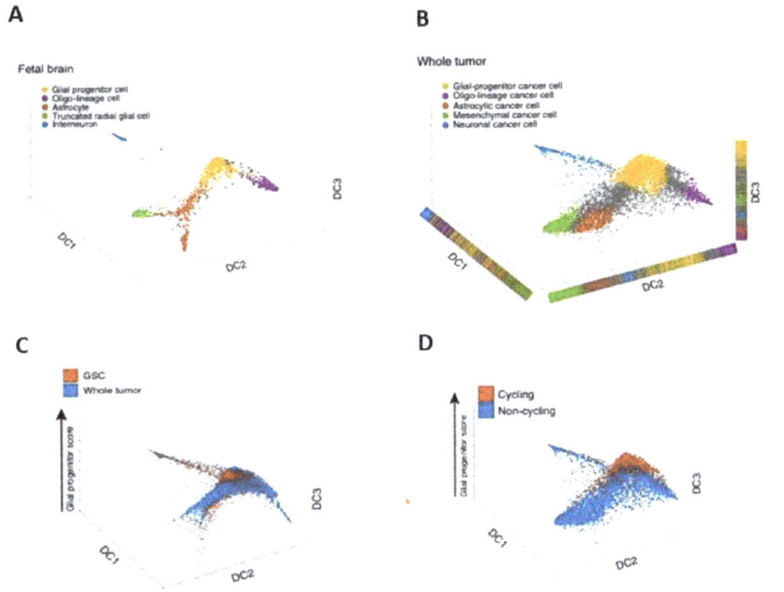

Figure 5. Developmental-related transcriptomic roadmaps in the human fetal brain and glioblastoma samples. Cell maps built with human prenatal telencephalic cells (**A**) and glioblastoma cells from whole-tumor samples (**B–D**) or enriched by CD133-positive selection; a population named glioma stem-like cells (GSC) in **C**. GPC cluster (orange, (**A**,**B**)), oligo-lineage cluster (violet, (**A**,**B**)), astrocyte cluster (coral, (**A**,**B**)), tRG cluster (green, (**A**,**B**)) and interneuron cluster (blue, (**A**,**B**)), but also a non-defined intermediate population (grey, (**B**); adapted from [74].

6. Towards Signaling-Specific Targeted Therapy in Gliomas

Significant obstacles hampering the development of effective cancer therapeutics include tumor heterogeneity and the persistence of GSCs that give rise to cancer recurrence. Most studies consider the GSC population to be uniform. However, recent single-cell transcriptomics studies have shown that GSCs display heterogeneity driven by a hierarchical developmental organization [74,81,84]. Some of these recent studies have evaluated chemoresistance and tumorigenicity in selected GSC populations in glioblastomas. In addition, identifying the signaling pathways that maintain tumor-initiating cell proliferation may provide therapeutic targets for inhibiting tumor growth. Identification of signaling pathway alterations between progenitor cancer cells and more differentiated cancer cells may yield meaningful new therapeutic targets.

The association of the genetic alterations in signaling pathway component genes such as *PDGFRα*, *EGFR* and *NF1* was explored in *IDH*wt glioblastomas, with each mutation being shown to favor a particular state [83]. However, these signaling pathway components are expressed broadly in different normal populations in the adult brain, making the targeted approach difficult. A more efficient approach should address signaling pathways related to more specific cell populations.

The proliferation and migration of the RG in the fetal brain occurs in response to different signals, parts of different signalling pathways, such as the FGF-MAPK, PI3K/PTEN/AKT, Hedgehog, N-cadherin/β-catenin, Notch, mTOR and Rho/Rho-kinase (ROCK) pathways [26,29]. Nowakowski et al. highlighted the gene enrichment at the single-cell level in oRG for *GLI2*, *NFAT2C*, and several regulators of the mTOR signalling pathway, as well as increased phosphorylation of the S6 ribosomal protein [73]. Bhaduri et al. explored the role of PTPRZ1 in oRG-like GSCs. PTPRZ1 and its ligand, PTN, have been previously identified as necessary for tumour invasion and viability and linked with the known effects of the

Rho/Rho-kinase signalling pathway in glioblastoma [92–94]. The authors used genetic and pharmacologic approaches and found that *PTPRZ1* and *PTN* double knockdown significantly reduced the migration of the oRG selected from fetal human brains. In order to relate these findings to invasive behaviour in glioblastoma, an in vitro invasion assay used tumour samples treated with either control shRNAs, *PTPRZ1* shRNAs, or Rock inhibitor. Both *PTPRZ1* knockdown and Rock pathway inhibition significantly decreased the invasive behaviour of oRG-like cells in an in vitro model using human brain organoids, suggesting a selective way of further addressing single members of the GSC population [84].

Couturier et al. focused on GP-like cells as the originators of the cancer cell hierarchy in glioblastoma [74]. These rapidly cycling progenitor cancer cells were seen as a prime cell population to target and were tested both in vitro and in vivo, in xenograft models. In GSC culture conditions, where all GSCs retain the ability to divide, GP-like cells were found to be the most resistant to chemotherapy. Investigating the scRNA-seq transcriptomic data, Couturier et al. identified several pathways with a significant enrichment in the GPC cluster, as compared to the astro-mesenchymal groups. Hits with significant and strong correlations were found in pathways previously established as relevant to GSC self-renewal and tumorigenicity, such as the WNT pathway and the EZH2 and FOXM1 genes, but also in pathways of previously unknown significance in glioblastoma. Of these, the E2F4 pathway was the most significant, and was thus selected for testing. While E2F4 expression in glioblastoma tissue has been previously shown [95], Couturier et al. provided the first description of its role in the GPC malignant population. The E2F gene family regulates the cell cycle and is important for progenitor cell survival [96]. It has been shown that E2F4 inhibition causes senescence of gastric cancer cells [97]. The effect of the small molecule inhibitor HLM006474, which prevents E2F4 binding to DNA, was tested in vitro and in vivo in IDHwt glioblastoma cells. Following HLM006474 treatment, the proliferation and survival of the GPC population in vitro was reduced significantly as compared to the neuronal and astro-mesenchymal populations, supporting its specific effect on the GP-like cell population in glioblastomas. After orthotopically xenografting glioblastoma cells, a significant reduction in tumor growth and improved survival in the HLM006474-treated mice was observed. In addition, no synergism or antagonistic effect was found between the HLM006474 and classical chemotherapy. In addition, mice xenografted with GP-like cancer cells developed tumors faster and exhibited a shorter survival time than mice engrafted with OPC-like cancer cells [74].

While targeting the most rapidly cycling and functionally aggressive progenitor cancer cell population may be an effective treatment approach, given the plasticity that can occur in the GSC population, separate targeting of all cell types within a cancer may need to be addressed in future for each GSC subtype. The signalling pathway components waiting for further exploration in selected cell populations from glioma include the oRG-like and GPC-like cells. Several combined targeted therapies could address each tumor and cell type in a personalized approach.

7. Conclusions and Outlook

Several transcriptomic-related studies have shown that the optimal classification of cell subtypes in a tissue or a tumour sample should imply a deeper knowledge of their origins and trajectories during development and maturation. The single cell transcriptomic approaches presented here provide convergent evidence for a robust description of cell type identity in both normal and pathological lineages, validating the neurodevelopment pathways as important players in tumour development. The developmental roadmaps generated from the transcriptomic studies still wait for additional work for validation, such as cell fate mapping in vitro and in vivo, which is necessary to uncover the exact position of these cells within the developmental hierarchy of the brain. However, the proposed models of cell hierarchies resemble a pattern that has also been shown in gliomas: GSCs generate daughter cells that subsequently differentiate into tumour bulk cells [93,98]. Using these tools, the cellular hierarchies proposed for the prenatal telencephalon can be linked with a diverse range of cerebral tumours, as we propose here for IDH-mutated glioma and IDHwt glioblastoma (Figure 6).

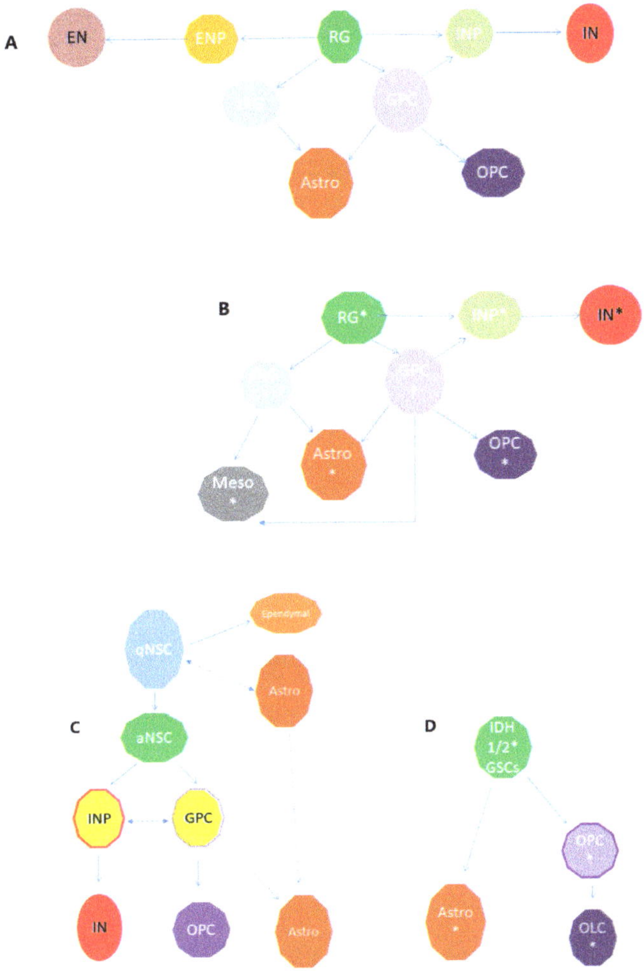

Figure 6. Towards a model of cell hierarchies in gliomas. Single-cell transcriptomics-based links proposed for the developing human telencephalon (**A**), adult human telencephalon (**C**), and the potential similarities with IDHwt glioblastomas (**B**) and IDH1/2 gliomas (**D**). The nodes in (**A**) represent the cell types from human fetal telencephalic samples corresponding to different types of radial glia (RG), excitatory neuron progenitors (ENPs), excitatory neurons (ENs), inhibitory neuron progenitors (INPs), inhibitory neurons (INs), truncated RG (tRG), astrocytes (Astros), glial progenitor cells (GPCs) and oligodendrocyte progenitor cells (OPCs; shown as clusters in Figure 2). Nodes in (**B**) marked with (*) represent cell types with IDHwt glioblastoma-related mutations. Mesenchymal-like cells (Mesos). The nodes in (**C**) represent the cell types in the adult subventricular zone (SZV) samples corresponding to different types of adult neural stem cells (NSCs), such as quiescent (qNSCs) and activated neural stem cells (aNSCs), inhibitory neuron progenitors (INPs), inhibitory neurons (INs), astrocytes (Astros), glial progenitor cells (GPCs) and oligodendrocyte progenitor cells (OPCs; shown as clusters in Figure 3). Nodes in (**D**) marked with (*) represent cell types with IDH1/2 glioma-related mutations. GSC: glioma stem-like cells, OLC: oligo-lineage cells.

The single-cell profiles from IDH-mutated glioma and IDHwt glioblastoma appeared to be different, suggesting that different progenitor populations maintain their cell pool while producing additional tumour cell types. IDH-A and IDH-O share the same developmental hierarchy, consisting in each case of three subpopulations of malignant cells: two nonproliferating populations differentiated along the astrocytic and oligodendrocytic lineages, and a proliferative population of GSCs that resembles NSCs.

For the IDHwt glioblastoma, different GSC populations were detected, which exist at the intersection of neuronal, glial and mesenchymal lineages. These GSCs correspond to different classes of stem cells and progenitors found in human fetal telencephalon, including oRG and GPCs [74,84], which support the high degree of heterogeneity in this tumour type. Single-cell analysis suggests that progenitor cancer cells have the potential to differentiate into all the cancer cell lineages identified. GSC cell type-specific xenograft models show evidence of both neuronal and glial lineage commitment.

The transcriptomic analysis of Bhaduri et al. indicates that RG-like glioblastoma cells, along with other progenitor populations, might serve as tumour propagating cells in glioblastoma. ORG are stem cells of the developing human brain which give rise to a transit-amplifying progenitor cell population, and further to neurons and glia in a temporal- and lineage-dependent fashion [84].

The roadmap built in parallel with the fetal human telencephalon as a training set using an equal number of cancer cells from each of the analyzed *IDH*wt [74] revealed that glioblastoma develops along conserved neurodevelopmental gene programs and contains a rapidly dividing progenitor population, which corresponds in development to a newly defined glial progenitor type. These GPCs share markers with NSCs, but also to OPCs, and could be clearly clustered only in the CD133-positive selected samples. These data also show that tumour initiating cells or GSCs are the cancer cell type with the highest rates of proliferation—they were identified in glioblastoma and phenotyped only after enrichment for stem cell-like populations using CD133-positive selection, followed by scRNA-seq [62,63,99,100]. These results suggest a model in which developmental programs are reactivated in IDHwt glioblastoma cells by specific mutations (Figure 6A,B).

Another possibility is that different adult NSCs in the neurogenic and gliogenic niches are affected directly by these mutations, and that these cells share phenotypes with the population found during development. This is more likely to be a model for IDH1/2 gliomas, which may originate from adult stem cells (Figure 6C,D). However, as scRNA-seq of the adult SVZ niche cells was performed only for the adult rodent SVZ, further investigations of the human SVZ are expected to confirm this pathway. Again, additional work such as cell fate mapping in vitro and in vivo will also be necessary for the adult NSC, transit-amplifying, and migrating cell populations.

Taken together, transcriptomic analysis in normal brain development reconciles glioblastoma development, suggesting possible origins for the glioblastoma hierarchy, and helping to identify cancer stem-like cell-specific targets. However, the transcriptomic phenotypes of the GSC in IDHwt glioblastomas and in IDH-mutated gliomas are not the same. This suggests a different cell of origin for these pathologies than in adult IDHwt glioblastoma and may underlie the disparate natural histories and treatment responses between these cancer types.

A better understanding of the spectrum and dynamics of cellular states in several types of glioma is critical for establishing faithful models and advancing therapeutic strategies that address the complexity of this disease. Further combining the single-cell transcriptomic profiles of normal and tumour cells from the same brain region can provide the basis to better define the potential origins of neural cells initiating different types of primary cerebral tumours, and the design of therapies targeting GSC phenotypes—a potentially novel avenue in the treatment of these currently incurable malignancies.

Single-cell transcriptomic studies have started to rewrite the knowledge regarding the composition of the brain and its tumors, but also of cell hierarchies and cell-specific signaling. They belong to an ongoing major collaborative effort directed to generating a complete description of cell types in the human brain (the National Institutes of Health (NIH) BRAIN Initiative Network, braininitiative.nih.gov, accessed on 9 November 2021), in the human neocortex (Allen Institute for Brain Science, www.allenbrain.org, accessed on 9 November 2021), the whole human body (the Human Cell Atlas, www.humancellatlas.org, accessed on 9 November 2021) [62,63,99,100], and the pathological brain, such as the IVY glioblastoma repository [89].

As complementary approaches, several newly released methodologies, such as in situ sequencing of fixed tissue—called spatial transcriptomics [101–104]—offer important information needed to interpret the results obtained from single-cell transcriptomics. They enrich the picture by addressing each single cell in its tissue context. On this line, single-cell analyses and spatial transcriptomics are expected to be together applied to better characterize the cells that constitute the nervous system and its tumors [100].

Funding: Grant PN-II-ID-PCE-3-1041, UEFISCDI, Romania.

Acknowledgments: We acknowledge the logistic support from the WTZ bilateral projects Austria-Romania between the Medical University of Innsbruck, Innsbruck, Austria, Victor Babeș National Institute of Pathology, Bucharest, Romania and the University of Medicine and Pharmacy, Craiova, Romania.

Conflicts of Interest: The authors declare no conflict of interest.

References

1. Azevedo, F.A.C.; Carvalho, L.R.B.; Grinberg, L.T.; Farfel, J.M.; Ferretti, R.E.L.; Leite, R.E.P.; Filho, W.J.; Lent, R.; Herculano-Houzel, S. Equal numbers of neuronal and nonneuronal cells make the human brain an isometrically scaled-up primate brain. *J. Comp. Neurol.* **2009**, *513*, 532–541. [CrossRef]
2. Lent, R.; Azevedo, F.A.C.; Andrade-Moraes, C.H.; Pinto, A.V.O. How many neurons do you have? Some dogmas of quantitative neuroscience under revision. *Eur. J. Neurosci.* **2012**, *35*, 1–9. [CrossRef] [PubMed]
3. Ostrom, Q.T.; Fahmideh, M.A.; Cote, D.J.; Muskens, I.S.; Schraw, J.M.; Scheurer, M.E.; Bondy, M.L. Risk factors for childhood and adult primary brain tumors. *Neuro. Oncol.* **2019**, *21*, 1357–1375. [CrossRef] [PubMed]
4. Louis, D.N.; Perry, A.; Reifenberger, G.; von Deimling, A.; Figarella-Branger, D.; Cavenee, W.K.; Ohgaki, H.; Wiestler, O.D.; Kleihues, P.; Ellison, D.W. The 2016 World Health Organization Classification of Tumors of the Central Nervous System: A summary. *Acta Neuropathol.* **2016**, *131*, 803–820. [CrossRef]
5. Wang, Q.; Hu, B.; Hu, X.; Kim, H.; Squatrito, M.; Scarpace, L.; DeCarvalho, A.C.; Lyu, S.; Li, P.; Li, Y.; et al. Tumor Evolution of Glioma-Intrinsic Gene Expression Subtypes Associates with Immunological Changes in the Microenvironment. *Cancer Cell* **2017**, *32*, 42–56.e6. [CrossRef]
6. Prager, B.C.; Xie, Q.; Bao, S.; Rich, J.N. Cancer Stem Cells: The Architects of the Tumor Ecosystem. *Cell Stem Cell* **2019**, *24*, 41–53. [CrossRef] [PubMed]
7. Singh, S.K.; Hawkins, C.; Clarke, I.D.; Squire, J.A.; Bayani, J.; Hide, T.; Henkelman, R.M.; Cusimano, M.D.; Dirks, P.B. Identification of human brain tumour initiating cells. *Nature* **2004**, *432*, 396–401. [CrossRef]
8. Gimple, R.C.; Bhargava, S.; Dixit, D.; Rich, J.N. Glioblastoma stem cells: Lessons from the tumor hierarchy in a lethal cancer. *Genes Dev.* **2019**, *33*, 591–609. [CrossRef] [PubMed]
9. Bakken, T.E.; Hodge, R.D.; Miller, J.A.; Yao, Z.; Nguyen, T.N.; Aevermann, B.; Barkan, E.; Bertagnolli, D.; Casper, T.; Dee, N.; et al. Single-nucleus and single-cell transcriptomes compared in matched cortical cell types. *PLoS ONE* **2018**, *13*. [CrossRef]
10. Krishnaswami, S.R.; Grindberg, R.V.; Novotny, M.; Venepally, P.; Lacar, B.; Bhutani, K.; Linker, S.B.; Pham, S.; Erwin, J.A.; Miller, J.A.; et al. Using single nuclei for RNA-seq to capture the transcriptome of postmortem neurons. *Nat. Protoc.* **2016**, *11*, 499–524. [CrossRef]
11. Lake, B.B.; Codeluppi, S.; Yung, Y.C.; Gao, D.; Chun, J.; Kharchenko, P.V.; Linnarsson, S.; Zhang, K. A comparative strategy for single-nucleus and single-cell transcriptomes confirms accuracy in predicted cell-type expression from nuclear RNA. *Sci. Rep.* **2017**, *7*. [CrossRef]
12. Finlay, B.L.; Darlington, R.B. Linked regularities in the development and evolution of mammalian brains. *Science* **1995**, *268*, 1578–1584. [CrossRef]
13. Zhao, Y.; Flandin, P.; Long, J.E.; Dela Cuesta, M.; Westphal, H.; Rubenstein, J.L.R. Distinct molecular pathways of development of telencephalic interneuron subtypes revealed through analysis of Lhx6 mutants. *J. Comp. Neurol.* **2008**, *510*, 79–99. [CrossRef] [PubMed]
14. Wonders, C.P.; Anderson, S.A. The origin and specification of cortical interneurons. *Nat. Rev. Neurosci.* **2006**, *7*, 687–696. [CrossRef] [PubMed]
15. Shen, Q.; Wang, Y.; Dimos, J.T.; Fasano, C.A.; Phoenix, T.N.; Lemischka, I.R.; Ivanova, N.B.; Stifani, S.; Morrisey, E.E.; Temple, S. The timing of cortical neurogenesis is encoded within lineages of individual progenitor cells. *Nat. Neurosci.* **2006**, *9*, 743–751. [CrossRef] [PubMed]
16. Turrero García, M.; Harwell, C.C. Radial glia in the ventral telencephalon. *FEBS Lett.* **2017**, *591*, 3942–3959. [CrossRef] [PubMed]

17. Lodato, S.; Arlotta, P. Generating Neuronal Diversity in the Mammalian Cerebral Cortex. *Annu. Rev. Cell Dev. Biol.* **2015**, *31*, 699–720. [CrossRef] [PubMed]
18. Cadwell, C.R.; Bhaduri, A.; Mostajo-Radji, M.A.; Keefe, M.G.; Nowakowski, T.J. Development and Arealization of the Cerebral Cortex. *Neuron* **2019**, *103*, 980–1004. [CrossRef]
19. Li, X.-J.; Du, Z.-W.; Zarnowska, E.D.; Pankratz, M.; Hansen, L.O.; Pearce, R.A.; Zhang, S.-C. Specification of motoneurons from human embryonic stem cells. *Nat. Biotechnol.* **2005**, *23*, 215–221. [CrossRef]
20. Karbanová, J.; Missol-Kolka, E.; Fonseca, A.-V.; Lorra, C.; Janich, P.; Hollerová, H.; Jászai, J.; Ehrmann, J.; Kolář, Z.; Liebers, C.; et al. The Stem Cell Marker CD133 (Prominin-1) Is Expressed in Various Human Glandular Epithelia. *J. Histochem. Cytochem.* **2008**, *56*, 977–993. [CrossRef]
21. Fietz, S.A.; Kelava, I.; Vogt, J.; Wilsch-Bräuninger, M.; Stenzel, D.; Fish, J.L.; Corbeil, D.; Riehn, A.; Distler, W.; Nitsch, R.; et al. OSVZ progenitors of human and ferret neocortex are epithelial-like and expand by integrin signaling. *Nat. Neurosci.* **2010**, *13*, 690–699. [CrossRef]
22. Lim, L.; Mi, D.; Llorca, A.; Marín, O. Development and Functional Diversification of Cortical Interneurons. *Neuron* **2018**, *100*, 294–313. [CrossRef]
23. Nowakowski, T.J.; Pollen, A.A.; Sandoval-Espinosa, C.; Kriegstein, A.R. Transformation of the Radial Glia Scaffold Demarcates Two Stages of Human Cerebral Cortex Development. *Neuron* **2016**, *91*, 1219–1227. [CrossRef]
24. Pollen, A.A.; Nowakowski, T.J.; Chen, J.; Retallack, H.; Sandoval-Espinosa, C.; Nicholas, C.R.; Shuga, J.; Liu, S.J.; Oldham, M.C.; Diaz, A.; et al. Molecular identity of human outer radial glia during cortical development. *Cell* **2015**, *163*, 55–67. [CrossRef] [PubMed]
25. Papadopoulos, N.; Lennartsson, J. The PDGF/PDGFR pathway as a drug target. *Mol. Asp. Med.* **2018**, *62*, 75–88. [CrossRef] [PubMed]
26. Li, Y.; Muffat, J.; Omer, A.; Bosch, I.; Lancaster, M.A.; Sur, M.; Gehrke, L.; Knoblich, J.A.; Jaenisch, R. Induction of Expansion and Folding in Human Cerebral Organoids. *Cell Stem Cell* **2017**, *20*, 385–396.e3. [CrossRef]
27. Wang, L.; Hou, S.; Han, Y.-G. Hedgehog signaling promotes basal progenitor expansion and the growth and folding of the neocortex. *Nat. Neurosci.* **2016**, *19*, 888–896. [CrossRef] [PubMed]
28. Penisson, M.; Ladewig, J.; Belvindrah, R.; Francis, F. Genes and Mechanisms Involved in the Generation and Amplification of Basal Radial Glial Cells. *Front. Cell. Neurosci.* **2019**, *13*, 381. [CrossRef]
29. Blackwood, C.A. Jagged1 is Essential for Radial Glial Maintenance in the Cortical Proliferative Zone. *Neuroscience* **2019**, *413*, 230–238. [CrossRef]
30. Kriegstein, A.; Noctor, S.; Martínez-Cerdeño, V. Patterns of neural stem and progenitor cell division may underlie evolutionary cortical expansion. *Nat. Rev. Neurosci.* **2006**, *7*, 883–890. [CrossRef]
31. Andrews, M.G.; Subramanian, L.; Kriegstein, A.R. mTOR signaling regulates the morphology and migration of outer radial glia in developing human cortex. *Elife* **2020**, *9*. [CrossRef]
32. Hébert, J.M.; Fishell, G. The genetics of early telencephalon patterning: Some assembly required. *Nat. Rev. Neurosci.* **2008**, *9*, 678–685. [CrossRef]
33. Hansen, D.V.; Lui, J.H.; Flandin, P.; Yoshikawa, K.; Rubenstein, J.L.; Alvarez-Buylla, A.; Kriegstein, A.R. Non-epithelial stem cells and cortical interneuron production in the human ganglionic eminences. *Nat. Neurosci.* **2013**, *16*, 1576–1587. [CrossRef]
34. Kwan, K.Y.; Šestan, N.; Anton, E.S. Transcriptional co-regulation of neuronal migration and laminar identity in the neocortex. *Development* **2012**, *139*, 1535–1546. [CrossRef] [PubMed]
35. Bayraktar, O.A.; Fuentealba, L.C.; Alvarez-Buylla, A.; Rowitch, D.H. Astrocyte Development and Heterogeneity. *Cold Spring Harb. Perspect. Biol.* **2015**, *7*, a020362. [CrossRef] [PubMed]
36. Reillo, I.; de Juan Romero, C.; García-Cabezas, M.Á.; Borrell, V. A Role for Intermediate Radial Glia in the Tangential Expansion of the Mammalian Cerebral Cortex. *Cereb. Cortex* **2011**, *21*, 1674–1694. [CrossRef]
37. Coletti, A.M.; Singh, D.; Kumar, S.; Shafin, T.N.; Briody, P.J.; Babbitt, B.F.; Pan, D.; Norton, E.S.; Brown, E.C.; Kahle, K.T.; et al. Characterization of the ventricular-subventricular stem cell niche during human brain development. *Development* **2018**, *145*, dev170100. [CrossRef]
38. Goldman, S.A.; Kuypers, N.J. How to make an oligodendrocyte. *Development* **2015**, *142*, 3983–3995. [CrossRef] [PubMed]
39. Bergles, D.E.; Richardson, W.D. Oligodendrocyte Development and Plasticity. *Cold Spring Harb. Perspect. Biol.* **2016**, *8*, a020453. [CrossRef]
40. Kessaris, N.; Fogarty, M.; Iannarelli, P.; Grist, M.; Wegner, M.; Richardson, W.D. Competing waves of oligodendrocytes in the forebrain and postnatal elimination of an embryonic lineage. *Nat. Neurosci.* **2006**, *9*, 173–179. [CrossRef]
41. Yeung, M.S.Y.; Zdunek, S.; Bergmann, O.; Bernard, S.; Salehpour, M.; Alkass, K.; Perl, S.; Tisdale, J.; Possnert, G.; Brundin, L.; et al. Dynamics of Oligodendrocyte Generation and Myelination in the Human Brain. *Cell* **2014**, *159*, 766–774. [CrossRef]
42. Bonaguidi, M.A.; Wheeler, M.A.; Shapiro, J.S.; Stadel, R.P.; Sun, G.J.; Ming, G.; Song, H. In Vivo Clonal Analysis Reveals Self-Renewing and Multipotent Adult Neural Stem Cell Characteristics. *Cell* **2011**, *145*, 1142–1155. [CrossRef]
43. Rivers, L.E.; Young, K.M.; Rizzi, M.; Jamen, F.; Psachoulia, K.; Wade, A.; Kessaris, N.; Richardson, W.D. PDGFRA/NG2 glia generate myelinating oligodendrocytes and piriform projection neurons in adult mice. *Nat. Neurosci.* **2008**, *11*, 1392–1401. [CrossRef] [PubMed]
44. Frisén, J. Neurogenesis and Gliogenesis in Nervous System Plasticity and Repair. *Annu. Rev. Cell Dev. Biol.* **2016**, *32*, 127–141. [CrossRef] [PubMed]

45. Kempermann, G. Activity Dependency and Aging in the Regulation of Adult Neurogenesis. *Cold Spring Harb. Perspect. Biol.* **2015**, *7*. [CrossRef] [PubMed]
46. Bergmann, O.; Spalding, K.L.; Frisén, J. Adult Neurogenesis in Humans. *Cold Spring Harb. Perspect. Biol.* **2015**, *7*, a018994. [CrossRef]
47. Khakh, B.S.; Sofroniew, M.V. Diversity of astrocyte functions and phenotypes in neural circuits. *Nat. Neurosci.* **2015**, *18*, 942–952. [CrossRef]
48. Magnusson, J.P.; Göritz, C.; Tatarishvili, J.; Dias, D.O.; Smith, E.M.K.; Lindvall, O.; Kokaia, Z.; Frisén, J. A latent neurogenic program in astrocytes regulated by Notch signaling in the mouse. *Science* **2014**, *346*, 237–241. [CrossRef]
49. Nakafuku, M.; Grande, A. Neurogenesis in the Damaged Mammalian Brain. In *Patterning and Cell Type Specification in the Developing CNS and PNS*; Elsevier: Amsterdam, The Netherlands, 2013; pp. 551–608.
50. Perez-Catalan, N.A.; Doe, C.Q.; Ackerman, S.D. The role of astrocyte-mediated plasticity in neural circuit development and function. *Neural Dev.* **2021**, *16*, 1. [CrossRef] [PubMed]
51. Lledo, P.-M.; Merkle, F.T.; Alvarez-Buylla, A. Origin and function of olfactory bulb interneuron diversity. *Trends Neurosci.* **2008**, *31*, 392–400. [CrossRef]
52. Bergmann, O.; Liebl, J.; Bernard, S.; Alkass, K.; Yeung, M.S.Y.; Steier, P.; Kutschera, W.; Johnson, L.; Landén, M.; Druid, H.; et al. The Age of Olfactory Bulb Neurons in Humans. *Neuron* **2012**, *74*, 634–639. [CrossRef]
53. Ernst, A.; Alkass, K.; Bernard, S.; Salehpour, M.; Perl, S.; Tisdale, J.; Possnert, G.; Druid, H.; Frisén, J. Neurogenesis in the striatum of the adult human brain. *Cell* **2014**, *156*, 1072–1083. [CrossRef]
54. Doetsch, F.; Caillé, I.; Lim, D.A.; García-Verdugo, J.M.; Alvarez-Buylla, A. Subventricular zone astrocytes are neural stem cells in the adult mammalian brain. *Cell* **1999**, *97*, 703–716. [CrossRef]
55. Doetsch, F. The glial identity of neural stem cells. *Nat. Neurosci.* **2003**, *6*, 1127–1134. [CrossRef]
56. Bakken, T.; Cowell, L.; Aevermann, B.D.; Novotny, M.; Hodge, R.; Miller, J.A.; Lee, A.; Chang, I.; McCorrison, J.; Pulendran, B.; et al. Cell type discovery and representation in the era of high-content single cell phenotyping. *BMC Bioinform.* **2017**, *18*. [CrossRef]
57. Tang, F.; Barbacioru, C.; Wang, Y.; Nordman, E.; Lee, C.; Xu, N.; Wang, X.; Bodeau, J.; Tuch, B.B.; Siddiqui, A.; et al. mRNA-Seq whole-transcriptome analysis of a single cell. *Nat. Methods* **2009**, *6*, 377–382. [CrossRef] [PubMed]
58. Lein, E.S.; Hawrylycz, M.J.; Ao, N.; Ayres, M.; Bensinger, A.; Bernard, A.; Boe, A.F.; Boguski, M.S.; Brockway, K.S.; Byrnes, E.J.; et al. Genome-wide atlas of gene expression in the adult mouse brain. *Nature* **2007**, *445*, 168–176. [CrossRef] [PubMed]
59. Habib, N.; Avraham-Davidi, I.; Basu, A.; Burks, T.; Shekhar, K.; Hofree, M.; Choudhury, S.R.; Aguet, F.; Gelfand, E.; Ardlie, K.; et al. Massively parallel single-nucleus RNA-seq with DroNc-seq. *Nat. Methods* **2017**, *14*, 955–958. [CrossRef]
60. Zhong, S.; Zhang, S.; Fan, X.; Wu, Q.; Yan, L.; Dong, J.; Zhang, H.; Li, L.; Sun, L.; Pan, N.; et al. A single-cell RNA-seq survey of the developmental landscape of the human prefrontal cortex. *Nature* **2018**, *555*, 524–528. [CrossRef] [PubMed]
61. Darmanis, S.; Sloan, S.A.; Zhang, Y.; Enge, M.; Caneda, C.; Shuer, L.M.; Gephart, M.G.H.; Barres, B.A.; Quake, S.R. A survey of human brain transcriptome diversity at the single cell level. *Proc. Natl. Acad. Sci. USA* **2015**, *112*, 7285–7290. [CrossRef]
62. Ecker, J.R.; Geschwind, D.H.; Kriegstein, A.R.; Ngai, J.; Osten, P.; Polioudakis, D.; Regev, A.; Sestan, N.; Wickersham, I.R.; Zeng, H. The BRAIN Initiative Cell Census Consortium: Lessons Learned toward Generating a Comprehensive Brain Cell Atlas. *Neuron* **2017**, *96*, 542–557. [CrossRef]
63. Regev, A.; Teichmann, S.A.; Lander, E.S.; Amit, I.; Benoist, C.; Birney, E.; Bodenmiller, B.; Campbell, P.; Carninci, P.; Clatworthy, M.; et al. The human cell atlas. *Elife* **2017**, *6*. [CrossRef]
64. Hodge, R.D.; Bakken, T.E.; Miller, J.A.; Smith, K.A.; Barkan, E.R.; Graybuck, L.T.; Close, J.L.; Long, B.; Johansen, N.; Penn, O.; et al. Conserved cell types with divergent features in human versus mouse cortex. *Nature* **2019**, *573*, 61–68. [CrossRef]
65. Lake, B.B.; Chen, S.; Sos, B.C.; Fan, J.; Kaeser, G.E.; Yung, Y.C.; Duong, T.E.; Gao, D.; Chun, J.; Kharchenko, P.V.; et al. Integrative single-cell analysis of transcriptional and epigenetic states in the human adult brain. *Nat. Biotechnol.* **2018**, *36*, 70–80. [CrossRef]
66. Velmeshev, D.; Schirmer, L.; Jung, D.; Haeussler, M.; Perez, Y.; Mayer, S.; Bhaduri, A.; Goyal, N.; Rowitch, D.H.; Kriegstein, A.R. Single-cell genomics identifies cell type–specific molecular changes in autism. *Science (80-.)* **2019**, *364*, 685–689. [CrossRef] [PubMed]
67. Lake, B.B.; Ai, R.; Kaeser, G.E.; Salathia, N.S.; Yung, Y.C.; Liu, R.; Wildberg, A.; Gao, D.; Fung, H.L.; Chen, S.; et al. Neuronal subtypes and diversity revealed by single-nucleus RNA sequencing of the human brain. *Science (80-.)* **2016**, *352*, 1586–1590. [CrossRef]
68. Kang, H.J.; Kawasawa, Y.I.; Cheng, F.; Zhu, Y.; Xu, X.; Li, M.; Sousa, A.M.M.; Pletikos, M.; Meyer, K.A.; Sedmak, G.; et al. Spatio-temporal transcriptome of the human brain. *Nature* **2011**, *478*, 483–489. [CrossRef] [PubMed]
69. Boldog, E.; Bakken, T.E.; Hodge, R.D.; Novotny, M.; Aevermann, B.D.; Baka, J.; Bordé, S.; Close, J.L.; Diez-Fuertes, F.; Ding, S.L.; et al. Transcriptomic and morphophysiological evidence for a specialized human cortical GABAergic cell type. *Nat. Neurosci.* **2018**, *21*, 1185–1195. [CrossRef]
70. Kiselev, V.Y.; Andrews, T.S.; Hemberg, M. Challenges in unsupervised clustering of single-cell RNA-seq data. *Nat. Rev. Genet.* **2019**, *20*, 273–282. [CrossRef] [PubMed]
71. Abdelaal, T.; Michielsen, L.; Cats, D.; Hoogduin, D.; Mei, H.; Reinders, M.J.T.; Mahfouz, A. A comparison of automatic cell identification methods for single-cell RNA sequencing data. *Genome Biol.* **2019**, *20*. [CrossRef]
72. Ranjan, B.; Schmidt, F.; Sun, W.; Park, J.; Honardoost, M.A.; Tan, J.; Arul Rayan, N.; Prabhakar, S. scConsensus: Combining supervised and unsupervised clustering for cell type identification in single-cell RNA sequencing data. *BMC Bioinform.* **2021**, *22*. [CrossRef]

73. Nowakowski, T.J.; Bhaduri, A.; Pollen, A.A.; Alvarado, B.; Mostajo-Radji, M.A.; Di Lullo, E.; Haeussler, M.; Sandoval-Espinosa, C.; Liu, S.J.; Velmeshev, D.; et al. Spatiotemporal gene expression trajectories reveal developmental hierarchies of the human cortex. *Science* **2017**, *358*, 1318–1323. [CrossRef] [PubMed]
74. Couturier, C.P.; Ayyadhury, S.; Le, P.U.; Nadaf, J.; Monlong, J.; Riva, G.; Allache, R.; Baig, S.; Yan, X.; Bourgey, M.; et al. Single-cell RNA-seq reveals that glioblastoma recapitulates a normal neurodevelopmental hierarchy. *Nat. Commun.* **2020**, *11*, 3406. [CrossRef] [PubMed]
75. Tasic, B.; Menon, V.; Nguyen, T.N.; Kim, T.K.; Jarsky, T.; Yao, Z.; Levi, B.; Gray, L.T.; Sorensen, S.A.; Dolbeare, T.; et al. Adult mouse cortical cell taxonomy revealed by single cell transcriptomics. *Nat. Neurosci.* **2016**, *19*, 335–346. [CrossRef]
76. Dulken, B.W.; Leeman, D.S.; Boutet, S.C.; Hebestreit, K.; Brunet, A. Single-Cell Transcriptomic Analysis Defines Heterogeneity and Transcriptional Dynamics in the Adult Neural Stem Cell Lineage. *Cell Rep.* **2017**, *18*, 777–790. [CrossRef] [PubMed]
77. Müller, S.; Cho, A.; Liu, S.J.; Lim, D.A.; Diaz, A. CONICS integrates scRNA-seq with DNA sequencing to map gene expression to tumor sub-clones. *Bioinformatics* **2018**, *34*, 3217–3219. [CrossRef]
78. Patel, A.P.; Tirosh, I.; Trombetta, J.J.; Shalek, A.K.; Gillespie, S.M.; Wakimoto, H.; Cahill, D.P.; Nahed, B.V.; Curry, W.T.; Martuza, R.L.; et al. Single-cell RNA-seq highlights intratumoral heterogeneity in primary glioblastoma. *Science* **2014**, *344*, 1396–1401. [CrossRef]
79. Tirosh, I.; Venteicher, A.S.; Hebert, C.; Escalante, L.E.; Patel, A.P.; Yizhak, K.; Fisher, J.M.; Rodman, C.; Mount, C.; Filbin, M.G.; et al. Single-cell RNA-seq supports a developmental hierarchy in human oligodendroglioma. *Nature* **2016**, *539*, 309–313. [CrossRef]
80. Liu, C.; Sage, J.C.; Miller, M.R.; Verhaak, R.G.W.; Hippenmeyer, S.; Vogel, H.; Foreman, O.; Bronson, R.T.; Nishiyama, A.; Luo, L.; et al. Mosaic Analysis with Double Markers Reveals Tumor Cell of Origin in Glioma. *Cell* **2011**, *146*, 209–221. [CrossRef]
81. Venteicher, A.S.; Tirosh, I.; Hebert, C.; Yizhak, K.; Neftel, C.; Filbin, M.G.; Hovestadt, V.; Escalante, L.E.; Shaw, M.L.; Rodman, C.; et al. Decoupling genetics, lineages, and microenvironment in IDH-mutant gliomas by single-cell RNA-seq. *Science* **2017**, *355*. [CrossRef]
82. Brennan, C.W.; Verhaak, R.G.W.; McKenna, A.; Campos, B.; Noushmehr, H.; Salama, S.R.; Zheng, S.; Chakravarty, D.; Sanborn, J.Z.; Berman, S.H.; et al. The Somatic Genomic Landscape of Glioblastoma. *Cell* **2013**, *155*, 462–477. [CrossRef]
83. Neftel, C.; Laffy, J.; Filbin, M.G.; Hara, T.; Shore, M.E.; Rahme, G.J.; Richman, A.R.; Silverbush, D.; Shaw, M.L.; Hebert, C.M.; et al. An Integrative Model of Cellular States, Plasticity, and Genetics for Glioblastoma. *Cell* **2019**, *178*, 835–849.e21. [CrossRef] [PubMed]
84. Bhaduri, A.; Di Lullo, E.; Jung, D.; Müller, S.; Crouch, E.E.; Espinosa, C.S.; Ozawa, T.; Alvarado, B.; Spatazza, J.; Cadwell, C.R.; et al. Outer Radial Glia-like Cancer Stem Cells Contribute to Heterogeneity of Glioblastoma. *Cell Stem Cell* **2020**, *26*, 48–63.e6. [CrossRef]
85. Alcantara Llaguno, S.R.; Wang, Z.; Sun, D.; Chen, J.; Xu, J.; Kim, E.; Hatanpaa, K.J.; Raisanen, J.M.; Burns, D.K.; Johnson, J.E.; et al. Adult Lineage-Restricted CNS Progenitors Specify Distinct Glioblastoma Subtypes. *Cancer Cell* **2015**, *28*, 429–440. [CrossRef] [PubMed]
86. Friedmann-Morvinski, D.; Bushong, E.A.; Ke, E.; Soda, Y.; Marumoto, T.; Singer, O.; Ellisman, M.H.; Verma, I.M. Dedifferentiation of Neurons and Astrocytes by Oncogenes Can Induce Gliomas in Mice. *Science* **2012**, *338*, 1080–1084. [CrossRef] [PubMed]
87. Weng, Q.; Wang, J.; Wang, J.; He, D.; Cheng, Z.; Zhang, F.; Verma, R.; Xu, L.; Dong, X.; Liao, Y.; et al. Single-Cell Transcriptomics Uncovers Glial Progenitor Diversity and Cell Fate Determinants during Development and Gliomagenesis. *Cell Stem Cell* **2019**, *24*, 707–723.e8. [CrossRef] [PubMed]
88. Zong, H.; Parada, L.F.; Baker, S.J. Cell of Origin for Malignant Gliomas and Its Implication in Therapeutic Development. *Cold Spring Harb. Perspect. Biol.* **2015**, *7*, a020610. [CrossRef]
89. Puchalski, R.B.; Shah, N.; Miller, J.; Dalley, R.; Nomura, S.R.; Yoon, J.-G.; Smith, K.A.; Lankerovich, M.; Bertagnolli, D.; Bickley, K.; et al. An anatomic transcriptional atlas of human glioblastoma. *Science* **2018**, *360*, 660–663. [CrossRef] [PubMed]
90. Linkous, A.; Balamatsias, D.; Snuderl, M.; Edwards, L.; Miyaguchi, K.; Milner, T.; Reich, B.; Cohen-Gould, L.; Storaska, A.; Nakayama, Y.; et al. Modeling Patient-Derived Glioblastoma with Cerebral Organoids. *Cell Rep.* **2019**, *26*, 3203–3211.e5. [CrossRef]
91. Ogawa, J.; Pao, G.M.; Shokhirev, M.N.; Verma, I.M. Glioblastoma Model Using Human Cerebral Organoids. *Cell Rep.* **2018**, *23*, 1220–1229. [CrossRef]
92. Fujikawa, A.; Sugawara, H.; Tanaka, T.; Matsumoto, M.; Kuboyama, K.; Suzuki, R.; Tanga, N.; Ogata, A.; Masumura, M.; Noda, M. Targeting PTPRZ inhibits stem cell-like properties and tumorigenicity in glioblastoma cells. *Sci. Rep.* **2017**, *7*, 5609. [CrossRef]
93. Qin, E.Y.; Cooper, D.D.; Abbott, K.L.; Lennon, J.; Nagaraja, S.; Mackay, A.; Jones, C.; Vogel, H.; Jackson, P.K.; Monje, M. Neural Precursor-Derived Pleiotrophin Mediates Subventricular Zone Invasion by Glioma. *Cell* **2017**, *170*, 845–859.e19. [CrossRef] [PubMed]
94. Shi, Y.; Ping, Y.-F.; Zhou, W.; He, Z.-C.; Chen, C.; Bian, B.-S.-J.; Zhang, L.; Chen, L.; Lan, X.; Zhang, X.-C.; et al. Tumour-associated macrophages secrete pleiotrophin to promote PTPRZ1 signalling in glioblastoma stem cells for tumour growth. *Nat. Commun.* **2017**, *8*, 15080. [CrossRef] [PubMed]
95. Donaires, F.S.; Godoy, P.R.D.V.; Leandro, G.S.; Puthier, D.; Sakamoto-Hojo, E.T. E2F transcription factors associated with up-regulated genes in glioblastoma. *Cancer Biomarkers* **2017**, *18*, 199–208. [CrossRef] [PubMed]
96. Chong, J.-L.; Wenzel, P.L.; Sáenz-Robles, M.T.; Nair, V.; Ferrey, A.; Hagan, J.P.; Gomez, Y.M.; Sharma, N.; Chen, H.-Z.; Ouseph, M.; et al. E2f1–3 switch from activators in progenitor cells to repressors in differentiating cells. *Nature* **2009**, *462*, 930–934. [CrossRef]

97. Dong, X.; Hu, X.; Chen, J.; Hu, D.; Chen, L.-F. BRD4 regulates cellular senescence in gastric cancer cells via E2F/miR-106b/p21 axis. *Cell Death Dis.* **2018**, *9*, 203. [CrossRef] [PubMed]
98. Garcia, N.A.; González-King, H.; Grueso, E.; Sánchez, R.; Martinez-Romero, A.; Jávega, B.; O'Connor, J.E.; Simons, P.J.; Handberg, A.; Sepúlveda, P. Circulating exosomes deliver free fatty acids from the bloodstream to cardiac cells: Possible role of CD36. *PLoS ONE* **2019**, *14*, e0217546. [CrossRef] [PubMed]
99. Bakken, T.E.; Miller, J.A.; Ding, S.L.; Sunkin, S.M.; Smith, K.A.; Ng, L.; Szafer, A.; Dalley, R.A.; Royall, J.J.; Lemon, T.; et al. A comprehensive transcriptional map of primate brain development. *Nature* **2016**, *535*, 367–375. [CrossRef]
100. Yuste, R.; Hawrylycz, M.; Aalling, N.; Aguilar-Valles, A.; Arendt, D.; Arnedillo, R.A.; Ascoli, G.A.; Bielza, C.; Bokharaie, V.; Bergmann, T.B.; et al. A community-based transcriptomics classification and nomenclature of neocortical cell types. *Nat. Neurosci.* **2020**, *23*, 1456–1468. [CrossRef] [PubMed]
101. Lein, E.; Borm, L.E.; Linnarsson, S. The promise of spatial transcriptomics for neuroscience in the era of molecular cell typing. *Science* **2017**, *358*, 64–69. [CrossRef]
102. Moffitt, J.R.; Hao, J.; Bambah-Mukku, D.; Lu, T.; Dulac, C.; Zhuang, X. High-performance multiplexed fluorescence in situ hybridization in culture and tissue with matrix imprinting and clearing. *Proc. Natl. Acad. Sci. USA* **2016**, *113*, 14456–14461. [CrossRef] [PubMed]
103. Battich, N.; Stoeger, T.; Pelkmans, L. Image-based transcriptomics in thousands of single human cells at single-molecule resolution. *Nat. Methods* **2013**, *10*, 1127–1136. [CrossRef] [PubMed]
104. Wang, X.; Allen, W.E.; Wright, M.A.; Sylwestrak, E.L.; Samusik, N.; Vesuna, S.; Evans, K.; Liu, C.; Ramakrishnan, C.; Liu, J.; et al. Three-dimensional intact-tissue sequencing of single-cell transcriptional states. *Science* **2018**, *361*. [CrossRef] [PubMed]

International Journal of
Molecular Sciences

MDPI

Article

IL-10 Signaling Elicited by Nivolumab-Induced Activation of the MAP Kinase Pathway Does Not Fully Contribute to Nivolumab-Modulated Heterogeneous T Cell Responses

Taylor A. Harper [1,†], Silvia M. Bacot [1], Christie Jane Fennell [1], Rebecca L. Matthews [1,‡], Christina Zhu [1], Peng Yue [1,§], Alexander Benton [1,‖], Devira Friedman [1,¶], Adovi Akue [2], Mark A. KuKuruga [2], Shiowjen Lee [2], Tao Wang [1,*] and Gerald M. Feldman [1,*]

1 Office of Biotechnology Products, Center for Drug Evaluation and Research, Food and Drug Administration, Silver Spring, MD 20993, USA; tah9h@virginia.edu (T.A.H.); Silvia.Bacot@fda.hhs.gov (S.M.B.); ChristieJane.Fennell@fda.hhs.gov (C.J.F.); rebecca.matthews@nih.gov (R.L.M.); Christina.Zhu@fda.hhs.gov (C.Z.); shandongyuepeng@hotmail.com (P.Y.); alexander.benton@okstate.edu (A.B.); dsf2141@barnard.edu (D.F.)

2 Office of Vaccines Research & Review, Center for Biologics Evaluation and Research, U.S. Food and Drug Administration, Silver Spring, MD 20993, USA; Adovi.Akue@fda.hhs.gov (A.A.); Mark.Kukuruga@fda.hhs.gov (M.A.K.); Shiowjen.Lee@fda.hhs.gov (S.L.)

* Correspondence: tao.wang@fda.hhs.gov (T.W.); Gerald.Feldman@fda.hhs.gov (G.M.F.); Tel.: +301-402-9392 (T.W.)

† Present Affiliation: Flow Cytometry Core Facility, School of Medicine, University of Virginia, Charlottesville, VA 22903, USA.

‡ Present Affiliation: Cancer ImmunoPrevention Laboratory, Vaccine, Immunity, and Cancer Directorate, Frederick National Laboratory for Cancer Research, Frederick, MD 21702, USA.

§ Present Affiliation: Department of Gastroenterology, Perelman School of Medicine, University of Pennsylvania, Philadelphia, PA 19104, USA.

‖ Present Affiliation: Department of Biochemistry, Oklahoma State University, Stillwater, OK 74075, USA.

¶ Present Affiliation: Barnard College, New York, NY 10027, USA.

Citation: Harper, T.A.; Bacot, S.M.; Fennell, C.J.; Matthews, R.L.; Zhu, C.; Yue, P.; Benton, A.; Friedman, D.; Akue, A.; KuKuruga, M.A.; et al. IL-10 Signaling Elicited by Nivolumab-Induced Activation of the MAP Kinase Pathway Does Not Fully Contribute to Nivolumab-Modulated Heterogeneous T Cell Responses. *Int. J. Mol. Sci.* **2021**, *22*, 11848. https://doi.org/10.3390/ijms222111848

Academic Editors: Anica Dricu and Phillip Darcy

Received: 23 August 2021
Accepted: 26 October 2021
Published: 31 October 2021

Abstract: Immune checkpoint inhibitor (ICI) therapy has revolutionized anti-cancer treatment for many late-stage cancer patients. However, ICI therapy has thus far demonstrated limited efficacy for most patients, and it remains unclear why this is so. Interleukin 10 (IL-10) is a cytokine that has been recognized as a central player in cancer biology with its ability to inhibit anti-tumor T cell responses. Recent studies suggest that IL-10 might also exert some intrinsic anti-tumor T cell responses, and clinical studies using recombinant IL-10 alone or in combination with ICI are underway. This paradoxical effect of IL-10 and its underlying mechanisms impacting ICI-modulated T cell responses remain poorly understood. In this study, using an in vitro mixed lymphocyte reaction assay, we found that treatment with ICIs such as the anti-programmed cell death receptor-1 (PD-1) mAb nivolumab elicits a strong expression of IL-10. While neutralization of IL-10 signaling with an anti-IL-10 specific mAb significantly decreases the production of IFN-γ by T cells in a cohort of donor cells, the opposite effect was observed in other donor cells. Similarly, neutralization of IL-10 signaling significantly decreases the expression of T cell activation markers Ki67 and CD25, as well as the production of Granzyme B in a cohort of donor cells, whereas the opposite effect was observed in others. Furthermore, we found that nivolumab and IL-10 differentially modulate the signal transducer and activator of transcription 3 (STAT3) and AKT serine–threonine kinase pathways. Finally, we found that nivolumab activates the mitogen-activated protein kinase (MAPK) pathway, which in turn is responsible for the observed induction of IL-10 production by nivolumab. These findings provide new insights into the mechanisms underlying anti-PD-1-modulated T cell responses by IL-10, which could lead to the discovery of novel combination treatments that target IL-10 and immune checkpoint molecules.

Keywords: nivolumab; interleukin 10; T cells; cytokines; STAT3 pathway; AKT serine–threonine kinase pathway; MAP kinase pathway

1. Introduction

Immune checkpoint inhibitors (ICIs) such as anti-programmed cell death receptor-1 (PD-1) and/or programmed cell death ligand-1 (PD-L1) therapeutic monoclonal antibodies reinvigorate the anti-tumor immunological activities by reversing immune checkpoint receptor-induced immunosuppressive effects [1–4]. The U.S. Food and Drug Administration has approved several anti-PD1 mAbs (nivolumab, pembrolizumab, cemiplimab, and dostarlimab) and anti-PD-L1 mAbs (atezolizumab, durvalumab, and avelumab) to treat cancer patients at a variety of stages due to their significant clinical benefits. However, most patients either do not respond to treatment or develop resistance to the ICI immunotherapy [5–8]. The mechanisms of unresponsiveness or resistance to this type of immunotherapy are poorly understood. Therefore, identifying factors that drive or prevent an effective T cell response to ICI immunotherapy is an urgent need for understanding these resistance mechanisms, and could lead to the discovery of novel effective combination therapies.

IL-10 has been recognized as one of the most important immunosuppressive cytokines, and accumulating evidence suggests that it has pleiotropic effects on immunoregulation and inflammation, as well as being one of the most critical modulators in anti-cancer immune responses [9–11]. IL-10 is widely expressed in the tumor microenvironment by tumor cells as well as various innate immune cells such as macrophages and dendritic cells. IL-10 is also highly expressed in $CD4^+$ cells, $CD8^+$ T cells, and regulatory T cells (Tregs) [12–14]. Historically, IL-10 has been thought to exert potent pro-tumor effects [15,16] mainly due to its immunosuppressive abilities [17,18]. However, recent studies also suggest that IL-10 provides significant anti-tumor effects in several mouse models. For example, tumor immune surveillance was shown to be decreased in IL-10 knockout mice, whereas transgenic overexpression of IL-10 protected mice from carcinogenesis. Furthermore, injection of PEGylated IL-10 into MMTV/HER2 transgenic mice led to tumor rejection that was dependent on activated CD8 T cells in an IFN-γ and Granzyme B-dependent manner [16].

Data from phase I/II clinical trials demonstrated that recombinant PEGylated IL-10 alone shows some anti-tumor effects against multiple cancer types including renal carcinoma, melanoma, and breast cancer [19]. In contrast, anti-IL-10 treatment increases the efficacy of an anticancer vaccine in a subset of patients [20]. Furthermore, early clinical trials demonstrated that recombinant IL-10 significantly inhibits T cell responses in both healthy donors [21] and kidney transplant patients receiving anti-CD3 mAb induction therapy [22] but does not have significant effects on autoimmune diseases such as rheumatoid arthritis and active Crohn's disease [23,24]. Despite all of these observations, the effects of IL-10 in cancer therapy remain inconclusive due to the lack of data from large-scale clinical trials, and the scant evidence gathered from human *in vitro* testing systems.

Anecdotal evidence from animal studies suggests that IL-10 might play a role in modulating anti-PD-1/PD-L1-induced T cell responses. In a mouse ovarian cancer model, treatment with an anti-PD-1 mAb significantly increases IL-10 levels in serum and ascites. The combination of anti-IL-10 with anti-PD-1 treatment in this model significantly inhibits tumor growth compared to treatment with either component alone [25]. Consistent with these observations, blocking IL-10 increases anti-tumor T cell activity and ICI responsiveness in a chronic lymphocytic leukemia mouse model [26]. Furthermore, in a phase II clinical trial, the serum levels of IL-10 prior to treatment have been shown to be associated with better efficacy in patients treated with nivolumab [27,28]. However, these studies have not investigated whether or how IL-10 directly affects anti-PD-1-induced T cell responses. More importantly, anti-mouse PD-1 mAb treatment does not affect tumor growth when used in preclinical tumor models, indicating that these animal models do not recapitulate the heterogeneous T-cell responses of human cancers treated with anti-PD-1/PD-L1 therapy [25]. Blocking IL-10 signaling enhances anti-PD-1 induced tumor antigen-specific $CD8^+$ T cell functions in metastatic melanoma patients [15]. In addition, treatment with recombinant IL-10 enhances nivolumab-induced anti-tumor activities in a small portion of these patients [29]. Nevertheless, these heterogeneous and paradoxical effects of IL-10 highlight the need to better understand the roles of IL-10 in anti-tumor responses, its impact on nivolumab-induced T cell responses, and its underlying mechanisms in eliciting these effects, especially in a human experimental system.

In the current study, using a well-established Mixed Lymphocyte Reaction (MLR) assay that has been used for the characterization of nivolumab and pembrolizumab in non-clinical studies [30–32], we show that treatment with either nivolumab or pembrolizumab significantly increases IL-10 production, concurrent with an increased production of the immune activation cytokine IFN-γ. Blocking IL-10 signaling induces a highly heterogeneous nivolumab-induced IFN-γ production and the expression of T cell activation markers Ki67 and CD25. Furthermore, combining nivolumab with IL-10 also impacts the expression of Granzyme B in a donor-dependent manner. Mechanistically, blocking IL-10 signaling and/or nivolumab activates various downstream signaling pathways of IL-10 and PD-1, including the STAT3 and AKT pathways. Finally, we demonstrate that nivolumab activates the MAPK pathway, leading to the increased expression of IL-10. Our study demonstrates that the induction of IL-10 by anti-PD-1 immunotherapy may be one of several possible mechanisms underlying resistance to anti-PD-1 immunotherapy.

2. Results

2.1. Treatment with Nivolumab or Pembrolizumab Induces IL-10 Cytokine Production

IL-10 has been recognized as one of the most potent and multifunctional immunoregulatory cytokines that has a profound effect on anti-cancer T cell responses. To analyze whether the expression of IL-10 is associated with anti-PD-1/PD-L1 immunotherapy, we performed a cytokine profiling analysis in a human allogenic MLR system using a Luminex assay detecting four cytokines: IL-10, IFN-γ, IL-2, and TNF-α. Nivolumab significantly increased production of IL-10 with 13 of 19 donor pairs showing a more than two-fold increase compared to controls (Figure 1A and Supplementary Table S1). Consistent with previous studies [33], concomitant analyses of immune activation cytokines indicated that nivolumab treatment also significantly increased the production of IFN-γ, IL-2, and TNF-α (Figure 1B and Supplementary Table S1), of which IFN-γ has been a key cytokine in IL-10 mediated anti-tumor responses [16]. Of note, while most donor pairs showed an increased level of all cytokines tested, the extent of the increase in levels of IL-10, as well as of IFN-γ, IL-2, and TNF-α varied among donor pairs.

We then analyzed the potential correlation between the increased IL-10 production and the production of other cytokines tested. We found that there is no significant correlation between the increased IL-10 production with the increased production of IFN-γ and IL-2, and only a slight correlation with the increased production of TNF-α (Supplementary Figure S1A). Interestingly, we found a significant correlation between increased IFN-γ and increased TNF-α production, whereas there is no significant correction of the increased IL-2 production with the increased IFN-γ and TNF-α production (Supplementary Figure S1B). The lack of correlation between changes in IL-2 production and production of IFN-γ and TNF-α might be due to IL-2 acting as an essential cytokine during T cell proliferation, in which IL-2 is constantly consumed by binding to IL-2 receptors on T cells.

To further confirm the effects of anti-PD-1 mAb on IL-10 production, we analyzed the effects of another anti-PD-1 therapeutic antibody, pembrolizumab, on cytokine production. Consistent with the results observed for nivolumab, pembrolizumab treatment also significantly increased production of IL-10 (Figure 1C and Supplementary Table S2), as well as the production of IFN-γ, IL-2, and TNF-α (Figure 1D and Supplementary Table S2). The levels of expression of these cytokines induced by pembrolizumab were comparable to those induced by nivolumab (Supplementary Figure S2). Similar to nivolumab treatment, the extent of the observed increase in the level of IL-10 also varied among donor pairs.

Together, our data demonstrate that both anti-PD-1 therapeutic antibodies nivolumab and pembrolizumab increase production of IL-10, as well as IFN-γ, IL-2, and TNF-α, although the extent of the cytokine responses to anti-PD-1 mAbs is significantly heterogeneous among donors.

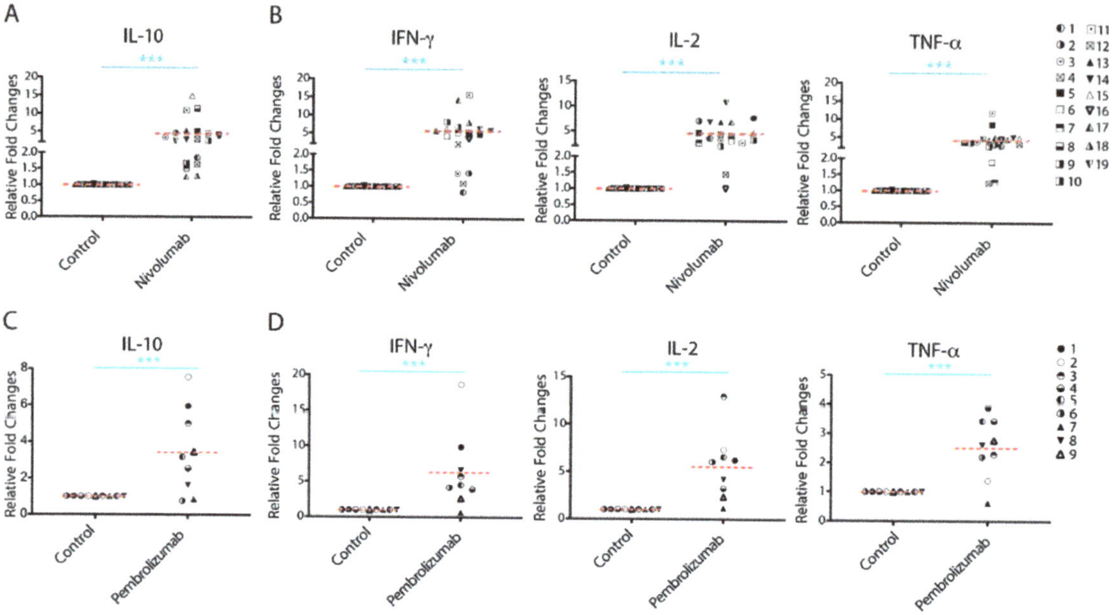

Figure 1. Treatment with nivolumab or pembrolizumab elicits potent interleukin-10 (IL-10) production from T cells. (**A**,**B**) Purified T cells were co-cultured with allogeneic matured monocyte-derived dendritic cells in the presence of nivolumab (20 µg/mL) for 5 days, after which the culture media was harvested for multiplex analysis of production of IL-10 (**A**), and interferon-γ (IFN-γ), IL-2, and tumor necrosis factor-α (TNF-α) (**B**). (**C**,**D**) Purified T cells were co-cultured with allogeneic monocyte-derived dendritic cells in the presence of pembrolizumab (20 µg/mL) for 5 days, after which the culture media was harvested for multiplex analysis of the production of IL-10 (**C**) and IFN-γ, IL-2 and TNF-α (**D**). Graphs show the relative fold changes over the no-antibody treatment controls. Each symbol represents data from one individual donor pair. Note: The representative symbols between Figure 1A,B (nivolumab treatment group) and Figure 1C,D (pembrolizumab treatment group) are not the same. Student's *t*-test, *** $p < 0.001$.

2.2. Anti-IL-10 mAb Exerts Heterogeneous Effects on Nivolumab-Induced Cytokine Production

To explore the biological outcome of anti-PD-1 treatment-induced IL-10 production, we investigated whether blockade of IL-10 signaling affects the production of other cytokines such as IFN-γ, IL-2, and TNF-α. In comparison to controls, anti-IL-10 alone did not significantly impact the production of IFN-γ, IL-2, and TNF-α (Figure 2A). However, anti-IL-10 treatment decreased the production of IFN-γ in three donor pairs (donor pairs 21, 26, and 27) and increased production of IFN-γ in two donor pairs (donor pairs 3 and 9) by more than two-fold in comparison to controls. Similarly, anti-IL-10 treatment decreased the production of IL-2 in one donor pair (donor pair 7) and the production of TNF-α in two donor pairs (donor pairs 3 and 29) by more than two-fold, respectively. Anti-IL-10 treatment increased the production of IL-2 in two donor pairs (donor pairs 3 and 9) and the production of TNF-α in two donor pairs (donor pairs 14 and 29) by more than two-fold, respectively. Of note, anti-IL-10 mAb treatment diminished subsequent IL-10 detection, suggesting that the anti-IL-10 blocking mAb interfered with the anti-IL-10 mAb detection antibody used in the Luminex assay (Figure 2B).

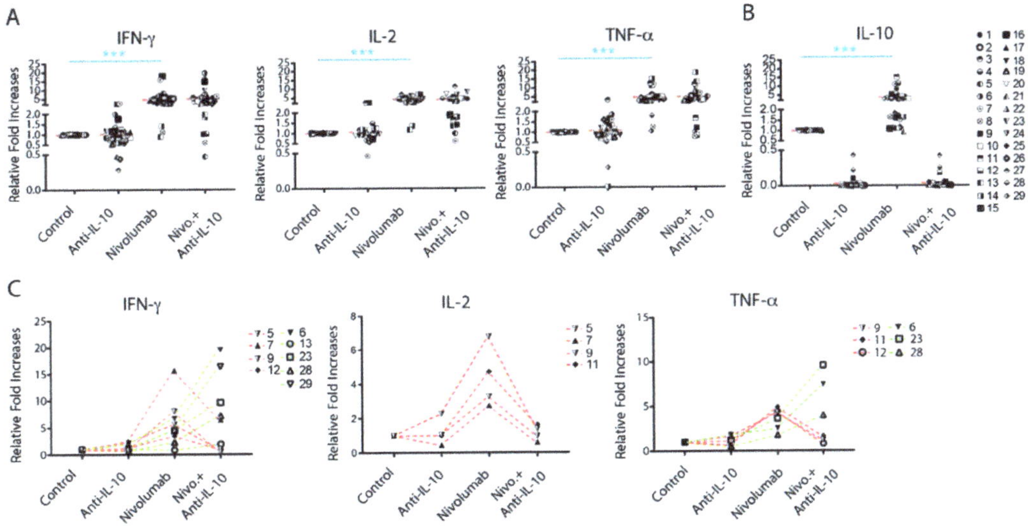

Figure 2. Anti-interleukin-10 (anti-IL-10) mAb induces heterogeneous cytokine responses. (**A**,**B**) Purified T cells were co-cultured with allogeneic matured monocyte-derived dendritic cells in the presence of nivolumab (20 μg/mL) with or without anti-IL-10 mAb (5 μg/mL) for 5 days, after which the culture media was harvested for multiplex analysis of the production of interferon-γ (IFN-γ), IL-2 and tumor necrosis factor-α (TNF-α), (**A**) and IL-10 (**B**). Graphics charting the relative fold change over the no-antibody treatment controls per donor pair. (**C**) Donor pairs with two-fold changes between nivolumab and nivolumab plus anti-IL10 mAb treatment are shown. Each symbol represents data from one individual donor pair. Student's *t*-test, *** $p < 0.001$.

Effects of anti-IL-10 mAb on nivolumab-induced cytokine production were also heterogeneous. Whereas anti-IL-10 mAb treatment on average did not have a significant impact on nivolumab-induced cytokine production (Figure 2A), anti-IL-10 mAb treatment significantly decreased IFN-γ production in four donor pairs and decreased IL-2 and IFN-γ production in three donor pairs by more than two-fold (Figure 2C). Anti-IL-10 treatment increased the nivolumab-induced production of IFN-γ in five tested donor pairs and TNF-α in three donor pairs by more than two-fold, respectively (Figure 2C). No donor pair showed a greater than two-fold increase in nivolumab-induced IL-2 production by anti-IL-10 mAb treatment (Figure 2C).

Since the level of PD-1 expression on T cells and the status of dendritic activation may potentially affect cytokine production modulated by nivolumab, we investigated the expression of PD-1 on both CD4$^+$ and CD8$^+$ T cells as well as the HLA-DR expression on dendritic cells. We found that anti-IL-10 mAb significantly, but modestly, decreased expression of PD-1 on CD4$^+$ and CD8$^+$ T cells (Supplementary Figure S3A). However, additional anti-IL-10 mAb treatment to nivolumab did not impact the expression of PD-1 in comparison to nivolumab treatment alone (Supplementary Figure S3A). Furthermore, using HLA-DR as a dendritic activation marker [34], treatment of nivolumab and/or anti-IL-10 mAb did not impact the expression of HLA-DR on dendritic cells (Supplementary Figure S3B). Of note, the variation of expression of PD-1 and HLA-DR between different donor pairs appears modest compared to that of cytokine production. Therefore, it is unlikely that the status of expression of PD-1 and dendritic cell activation contributes to the changes in cytokine production modulated by nivolumab and/or anti-IL-10 mAb.

In summary, anti-IL-10 mAb treatment showed a heterogeneous effect on nivolumab-induced cytokine production.

2.3. Anti-IL-10 mAb Exerts Heterogeneous Effects on Nivolumab-Induced T Cell Activation

To further define the potential role of IL-10 in T cell functions, we tested whether treatment with anti-IL-10 mAb affects the expression of T cell activation markers nuclear protein Ki67 (Ki67) and CD25. On average, treatment with anti-IL-10 mAb alone did not significantly affect expression of Ki67 and CD25 (Figure 3A,C). However, anti-IL-10 mAb treatment decreased Ki67 expression more than two-fold in two donor pairs for both CD4+ (donor pairs 6 and 7) and CD8+ (donor pairs 5 and 6) T cells (Figure 3A,C). Anti-IL-10 mAb treatment decreased expression of CD25 more than two-fold for both CD4+ (donor pairs 5 and 6) and CD8+ (donor pair 6) T cells (Figure 3A,C). No donor pairs showed an increased expression of Ki67 and CD25 for both CD4+ and CD8+ T cells by more than two-fold (Figure 3A,C).

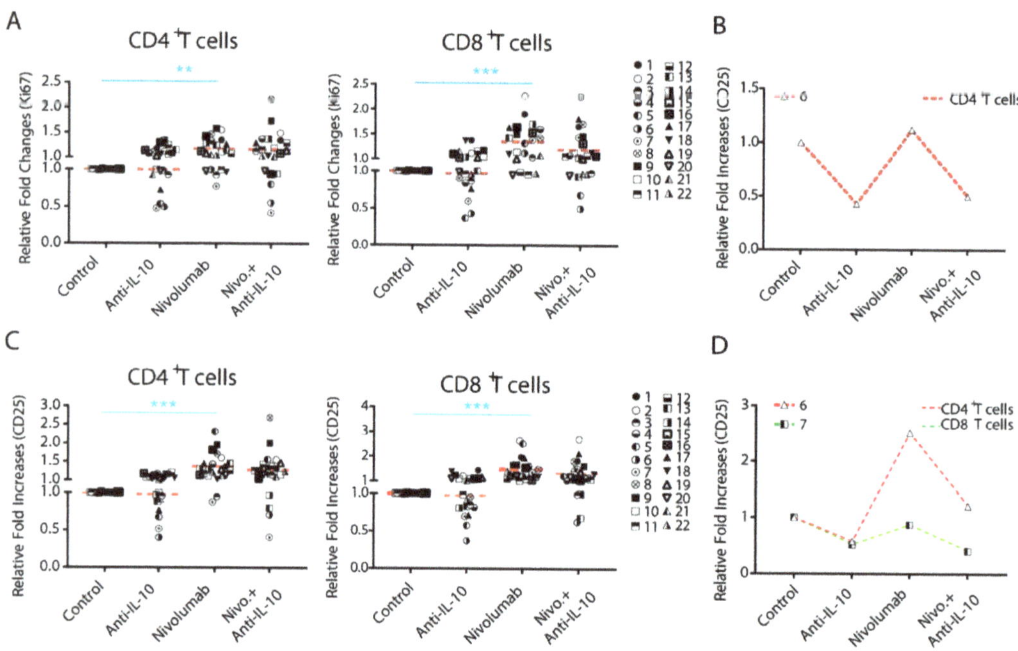

Figure 3. Anti-interleukin-10 (anti-IL-10) mAb induces heterogeneous expression of nuclear protein Ki67 (Ki67) and CD25 on CD4+ and CD8+ T cells. Purified T cells were co-cultured with allogeneic monocyte-derived dendritic cells in the presence of nivolumab (20 μg/mL) with or without anti-IL-10 mAb (5 μg/mL) for 5 days, after which cells were harvested for flow cytometry analysis of Ki67 expression. (**A**,**B**) Donor pairs showing that anti-IL-10 mAb decreased nivolumab-modulated expression of Ki67 in CD4+ and CD8+ T cells are shown (**A**). Donor pairs with two-fold changes between nivolumab and nivolumab plus anti-IL10 mAb treatment are shown (**B**). (**C**,**D**) Purified T cells were co-cultured with allogeneic monocyte-derived dendritic cells in the presence of nivolumab (20 μg/mL) with or without anti-IL-10 mAb (5 μg/mL) for 5 days, after which cells were harvested for flow cytometry analysis of CD25. Graphs show the relative fold change over the no-antibody treatment controls (**C**). Donor pairs showing that anti-IL-10 mAb decreased nivolumab-modulated expression of CD25 in CD4+ and CD8+ T cells are shown (**D**). Each symbol represents data from one individual donor pair. Student's *t*-test, ** $p < 0.01$.*** $p < 0.001$.

We next analyzed the effect of anti-IL-10 mAb treatment on nivolumab-modulated Ki67 and CD25 expression. on average, anti-IL-10 mAb treatment did not impact the expression of Ki67 and CD25 induced by nivolumab. However, the effect of anti-IL-10 mAb appears heterogeneous (Figure 3A,C). Furthermore, anti-IL-10 mAb treatment decreased expression of nivolumab-modulated Ki67 in one donor pair for CD4+ T cells by more than two fold, but did not have an impact on CD8+ T cells. Anti-IL-10 mAb treatment decreased the nivolumab-modulated expression of CD25 in one donor pair for $CD4^+$ T cells (donor pair 6) and one donor pair for $CD8^+$ T cells (donor pair 7) by more than two-fold, respectively.

Together, our data suggest that anti-IL-10 mAb displays an inhibitory effect on the expression of Ki67 and CD25 in a small subset of donor pairs, and does not have an additive effect in any of the donor pairs.

2.4. Anti-IL-10 mAb Exerts Heterogeneous Effects on Granzyme B Expression in T Cells

The expression of Granzyme B (GzmB) has been shown to be a major mechanism for IL-10 mediated anti-tumor T cell responses in a mouse tumor model [16]. We examined whether blockade of IL-10 affects the expression of Granzyme B, a functional marker for T cells. We found that anti-IL-10 mAb treatment alone on average did not significantly affect the expression of Granzyme B in either $CD4^+$ or $CD8^+$ T cells (Figure 4A). However, anti-IL-10 mAb treatment decreased the expression of Granzyme B in two donor pairs for $CD4^+$ T cells (donor pairs 2 and 13) and two donor pairs for $CD8^+$ T cells (donor pairs 1 and 3) (Figure 4A).

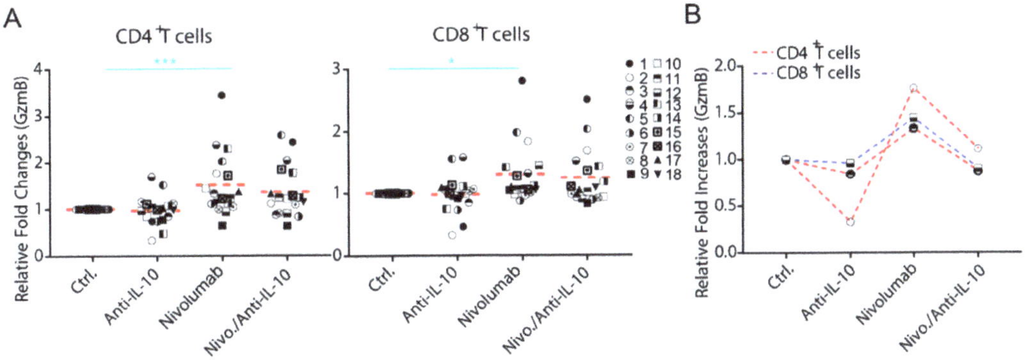

Figure 4. Anti-interleukin-10 (anti-IL-10) mAb induces heterogeneous expression of Granzyme B (GzmB) in $CD4^+$ and $CD8^+$ T cells. (**A**,**B**) Purified T cells were co-cultured with allogeneic monocyte-derived dendritic cells in the presence of nivolumab (20 μg/mL) with or without anti-IL-10 mAb (5 μg/mL) for 5 days, after which cells were harvested for flow cytometry analysis of Granzyme B expression. (**A**) Donor pairs showing that anti-IL-10 mAb decreased nivolumab-modulated expression of Granzyme B in $CD4^+$ and $CD8^+$ T cells are shown. Graphs show the relative fold change over the no-antibody treatment controls. (**B**) Donor pairs with two-fold changes between nivolumab and nivolumab plus anti-IL10 mAb treatment are shown. Each symbol represents data from an individual donor pair. Student's *t*-test, * $p < 0.05$, *** $p < 0.001$.

Whereas anti-IL-10 treatment did not significantly affect the expression of Granzyme B modulated by nivolumab, the effect of anti-IL-10 mAb appears heterogeneous (Figure 4A). Furthermore, anti-IL-10 treatment decreased nivolumab-modulated expression of Granzyme B in two donor pairs for $CD4^+$ T cell by more than two-fold (donor pairs 9 and 12) and did not impact the expression of Granzyme B in any of the donor pairs for $CD8^+$ T cells (Figure 4B).

Together, our data suggest that anti-IL-10 mAb displays an inhibitory effect on the expression of Granzyme B, in a small subset of donors, and does not have an additive effect on nivolumab-mediated Granzyme B expression in any of them.

2.5. Anti-IL-10 mAb and/or Nivolumab Display Heterogeneous Effects on Activation of the AKT and STAT3 Pathways

To determine the mechanism(s) by which nivolumab-induced IL-10 production affects the nivolumab-induced T cell responses, we examined the effect of anti-IL-10 mAb and/or nivolumab on the activation of the PD-1 downstream signaling pathways such as AKT [35–37], as well as the IL-10 downstream signaling pathway STAT3 that has been shown to be a critical pathway modulating IL-10 mediated anti-tumor T cell responses [16].

Western Blot analyses indicated that the patterns of activation of these pathways by anti-IL-10 mAb and/or nivolumab were highly heterogeneous (Figure 5 and Supplementary Figure S4). For donor pair 1, treatments of anti-IL-10 and/or nivolumab did not impact the expression of phospho-AKT or phospho-STAT3. However, for donor pair 2, anti-IL-10 mAb treatment slightly increased the expression of phospho-STAT3 but not the expression of phospho-AKT. nivolumab, however, increased expression of both phospho-AKT and phospho-STAT3 in this donor pair. Additional anti-IL-10 mAb did not have an impact on nivolumab-modulated regulation of these pathways. Lastly, for donor 3, anti-IL-10 mAb treatment slightly decreased the expression of phospho-STAT3 but did not have an impact on the expression of phospho-AKT. Nivolumab significantly increased the expression of phospho-AKT, and slightly increased the expression of phospho-STAT3. Anti-IL-10 mAb reversed nivolumab-induced expression of phospho-AKT but did not have an impact on the expression of phospho-STAT3. In summary, our data suggest that anti-IL-10 mAb and/or nivolumab exert heterogeneous responses on activation of the AKT and STAT3 signaling pathways.

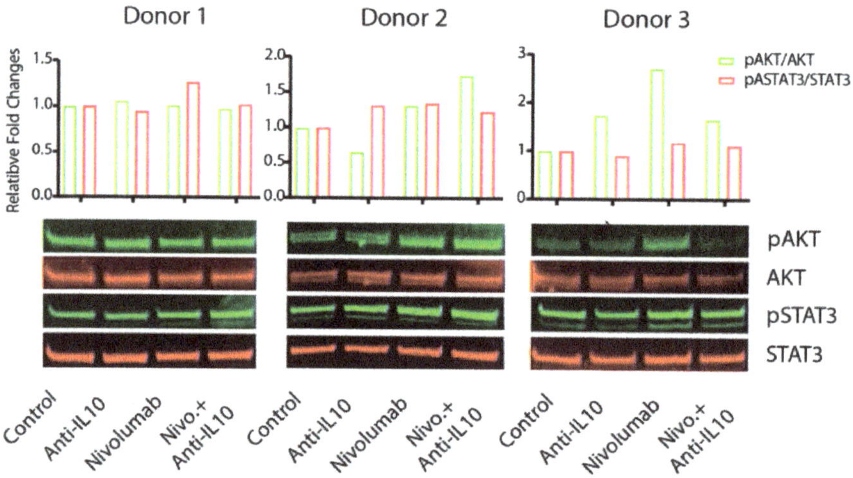

Figure 5. Anti-interleukin-10 (anti-IL-10) and nivolumab differentially affect activation of the AKT and signal transducer and activator of transcription 3 (STAT3) signaling pathways. Purified T cells were co-cultured with allogeneic monocyte-derived dendritic cells in the presence of nivolumab (20 µg/mL) with or without anti-IL-10 mAb (5 µg/mL) for 5 days, after which cells were harvested for WB analyses to determine activation of the AKT and STAT3 signaling pathways. Total AKT and total STAT3 were used as loading controls. Upper panel is the quantitative data of the Western blot results as conducted using ImageJ software. The activation of the AKT and STAT3 pathways is determined by the band intensity of pAKT (or pSTAT3) relative to the band intensity of total AKT (or STAT3). Data are from three donors, representative of at least eleven donor pairs.

2.6. *Nivolumab-Induced IL-10 Production Depends on the Activation of the MAPK Kinase Pathway*

Many factors have been shown to regulate the expression of IL-10 in T cells, but the activation of the mitogen-activated protein kinase (MAPK) pathway has been shown to be a dominant signaling pathway that regulates the expression of IL-10 in T cells [14]. Conversely, engagement of PD-L1 with PD-1 on T cells has been shown to significantly inhibit the activation of the MAPK pathway [38,39].

To further elucidate the mechanistic effect of IL-10 on nivolumab-modulated T cell responses, we first examined whether nivolumab can activate the MAPK signaling pathway in T cells. We found that nivolumab significantly increased the expression of phospho-ERK, an activation marker for the MAPK pathway in T cells, in most donors (Figure 6A and Supplementary Figure S5). Interestingly, anti-IL-10 mAb treatment also increased the expression of phospho-ERK in two donor pairs (donor pair 2 and 3), and anti-IL-10 mAb treatment exerted an additive effect on nivolumab-induced expression of phospho-ERK in donor pair 2 but did not have significant effects on donor pairs 1 or 3 (Figure 6A). We then analyzed whether nivolumab-induced activation of the MAPK pathway results in the nivolumab-induced IL-10 production by trametinib, a MEK inhibitor which has been shown to effectively and specifically block activation of the MAPK pathway in T cells [40]. We found that trametinib treatment alone significantly decreased IL-10 production and inhibited nivolumab-induced IL-10 production (Figure 6B). Furthermore, trametinib treatment abolished the expression of phospho-ERK and reversed the nivolumab-upregulated expression of phospho-ERK (Figure 6C and Supplementary Figure S6). Of note, the abolished activation of the MAPK pathway by trametinib did not induce cell death (Supplementary Figure S7), indicating that trametinib-induced inhibition of cytokine production is likely due to its inhibitory effect on cell proliferation, [33] rather than cell death.

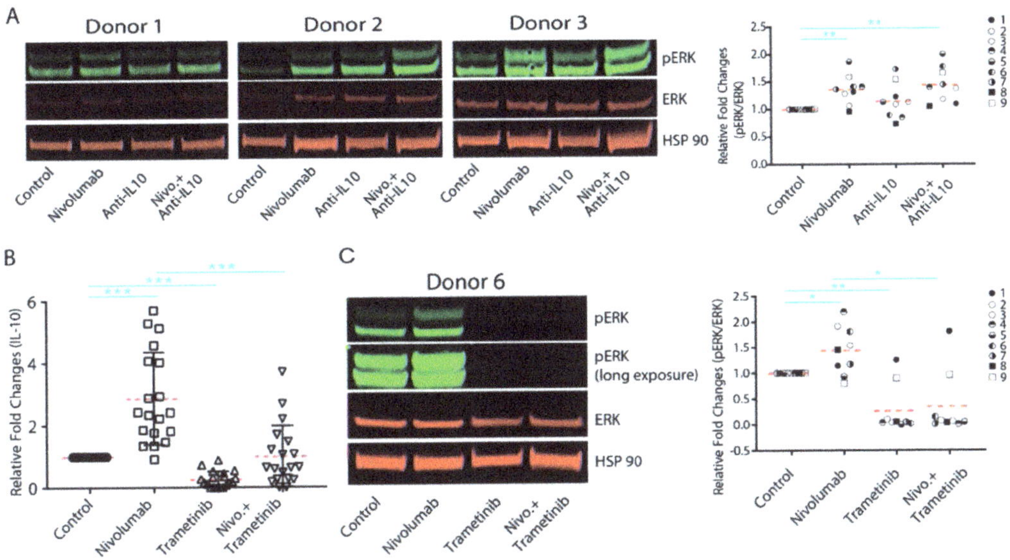

Figure 6. Nivolumab-mediated upregulation of interleukin-10 (IL-10) production depends on the activation of the mitogen-activated protein kinase (MAPK) pathway. (**A**,**B**,**C**) Purified T cells were co-cultured with allogeneic monocyte-derived dendritic cells in the presence of nivolumab (20 μg/mL) with or without anti-IL-10 mAb (5 μg/mL) for 5 days. (**A**) Cells were harvested for WB analyses for the MAPK signaling pathway. Shown are three donor pairs representative of 9 donor pairs. HSP90 and total ERK were used as loading controls (left panel). Quantitation of the Western blot results (combined with donor 4 to 9 is presented in Figure S5) was conducted using ImageJ software. The activation of the ERK pathway is determined by the band intensity of pERK divided by the band intensity of total ERK (right panel). Student's t test, ** $p < 0.01$. (**B**) Purified T cells were co-cultured with allogeneic monocyte-derived dendritic cells in the presence of nivolumab (20 μg/mL) with or without trametinib (0.2 μg/mL) for 5 days. Cell culture media were harvested for Luminex analyses of IL-10 production ($n = 19$). (**C**) Purified T cells were co-cultured with allogeneic monocyte-derived dendritic cells in the presence of nivolumab (20 μg/mL) with or without trametinib (0.2 μg/mL) for 5 days. Cells were harvested for WB analyses for the MAPK signaling pathway. Shown is a Western Blot of one donor pair representative of nine donor pairs (left panel). Quantitation of the Western blot results (combined with donor 2 to 9 is presented in Figure S6) was conducted using ImageJ software. The activation of the ERK pathway is determined by the band intensity of pERK divided by the band intensity of total ERK (right panel). Student's *t*-test, * $p < 0.05$, ** $p < 0.01$, *** $p < 0.001$.

Together, our data indicate that anti-IL-10 and/or nivolumab activates the MAPK pathway in a donor-dependent manner and that nivolumab-induced IL-10 production depends on activation of the MAPK pathway.

3. Discussion

One of many challenges for ICI immunotherapy is the low response rate coupled with patients who develop resistance to the therapy. The mechanisms underlying these challenges are not fully understood [6,8,41–44]. IL-10 is one of the most important immunoregulatory cytokines that regulates the T cell response via modulating multiple signaling pathways [12,14,15,25,45]. In the present study, we describe that nivolumab induces potent IL-10 secretion in T cells due to activation of the MAPK pathway. The addition of anti-IL-10 mAb results in a heterogeneous T cell response and modulates many downstream signaling pathways of nivolumab and IL-10 in a donor-dependent manner. Our findings facilitate a better understanding of the mechanisms by which IL-10 modulates nivolumab-induced T cell functions which may lead to potential novel combination therapies that would improve the efficacy of, or overcome the resistance to, ICI mediated immunotherapy.

Our data show that the levels of IL-10 induced by nivolumab or pembrolizumab are highly heterogeneous. Consistent with this, our recent studies also showed that nivolumab induces IL-10 production using a PBMC-based assay [45]. This is consistent with a recent report that elevated IL-10 levels caused by anti-PD-1 mAb treatment were also found in an animal tumor model [25]. These data suggest that IL-10 may confer a heterogeneous T cell response to either nivolumab or pembrolizumab, as is observed in the clinic. Indeed, although blockade of IL-10 signaling does not significantly impact the nivolumab-modulated T cell responses, our data demonstrate that these responses are induced in a donor-dependent manner when individual donor pair data were analyzed. In terms of cytokine responses, anti-IL-10 either increased or decreased production of IFN-γ and TNF-α in a subset of donor pairs, which is also in line with the clinical data (Figure 2C). However, whereas anti-IL-10 mAb increases the level of IL-2 production, there is no evidence showing that anti-IL-10 mAb increases nivolumab-induced IL-2 production (Figure 2C). Similar to IL-2, anti-IL-10 mAb only decreases and does not increase, expression of other T cell activation markers induced by nivolumab, such as Ki67, CD25 (Figure 3B,C), and Granzyme B (Figure 4B). Interestingly, blockade of IL-10 signaling alone also exerts similar heterogeneous T cell responses. Therefore, our data reflect multiple aspects observed in clinical studies and may partially explain why some cancer patients respond to PEGylated IL-10 treatment or IL-10 in combination with anti-PD-1 immunotherapy, while others do not. However, the clinical data of PEGylated IL-10 treatment with or without anti-PD-1 immunotherapy should be interpreted cautiously due to the limited patient sample size and because this trial was designed as a single-arm trial [29]. Future studies using large-scale randomized clinical trials will validate whether PEGylated IL-10 treatment does indeed enhance the efficacy of anti-PD-1 immunotherapy. Of note, and consistent with our previous studies, the cytokine responses are more sensitive to nivolumab treatment compared to other T cell activation markers analyzed by flow cytometry [33].

The engagement of IL-10 with its receptor activates multiple signaling pathways, particularly the JAK-STAT3 pathway [12,13,46], similarly, nivolumab induces the activation of the AKT pathway, resulting in the alteration of T cell responses. Our data show that the activation of several major downstream signaling pathways by anti-IL-10 and/or nivolumab is also heterogeneous. While some donor pairs showed that anti-IL-10 mAb and/or nivolumab alters the activation of the STAT3 and AKT signaling pathways (Figure 5 and Supplementary Figure S4), other donor pairs showed different activation patterns (Figure 5 and Supplementary Figure S4) indicating that neither these two known pathways alone contribute to the heterogeneous T cell responses. These data also indicate the possibility that other pathways might play a role in the heterogeneous T responses. For example, IL-10 has been shown to regulate activation of the NF-κB pathway, [13,47] Additionally, the expression levels of IL-10 receptors on T cells between different donors might also contribute to the heterogeneous responses modulated by IL-10. Future work to establish a direct connection between the activation of IL-10 and PD-1 signaling pathways and T cell responses is warranted.

Our study demonstrates that activation of the MAPK pathway not only regulates the nivolumab-induced signaling pathways, but it is also a consequence of activation of nivolumab downstream signaling pathways. Unlike activation of the AKT and STAT3 pathways, upregulation by nivolumab occurs in most donor pairs (Figure 6A,C and Supplementary Figures S4 and S5), and blocking of activation of the MAPK pathway significantly decreases IL-10 production (Figure 6 and Supplementary Figure S6), indicating that activation of the MAPK pathway might play a potential role in regulating the heterogeneous T cell responses caused by anti-IL-10 mAb and/or ICI immunotherapy. Collectively, our study suggests that the differential activation of multiple signaling pathways likely contributes to the heterogeneous T responses modulated by anti-IL-10 mAb and/or nivolumab treatment. Future work to identify the relevant signaling pathways modulating the effects of nivolumab and/or anti-IL-10 mAb on T cell responses will facilitate a greater understanding of the mechanism of action that could guide the identification of biomarker(s) for the clinical application of combination therapy of ICI and IL-10.

Although our study demonstrates the important roles of IL-10 and delineates the signaling pathways associated with anti-IL-10 and/or nivolumab on modulating nivolumab-induced T cell responses, there are several important aspects that need to be further investigated. (1) The serum levels of IL-10 prior to treatment have been shown to be associated with better outcomes for patients treated with nivolumab [27]. The reasons for this are likely due to the immunostimulatory effects of IL-10 that prevail over its immunosuppressive functions in these patients with high endogenous levels of IL-10. Nonetheless, additional IL-10 treatment might not be beneficial for these patients, and in some cases, might even potentially suppress T cell responses as demonstrated in our functional assessment in which many donor pairs show increased cytokine production by anti-IL-10 treatment in the absence or presence of nivolumab (Figure 2B,D). Therefore, it is critical to identify predictive biomarkers in patients that respond to IL-10 monotherapy alone, as well as in combination with nivolumab [15,48]. (2) The experimental system we used in this study focuses on $CD4^+$ and $CD8^+$ T cell functions. IL-10 not only plays an important role in modulating T cell functions but also plays critical roles in modulating anti-tumor or pro-tumor effects by other immune cells that are abundant in the tumor microenvironment. These immune cells include different types of helper T cells, other types of immune cells such as macrophages, NK cells, NKT cells, myeloid-derived suppressor cells (MDSC), etc., all of which have been shown to play critical roles in anti-cancer immune responses [6,49–52]. The development of an experimental system that contains these immune cells, as well as tumor cells, to better reflect the true complexity of the tumor microenvironment will be key to further delineating the mechanism of IL-10 in regulating ICI immunotherapy. Our ongoing studies of the roles of IL-10 and/or nivolumab in regulating the function of other types of helper T cells such as regulatory T cells and other types of cells such as NK cells among others are an effort to shed more light on the roles and mechanisms of IL-10 in anti-cancer immune responses. (3) Although we demonstrate that the involvement of several signaling pathways pertains to the roles of IL-10 in nivolumab-mediated T cell responses, we did not identify the specific pathway or pathways that determine the fate of heterogeneous T cell responses modulated by anti-IL-10 and/or nivolumab. Since the T cell responses modulated by anti-IL-10 and/or nivolumab are highly heterogeneous and may spatially and temporally be controlled through convergence or divergence of multiple downstream signaling pathways, future work is needed to identify the key critical signaling pathways by unbiased approaches such as next-generation sequencing (NGS) or proteomics in addition to exploring other signaling pathways modulated by PD-1 such as the SHP2 and TCR signaling pathways [35,36,53]. These studies will not only reveal additional mechanisms critically important for future clinical application, but they will also facilitate finding novel combination therapies that are more specific to targeting IL-10 or its downstream pathways, and thus able to exclusively enhance the immunostimulatory anti-tumor effects while excluding the immunosuppressive functions of IL-10.

In conclusion, our study not only illustrates how nivolumab-induced IL-10 production may shape the T cell responses via modulating multiple downstream signaling pathways, but also provides a rationale for the combination of targeting IL-10 and PD-1 for cancer patients. Future work on dissecting the mechanisms using more comprehensive approaches on how IL-10 and/or nivolumab modulate anti-cancer immune responses in the context of the tumor microenvironment, will lead to identifying biomarkers and discovering more effective anti-cancer immunotherapies targeting IL-10 and PD-1.

4. Materials and Methods

4.1. PBMCs and Monocyte Derived Dendritic Cells

This study—using human peripheral blood mononuclear cells (PBMCs) and monocytes—was reviewed and approved by both the National Institutes of Health and the Food and Drug Administration Internal Review Boards. The demographic information of donors is listed in Supplementary Table S1. Human peripheral blood mononuclear cells (PBMCs) were obtained from healthy human volunteers by leukophoresis and purified from whole blood by ficoll-hypaque sedimentation. Monocytes were purified from PBMCs by countercurrent centrifugal elutriation [54]. PBMCs and monocytes were frozen at concentrations of 50×10^6 and 20×10^6, respectively, in freezing medium containing 90% fetal bovine serum (Cat#BS3032, Valley Biomedical, Winchester, VA, USA) and 10% DMSO (Cat#D8418, Sigma-Aldrich, St. Louis, MO, USA). T cells were enriched from PBMCs using the RoboSep cell isolation platform (#19051, StemCell Technologies, Vancouver, Canada). Isolated monocytes were differentiated into dendritic cells by culturing at a concentration of 1×10^6 cells per mL in complete RPMI1640 (Cat#11875-093, Thermo Fisher Scientific, Waltham, MA, USA) containing 10% human AB serum (Cat#HP1022, Valley Biomedical), 1% Penn/Strep (#15140, Thermo Fisher Scientific), 1% HEPES (Cat#15630-080, Thermo Fisher Scientific), 0.1 mM 2-Mercaptoehanol (Cat#M6250, Sigma-Aldrich), 50 ng/mL GM-CSF (Cat#215-GM/CF, R&D Systems, Minneapolis, MN, USA), and 20 ng/mL IL-4 (Cat#204-IL/CF, R&D Systems) for 7 days as previous described [33]. On day 6 of culture, dendritic cells were matured with 100 ng/mL LPS (Cat#L2880, Sigma-Aldrich) for 24 h. All donor pairs used in this manuscript are listed in Table S1.

4.2. Mixed Lymphocyte Reaction

For the mixed lymphocyte reaction (MLR), monocyte-derived dendritic cells were harvested and resuspended at a concentration of 2×10^5 cells/mL in complete RPMI medium (Cat#11875-093, Invitrogen) containing 5% human AB serum, 1% Penn/Strep, 1% HEPES, and 0.1 mM 2-Mercaptoethanol at 1×10^5 cells/mL. Allogeneic T-cells were co-cultured with the matured dendritic cells at a concentration of 1×10^6 cells/mL in 96 well plates (Cat#3595, Corning, NY 14831 USA) for 5 days in the presence or absence of nivolumab and pembrolizumab (20 µg/mL, via McKesson Specialty Health, Scottsdale, Arizona), and functional anti-IL-10 monoclonal neutralizing antibody (Cat#AHC0103, clone JES3-9D7, Thermo Fisher Scientific) (5 µg/mL). After 5 days, cell culture supernatants were harvested for Luminex cytokine analysis, and cells were harvested for flow cytometry analysis as well as for Western blot protein analysis.

4.3. Cytokine Luminex Assay

Cytokine assays on cell culture supernatants were performed in duplicate using multiplex bead-based kits (eBioscience, San Diego, CA, USA) for the indicated cytokines as per the manufacturer's instructions. The fluorescence of beads was measured using a BioPlex 200 analyzer (Bio-Rad Laboratories, Hercules, CA, USA). Cytokine data analysis was performed using the BioPlex Manager software (v. 6.2, BioHercules, CA). Concentrations were determined using a 5-parametric logistic nonlinear regression curve-fitting algorithm.

4.4. Flow Cytometry

Cells were harvested and stained with Live/Dead Aqua (Cat#L34966, Invitrogen, Eugene, OR, USA), prepared according to the manufacturer's instructions. Once cells were washed in PBS containing 5% fetal bovine serum and 0.1% sodium azide (Cat#26628-22-8, Sigma-Aldrich, St. Louis, MO, USA), they were fixed and permeabilized using eBioscience FoxP3 Fixation/Permeabilization kit (Cat#00-5521-00, Thermo Fisher Scientific). Cells were then stained for both extracellular and intracellular antigens using fluorescence-conjugated antibodies to human CD3 (BV711 Clone UCHT1, Cat#563725, BD Biosciences, San Jose, CA 95131, USA), CD4 (FITC Clone RPA-T4, Cat#561842, BD Biosciences), CD8 (BV650 Clone RPA-T8, Cat#563821, BD Biosciences), CD25 (BV605 Clone 2A3, Cat#562660, BD Biosciences), Granzyme B (BV421 Clone GB11, Cat#563389, BD Biosciences), PD1 (BV421 Clone MIH4, Cat#562323, BD Biosciences), Ki67 (Alexa Fluor 700 Clone B56, Cat#561277, BD Biosciences), and HLA DR (AF700 Clone G46-6 Cat# 560743 BD Biosciences). Flow cytometry analysis was performed using a 5-laser BD LSR Fortessa™ flow cytometry system, and data were analyzed with FlowJo software (v. 10.7.1, BD Biosciences). For analysis of expression of HLA-DR on dendritic cells, CD3-negative cells were gated for dendritic cells.

4.5. Western Blot Analysis

T cells were gently harvested from the dendritic and T cell co-cultures that were treated with nivolumab and/or anti-IL-10 mAb for Figure 5, or MEK inhibitor trametinib (GSK1120212, 0.2 μM, Cat# HY-10999, Selleckchem, Houston, TX, USA) and/or nivolumab for Figure 6 for five days. Cells were washed twice with cold PBS and cell pellets were lysed using NuPAGE LDS sample buffer (Cat#NP0008, Invitrogen). Lysates were separated on 4–12% Tris-glycine gels (Invitrogen) and were transferred to nitrocellulose membranes. Anti phospho-STAT3 (Cat#9145S), -STAT3 (Cat#9139S), -phospho-AKT (Cat#4060L), -AKT (Cat#9272S), -phospho-ERK (Cat# 9107) and -ERK (Cat#9107) Abs were used as primary antibodies (Cell Signaling Technology). HSP90 (Cat#4877S) was used as a loading control (Cell Signaling Technology). Membranes were blocked with Odyssey Blocking Buffer (Cat#927-60001, LI-COR Biosciences, Lincoln, NE, USA) and incubated overnight at 4 °C with the respective antibody as per the manufacturer's recommendations. Donkey anti-mouse (680nm Cat#926-68072, LI-COR Biosciences) or donkey anti-rabbit IRDye (800nm Cat#926-32213, 680nm Cat#926-68023, LI-COR Biosciences at 1:10,000) were used as secondary antibodies and incubated for one hour at room temperature. Following washing, the blots were scanned and analyzed using the Odyssey Classic Imaging Detection System (LI-COR Biosciences). Quantification of WB data was performed with ImageJ software (v. 1.83e). The bands of interest were manually circled. The area and average intensity of the circled bands was calculated by the software. Similar circles in the adjacent area without bands were selected as controls. The intensity of bands was calculated by subtracting the control from the intensity of the phospho- or total targeted protein [55].

4.6. Statistical Analysis

Statistical analysis was performed using GraphPad Prism software (v. 9, GraphPad Software, San Diego, CA, USA). A two-tailed Student's t-test was used to analyze differences in cytokine production and expression of T cell activation markers among the experimental groups. Regression analysis was performed to determine the correlation between cytokine production after treatment. A p-value of <0.05 was considered to be statistically significant.

Supplementary Materials: Supplementary materials can be found at https://www.mdpi.com/article/10.3390/ijms222111848/s1.

Author Contributions: T.A.H., S.M.B., C.J.F., R.L.M., C.Z., P.Y., A.B. and D.F. performed all of the experiments; A.A. and M.A.K. helped set up flow cytometry and analyze the results; S.L. helped perform statistical analysis of data; T.W. conceived the project, analyzed the results, and wrote the manuscript; G.M.F. helped design the experiments and helped write and edit the manuscript. All authors have read and agreed to the published version of the manuscript.

Funding: This project was supported in part by the appointment of T.a.H., C.J.F. and R.L.M., C.Z., A.B., D.F. to the Research Participation Program at the Center for Drug Evaluation and Research, administered by the Oak Ridge Institute for Science and Education through an interagency agreement between the U.S. Department of Energy and the U.S. Food and Drug Administration. The project is also supported by a grant from the Office of the Chief Scientist, the U.S. Food and Drug Administration. This research received no external funding.

Institutional Review Board Statement: The study was conducted according to the guidelines of the Declaration of Helsinki and approved by the Institutional Review Board (or Ethics Committee) of both the National Institutes of Health (the Human Research Protection Program, or HRPP, protocol code 99CC0168) and the Food and Drug Administration (Research Involving Human Subjects Committee, or RIHSC, protocol code 03-117D) internal review boards.

Informed Consent Statement: All donors provided written informed consent prior to venipuncture. All donors were anonymized and no possibility of identifying donors exists.

Data Availability Statement: The datasets used and/or analyzed for this study will be made available upon reasonable request to the corresponding authors.

Conflicts of Interest: The authors declare no conflict of interest.

Disclaimer: The views in this paper have not been formally disseminated by the U.S. Food and Drug Administration and should not be construed to represent Agency opinion or policy.

References

1. Brahmer, J.R.; Pardoll, D.M. Immune checkpoint inhibitors: Making immunotherapy a reality for the treatment of lung cancer. *Cancer Immunol. Res.* **2013**, *1*, 85–91. [CrossRef] [PubMed]
2. Hodi, F.S.; O'Day, S.J.; McDermott, D.F.; Weber, R.W.; Sosman, J.A.; Haanen, J.B.; Gonzalez, R.; Robert, C.; Schadendorf, D.; Hassel, J.C.; et al. Improved survival with ipilimumab in patients with metastatic melanoma. *N. Engl. J. Med.* **2010**, *363*, 711–723. [CrossRef] [PubMed]
3. Makkouk, A.; Weiner, G.J. Cancer immunotherapy and breaking immune tolerance: New approaches to an old challenge. *Cancer Res.* **2015**, *75*, 5–10. [CrossRef] [PubMed]
4. Pardoll, D.M. The blockade of immune checkpoints in cancer immunotherapy. *Nat. Rev. Cancer* **2012**, *12*, 252–264. [CrossRef]
5. Ribas, A.; Wolchok, J.D. Cancer immunotherapy using checkpoint blockade. *Science* **2018**, *359*, 1350–1355. [CrossRef]
6. Sharma, P.; Hu-Lieskovan, S.; Wargo, J.A.; Ribas, A. Primary, Adaptive, and Acquired Resistance to Cancer Immunotherapy. *Cell* **2017**, *168*, 707–723. [CrossRef]
7. Jenkins, R.W.; Barbie, D.A.; Flaherty, K.T. Mechanisms of resistance to immune checkpoint inhibitors. *Br. J. Cancer* **2018**, *118*, 9–16. [CrossRef]
8. Pitt, J.M.; Vetizou, M.; Daillere, R.; Roberti, M.P.; Yamazaki, T.; Routy, B.; Lepage, P.; Boneca, I.G.; Chamaillard, M.; Kroemer, G.; et al. Resistance Mechanisms to Immune-Checkpoint Blockade in Cancer: Tumor-Intrinsic and -Extrinsic Factors. *Immunity* **2016**, *44*, 1255–1269. [CrossRef]
9. Ouyang, W.; O'Garra, A. IL-10 Family Cytokines IL-10 and IL-22: From Basic Science to Clinical Translation. *Immunity* **2019**, *50*, 871–891. [CrossRef] [PubMed]
10. Oft, M. IL-10: Master switch from tumor-promoting inflammation to antitumor immunity. *Cancer Immunol. Res.* **2014**, *2*, 194–199. [CrossRef]
11. Teng, M.W.; Darcy, P.K.; Smyth, M.J. Stable IL-10: A new therapeutic that promotes tumor immunity. *Cancer Cell* **2011**, *20*, 691–693. [CrossRef]
12. Mosser, D.M.; Zhang, X. Interleukin-10: New perspectives on an old cytokine. *Immunol. Rev.* **2008**, *226*, 205–218. [CrossRef]
13. Moore, K.W.; Malefyt, R.d.W.; Coffman, R.L.; O'Garra, A. Interleukin-10 and the Interleukin-10 Receptor. *Annu. Rev. Immunol.* **2001**, *19*, 683–765. [CrossRef]
14. Saraiva, M.; O'Garra, A. The regulation of IL-10 production by immune cells. *Nat. Rev. Immunol.* **2010**, *10*, 170–181. [CrossRef]
15. Sun, Z.; Fourcade, J.; Pagliano, O.; Chauvin, J.M.; Sander, C.; Kirkwood, J.M.; Zarour, H.M. IL10 and PD-1 Cooperate to Limit the Activity of Tumor-Specific CD8+ T Cells. *Cancer Res.* **2015**, *75*, 1635–1644. [CrossRef] [PubMed]
16. Mumm, J.B.; Emmerich, J.; Zhang, X.; Chan, I.; Wu, L.; Mauze, S.; Blaisdell, S.; Basham, B.; Dai, J.; Grein, J.; et al. IL-10 elicits IFNgamma-dependent tumor immune surveillance. *Cancer Cell* **2011**, *20*, 781–796. [CrossRef]
17. Dennis, K.L.; Blatner, N.R.; Gounari, F.; Khazaie, K. Current status of interleukin-10 and regulatory T-cells in cancer. *Curr. Opin. Oncol.* **2013**, *25*, 637–645. [CrossRef] [PubMed]
18. Beyer, M.; Schultze, J.L. Regulatory T cells in cancer. *Blood* **2006**, *108*, 804–811. [CrossRef] [PubMed]

19. Naing, A.; Papadopoulos, K.P.; Autio, K.A.; Ott, P.A.; Patel, M.R.; Wong, D.J.; Falchook, G.S.; Pant, S.; Whiteside, M.; Rasco, D.R.; et al. Safety, Antitumor Activity, and Immune Activation of Pegylated Recombinant Human Interleukin-10 (AM0010) in Patients With Advanced Solid Tumors. *J. Clin. Oncol.* **2016**, *34*, 3562–3569. [CrossRef] [PubMed]
20. Llopiz, D.; Ruiz, M.; Silva, L.; Sarobe, P. Enhancement of Antitumor Vaccination by Targeting Dendritic Cell-Related IL-10. *Front. Immunol.* **2018**, *9*. [CrossRef] [PubMed]
21. Chernoff, A.E.; Granowitz, E.V.; Shapiro, L.; Vannier, E.; Lonnemann, G.; Angel, J.B.; Kennedy, J.S.; Rabson, A.R.; Wolff, S.M.; Dinarello, C.A. A randomized, controlled trial of IL-10 in humans. Inhibition of inflammatory cytokine production and immune responses. *J. Immunol.* **1995**, *154*, 5492–5499. [PubMed]
22. Wissing, K.M.; Morelon, E.; Legendre, C.; De Pauw, L.; LeBeaut, A.; Grint, P.; Maniscalki, M.; Ickx, B.; Vereerstraeten, P.; Chatenoud, L.; et al. A pilot trial of recombinant human interleukin-10 in kidney transplant recipients receiving OKT3 induction therapy. *Transplantation* **1997**, *64*, 999–1006. [CrossRef]
23. Schreiber, S.; Fedorak, R.N.; Nielsen, O.H.; Wild, G.; Williams, C.N.; Nikolaus, S.; Jacyna, M.; Lashner, B.A.; Gangl, A.; Rutgeerts, P.; et al. Safety and efficacy of recombinant human interleukin 10 in chronic active Crohn's disease. Crohn's Disease IL-10 Cooperative Study Group. *Gastroenterology* **2000**, *119*, 1461–1472. [CrossRef] [PubMed]
24. Smeets, T.J.; Kraan, M.C.; Versendaal, J.; Breedveld, F.C.; Tak, P.P. Analysis of serial synovial biopsies in patients with rheumatoid arthritis: Description of a control group without clinical improvement after treatment with interleukin 10 or placebo. *J. Rheumatol.* **1999**, *26*, 2089–2093. [PubMed]
25. Lamichhane, P.; Karyampudi, L.; Shreeder, B.; Krempski, J.; Bahr, D.; Daum, J.; Kalli, K.R.; Goode, E.L.; Block, M.S.; Cannon, M.J.; et al. IL10 Release upon PD-1 Blockade Sustains Immunosuppression in Ovarian Cancer. *Cancer Res.* **2017**, *77*, 6667–6678. [CrossRef]
26. Rivas, J.R.; Liu, Y.; Alhakeem, S.S.; Eckenrode, J.M.; Marti, F.; Collard, J.P.; Zhang, Y.; Shaaban, K.A.; Muthusamy, N.; Hildebrandt, G.C.; et al. Interleukin-10 suppression enhances T-cell antitumor immunity and responses to checkpoint blockade in chronic lymphocytic leukemia. *Leukemia* **2021**, *35*, 3188–3200. [CrossRef]
27. Yamazaki, N.; Kiyohara, Y.; Uhara, H.; Iizuka, H.; Uehara, J.; Otsuka, F.; Fujisawa, Y.; Takenouchi, T.; Isei, T.; Iwatsuki, K.; et al. Cytokine biomarkers to predict antitumor responses to nivolumab suggested in a phase 2 study for advanced melanoma. *Cancer Sci.* **2017**, *108*, 1022–1031. [CrossRef]
28. Giunta, E.F.; Barra, G.; De Falco, V.; Argenziano, G.; Napolitano, S.; Vitale, P.; Zanaletti, N.; Terminiello, M.; Martinelli, E.; Morgillo, F.; et al. Baseline IFN-γ and IL-10 expression in PBMCs could predict response to PD-1 checkpoint inhibitors in advanced melanoma patients. *Sci. Rep.* **2020**, *10*, 17626. [CrossRef]
29. Naing, A.; Wong, D.J.; Infante, J.R.; Korn, W.M.; Aljumaily, R.; Papadopoulos, K.P.; Autio, K.A.; Pant, S.; Bauer, T.M.; Drakaki, A.; et al. Pegilodecakin combined with pembrolizumab or nivolumab for patients with advanced solid tumours (IVY): A multicentre, multicohort, open-label, phase 1b trial. *Lancet Oncol.* **2019**, *20*, 1544–1555. [CrossRef]
30. Selby, M.J.; Engelhardt, J.J.; Johnston, R.J.; Lu, L.S.; Han, M.; Thudium, K.; Yao, D.; Quigley, M.; Valle, J.; Wang, C.; et al. Preclinical Development of Ipilimumab and Nivolumab Combination Immunotherapy: Mouse Tumor Models, In Vitro Functional Studies, and Cynomolgus Macaque Toxicology. *PLoS ONE* **2016**, *11*, e0161779. [CrossRef]
31. Wang, C.; Thudium, K.B.; Han, M.; Wang, X.T.; Huang, H.; Feingersh, D.; Garcia, C.; Wu, Y.; Kuhne, M.; Srinivasan, M.; et al. In vitro characterization of the anti-PD-1 antibody nivolumab, BMS-936558, and in vivo toxicology in non-human primates. *Cancer Immunol. Res.* **2014**, *2*, 846–856. [CrossRef] [PubMed]
32. Lázár-Molnár, E.; Scandiuzzi, L.; Basu, I.; Quinn, T.; Sylvestre, E.; Palmieri, E.; Ramagopal, U.A.; Nathenson, S.G.; Guha, C.; Almo, S.C. Structure-guided development of a high-affinity human Programmed Cell Death-1: Implications for tumor immunotherapy. *EBioMedicine* **2017**, *17*, 30–44. [CrossRef]
33. Yue, P.; Harper, T.; Bacot, S.M.; Chowdhury, M.; Lee, S.; Akue, A.; Kukuruga, M.A.; Wang, T.; Feldman, G.M. BRAF and MEK inhibitors differentially affect nivolumab-induced T cell activation by modulating the TCR and AKT signaling pathways. *OncoImmunology* **2019**, *8*, e1512456. [CrossRef]
34. Schnorr, J.-J.; Xanthakos, S.; Keikavoussi, P.; Kämpgen, E.; ter Meulen, V.; Schneider-Schaulies, S. Induction of maturation of human blood dendritic cell precursors by measles virus is associated with immunosuppression. *Proc. Natl. Acad. Sci. USA* **1997**, *94*, 5326. [CrossRef]
35. Sharpe, A.H.; Pauken, K.E. The diverse functions of the PD1 inhibitory pathway. *Nat. Rev. Immunol.* **2018**, *18*, 153–167. [CrossRef] [PubMed]
36. Hui, E.; Cheung, J.; Zhu, J.; Su, X.; Taylor, M.J.; Wallweber, H.A.; Sasmal, D.K.; Huang, J.; Kim, J.M.; Mellman, I.; et al. T cell costimulatory receptor CD28 is a primary target for PD-1-mediated inhibition. *Science* **2017**, *355*, 1428–1433. [CrossRef] [PubMed]
37. Boussiotis, V.A. Molecular and Biochemical Aspects of the PD-1 Checkpoint Pathway. *N. Engl. J. Med.* **2016**, *375*, 1767–1778. [CrossRef]
38. Bardhan, K.; Anagnostou, T.; Boussiotis, V.A. The PD1:PD-L1/2 Pathway from Discovery to Clinical Implementation. *Front. Immunol.* **2016**, *7*. [CrossRef]
39. Dermani, F.K.; Samadi, P.; Rahmani, G.; Kohlan, A.K.; Najafi, R. PD-1/PD-L1 immune checkpoint: Potential target for cancer therapy. *J. Cell. Physiol.* **2019**, *234*, 1313–1325. [CrossRef]

40. Dushyanthen, S.; Teo, Z.L.; Caramia, F.; Savas, P.; Mintoff, C.P.; Virassamy, B.; Henderson, M.A.; Luen, S.J.; Mansour, M.; Kershaw, M.H.; et al. Agonist immunotherapy restores T cell function following MEK inhibition improving efficacy in breast cancer. *Nat. Commun.* **2017**, *8*, 606. [CrossRef]
41. Gide, T.N.; Wilmott, J.S.; Scolyer, R.A.; Long, G.V. Primary and Acquired Resistance to Immune Checkpoint Inhibitors in Metastatic Melanoma. *Clin. Cancer Res.* **2018**, *24*, 1260–1270. [CrossRef] [PubMed]
42. Syn, N.L.; Teng, M.W.L.; Mok, T.S.K.; Soo, R.A. De-novo and acquired resistance to immune checkpoint targeting. *Lancet Oncol.* **2017**, *18*, e731–e741. [CrossRef]
43. Wei, S.C.; Duffy, C.R.; Allison, J.P. Fundamental Mechanisms of Immune Checkpoint Blockade Therapy. *Cancer Discov.* **2018**, *8*, 1069–1086. [CrossRef] [PubMed]
44. Ouyang, W.; Rutz, S.; Crellin, N.K.; Valdez, P.A.; Hymowitz, S.G. Regulation and functions of the IL-10 family of cytokines in inflammation and disease. *Annu. Rev. Immunol.* **2011**, *29*, 71–109. [CrossRef]
45. Bacot, S.M.; Harper, T.A.; Matthews, R.L.; Fennell, C.J.; Akue, A.; KuKuruga, M.A.; Lee, S.; Wang, T.; Feldman, G.M. Exploring the Potential Use of a PBMC-Based Functional Assay to Identify Predictive Biomarkers for Anti-PD-1 Immunotherapy. *Int. J. Mol. Sci.* **2020**, *21*, 9023. [CrossRef] [PubMed]
46. Walter, M.R. The molecular basis of IL-10 function: From receptor structure to the onset of signaling. *Curr. Top. Microbiol. Immunol.* **2014**, *380*, 191–212. [CrossRef]
47. Pestka, S.; Krause, C.D.; Sarkar, D.; Walter, M.R.; Shi, Y.; Fisher, P.B. Interleukin-10 and Related Cytokines and Receptors. *Annu. Rev. Immunol.* **2004**, *22*, 929–979. [CrossRef]
48. Ganesan, S.; Mehnert, J. Biomarkers for Response to Immune Checkpoint Blockade. *Annu. Rev. Cancer Biol.* **2020**, *4*, 331–351. [CrossRef]
49. Chaudhry, A.; Samstein, R.M.; Treuting, P.; Liang, Y.; Pils, M.C.; Heinrich, J.-M.; Jack, R.S.; Wunderlich, F.T.; Brüning, J.C.; Müller, W.; et al. Interleukin-10 Signaling in Regulatory T Cells Is Required for Suppression of Th17 Cell-Mediated Inflammation. *Immunity* **2011**, *34*, 566–578. [CrossRef]
50. Kumar, V.; Patel, S.; Tcyganov, E.; Gabrilovich, D.I. The Nature of Myeloid-Derived Suppressor Cells in the Tumor Microenvironment. *Trends Immunol.* **2016**, *37*, 208–220. [CrossRef]
51. Parker, K.H.; Beury, D.W.; Ostrand-Rosenberg, S. Chapter Three—Myeloid-Derived Suppressor Cells: Critical Cells Driving Immune Suppression in the Tumor Microenvironment. In *Advances in Cancer Research*; Wang, X.-Y., Fisher, P.B., Eds.; Academic Press: Cambridge, MA, USA, 2015; Volume 128, pp. 95–139.
52. Sato, T.; Terai, M.; Tamura, Y.; Alexeev, V.; Mastrangelo, M.J.; Selvan, S.R. Interleukin 10 in the tumor microenvironment: A target for anticancer immunotherapy. *Immunol. Res.* **2011**, *51*, 170–182. [CrossRef] [PubMed]
53. Salmond, R.J.; Alexander, D.R. SHP2 forecast for the immune system: Fog gradually clearing. *Trends Immunol.* **2006**, *27*, 154–160. [CrossRef] [PubMed]
54. Wahl, L.M.; Katona, I.M.; Wilder, R.L.; Winter, C.C.; Haraoui, B.; Scher, I.; Wahl, S.M. Isolation of human mononuclear cell subsets by counterflow centrifugal elutriation (CCE). I. Characterization of B-lymphocyte-, T-lymphocyte-, and monocyte-enriched fractions by flow cytometric analysis. *Cell Immunol.* **1984**, *85*, 373–383. [CrossRef]
55. Yue, P.; Zhang, Y.; Mei, K.; Wang, S.; Lesigang, J.; Zhu, Y.; Dong, G.; Guo, W. Sec3 promotes the initial binary t-SNARE complex assembly and membrane fusion. *Nat. Commun.* **2017**, *8*, 14236. [CrossRef] [PubMed]

International Journal of *Molecular Sciences*

MDPI

Review

Growth Factors, PI3K/AKT/mTOR and MAPK Signaling Pathways in Colorectal Cancer Pathogenesis: Where Are We Now?

Constantin Stefani [1], Daniela Miricescu [2,*], Iulia-Ioana Stanescu-Spinu [2], Remus Iulian Nica [3], Maria Greabu [2,*], Alexandra Ripszky Totan [2] and Mariana Jinga [4]

1 Department of Family Medicine and Clinical Base, "Dr. Carol Davila' Central Military Emergency University Hospital, 051075 Bucharest, Romania; constantin.stefani@umfcd.ro
2 Department of Biochemistry, Faculty of Dental Medicine, Carol Davila University of Medicine and Pharmacy, 8 Eroii Sanitari Blvd, 050474 Bucharest, Romania; iulia.stanescu@umfcd.ro (I.-I.S.-S.); alexandra.totan@umfcd.ro (A.R.T.)
3 Surgery 2, 'Dr. Carol Davila' Central Military Emergency University Hospital, 051075 Bucharest, Romania; nica.remus@yahoo.com
4 Department of Gastroenterology, 'Dr. Carol Davila' Central Military Emergency University Hospital, 051075 Bucharest, Romania; mariana.jinga@umfcd.ro
* Correspondence: daniela.miricescu@umfcd.ro (D.M.); mariagreabu@umfcd.ro (M.G.)

Abstract: Colorectal cancer (CRC) is a predominant malignancy worldwide, being the fourth most common cause of mortality and morbidity. The CRC incidence in adolescents, young adults, and adult populations is increasing every year. In the pathogenesis of CRC, various factors are involved including diet, sedentary life, smoking, excessive alcohol consumption, obesity, gut microbiota, diabetes, and genetic mutations. The CRC tumor microenvironment (TME) involves the complex cooperation between tumoral cells with stroma, immune, and endothelial cells. Cytokines and several growth factors (GFs) will sustain CRC cell proliferation, survival, motility, and invasion. Epidermal growth factor receptor (EGFR), Insulin-like growth factor -1 receptor (IGF-1R), and Vascular Endothelial Growth Factor -A (VEGF-A) are overexpressed in various human cancers including CRC. The phosphatidylinositol 3-kinase (PI3K)/protein kinase B (AKT)/mammalian target of rapamycin (mTOR) and all the three major subfamilies of the mitogen-activated protein kinase (MAPK) signaling pathways may be activated by GFs and will further play key roles in CRC development. The main aim of this review is to present the CRC incidence, risk factors, pathogenesis, and the impact of GFs during its development. Moreover, the article describes the relationship between EGF, IGF, VEGF, GFs inhibitors, PI3K/AKT/mTOR-MAPK signaling pathways, and CRC.

Keywords: colorectal cancer; growth factors; PI3K/AKT/mTOR; MAPK; growth factor inhibitors

Citation: Stefani, C.; Miricescu, D.; Stanescu-Spinu, I.-I.; Nica, R.I.; Greabu, M.; Totan, A.R.; Jinga, M. Growth Factors, PI3K/AKT/mTOR and MAPK Signaling Pathways in Colorectal Cancer Pathogenesis: Where Are We Now?. *Int. J. Mol. Sci.* **2021**, *221*, 260. https://doi.org/10.3390/ijms221910260

Academic Editor: Anica Dricu

Received: 12 August 2021
Accepted: 20 September 2021
Published: 23 September 2021

1. Introduction

If, in 1950, colorectal cancer (CRC) was a rare malignancy, today, it became a predominant form worldwide [1]. After breast, lung, and prostate cancer, CRC is the fourth most common cause of cancer [2] and aggressive malignancy [3]. In the United States, it is the second-leading cause of death [2]. Every year, more than 1.2 million patients are diagnosed with CRC, and more than 600,000 lose the battle with this disease [4]. Worldwide, CRC is the third most common cancer, and the incidence is increasing with age [5–7]. In Europe, around 11% of CRC cases are attributed to overweight and obesity, especially visceral fat or abdominal obesity. The epidemiologic studies reported an incidence of 30–70% increased risk of CRC in obese men [8]. The most common CRC subsets are colon, proximal colon, distal colon, and rectum [9]. Since 1994, the CRC incidence in adolescents and young adults under 45 years has been increasing every year [10–13]. The statistical data published in 2014 revealed that 26% of proximal colon cancers were diagnosed in women younger than 50 years, while 56% of the cases were registered in women aged 80 years and older [14]. Compared with older CRC patients, early-onset CRC is a heterogenous group that is distinct from the clinical, pathologic, and molecular points of view [15]. Therefore, an increased incidence was observed between 49 and 50 years [16]. Kim SE reported that women are more prone to developing right-sided (proximal) colon cancer compared with men. Proximal colon cancer is a more aggressive form versus the left-sided (distal) form [17]. Depending on the mutation origin, CRC carcinomas are classified as sporadic (70%), inherited (5%), and familial (25%). Unfortunately, metastatic CRC (mCRC) is often incurable in most cases, representing 13% of all diagnosed cancers [18,19], with an overall survival rate of 13% [18,20]. Corroborating all the information received from genomic, epigenomic, transcriptomic, and microenvironment levels, CRC has molecular heterogeneity. Moreover, genomic events accumulated during carcinogenesis remain the leaders of cancer progression in the metastatic stage [21]. For early CRC, the 5-year survival rate is ~90%, but this rate decreases to 15% in the case of mCRC [22].

2. Risk Factors in CRC

Both environmental and genetic factors are involved in the etiology of CRC [23]. More than 80% of CRC cases are sporadic, as patients do not present a family history [23]. Therefore, the majority of the CRC cases (>90%) can be prevented if they are tested and screened early [24]. Several modifiable risk factors are involved in CRC pathogenesis such as diet, obesity, sedentary life, smoking, and moderate-to-heavy alcohol consumption [25]. Diet plays a pivotal role in CRC development [26,27], the consumption of unhealthy food being a significant factor in CRC development [28]. Moreover, a diet rich in red meats, processed meats, saturated animal fats, spicy foods, refined carbohydrates are associated with increased CRC development [27]. The International Agency for Research on Cancer (WHO-IARC) classified the consumption of processed meat as "carcinogenic to humans". Several compounds present in red (haem iron) and/or processed meat (nitrates and nitrites) as well as those formed during cooking will react with colorectal mucosa and promote carcinogenesis [29]. Experimental studies, performed on rodent models, reported that red meat haem iron induces lipid oxidation with the formation of 4-hydroxynonenal (HNE) from n-6 fatty acids. Aldehydes' synthesis is correlated in rats with preneoplastic lesions. In vitro, it has been observed that haem iron and aldehydes can enhance cellular inflammatory processes and cellular permeability, as well as promoting cellular DNA damage [30]. The process of meat cooking can incorporate or develop mutagens and carcinogens, which have been shown to enhance carcinogenesis. During high-temperature or open-flame meat cooking, heterocyclic amines (HCAs) and polycyclic aromatic hydrocarbons (PAHs) are formed. In meat, the most common PAH compound is benzo(a)pyrene. Cytochrome P450 enzymes activate these pro-carcinogens, which will be further converted in several metabolic pathways [31]. Moreover, N-nitroso compounds (NOC) obtained by the interaction between nitrogen oxides or nitrite with secondary amines and N-alkillamides have CRC carcinogenic properties [32]. In addition, the consumption of red meat and other animal products is conducive to trimethylamine N-oxide (TMAO) synthesis, a gut microbiota-derived metabolite of choline and L-carnitine, associated with an increased risk of CRC, cardiovascular disease, and diabetes. The correlation between TMAO and cancer is performed via inflammation, OS, DNA damage, and protein folding disruption [33].

The results from the epidemiologic and experimental studies performed in the last few decades revealed that calcium, fibers, milk, and whole grains decrease the CRC incidence, while red and processed meat increase the risk [26]. While the Western society prefers to eat red and processed meat associated with an increased cancer incidence, the Mediterranean diet is correlated with a decreased cancer incidence [34]. Smoking and a sedentary lifestyle are major risk factors for early-onset CRC [35]. Smoking cigarettes generates more than 7000 toxic chemicals, with at least 70 known carcinogens that can affect the entire human body. Carcinogens from the cigarette smoke (nitrosamines, heterocyclic amines, benzene, and polycyclic aromatic hydrocarbons) directly interact with the colorectal mucosa in two ways—by direct ingestion or through the bloodstream. Overall, smoking has a direct oncogenic effect, being correlated with CRC adenoma [36]. Moreover, passive smoking is an independent risk factor for CRC neoplasia in non- and former smokers [37].

In addition, it seems that physical activity after CRC diagnosis may reduce the risk of mortality by 38% [38]. In 2015, Baena R and co-workers published the results of the epidemiologic studies from EMBASE and PubMed-NCBI, carried out since November 2014, and revealed that obesity increases the risk of CRC by 19%, while regular physical activity reduces this risk by 24%. In addition, fish, fibers, and milk consumption reduce the risk of colon cancer [39]. Among students, the most important factors for CRC development are smoking (90.5%), excessive alcohol consumption (87.4%), family history of cancer (84.2%), and obesity (82.6%) [40]. The results of a prospective study regarding the effect of diet on CRC development were published in 2020. The study was conducted over a period of 4 years (2006–2010) and included men and women aged 40–69 years. The study revealed that consumption of 76 g/d red and processed meat and alcohol consumption increase the risk of CRC, while fibers from bread and breakfast cereals were associated with a decreased risk [41]. Ethanol is metabolized to acetaldehyde by alcohol dehydrogenases (ADH), catalase, or cytochrome P450 2E1. Aldehyde dehydrogenase further oxidases ethanal to acetate, a Group 1 carcinogen for humans. In the stomach and colon, the ethanal level is influenced by gastric colonization, by *Helicobacter pylori*, or by colonic enzymes. Furthermore, alcohol generates reactive oxygen species (ROS), leading to DNA damage and activating signaling pathways involved in inflammation, metastasis, and angiogenesis [42].

Diabetes is another risk factor for CRC [43]. An elevated body weight associated with a sedentary lifestyle plays an important role in CRC pathogenesis [43]. A link between insulin resistance (IR), hyperinsulinemia and cancer, and changes in the expression of insulin receptors and insulin growth factor (IGF) system, including IGF-I, IGF-II, has been observed. When insulin binds to IGF-1 receptor (IGF-1R) with low affinity, cell proliferation is stimulated via phosphatidylinositol 3 kinase (PI3K)/protein kinase B (AKT)/mammalian target of rapamycin (mTOR) signaling pathway [44]. Therefore, an IGF-I serum level within the upper part of the normal range has been associated with an increased risk of cancer development. In tumor cells, including in CRC and liver cancer, fetal isoforms of the IR have been observed to be increased. Leptin, a hormone produced by the adipose tissue stimulates cell growth, migration, and cytokines production by macrophages. Moreover, leptin activates proangiogenic factors, being also involved in tumor development [44]. Some cancer cells, such as those from human CRC, can locally produce IGF-II, triggering tumor proliferation and further metastatic effects [44].

Leptin and adiponectin are involved in cancer cell proliferation, invasion, and metastasis by the activation of the Janus kinase (JAKs)/signal transducer and activator transcription proteins (STATs), mitogen-activated protein kinase (MAPK), PI3K, mTOR, and the AMP-activated protein kinase (5′AMPK) signaling pathways and induce multiple dysregulations, including those of Cyclooxygenase 2 (COX-2) and mRNA expression [45].

The adipose tissue can produce pro-inflammatory cytokines (Interleukins-ILs, IL-8, IL-6, and IL-2), enzymes (lactate dehydrogenase-LDH) and tumor necrosis factor alpha (TNF-α). The lipid peroxidation process leads to 4-hydroxynonenal (4-HNE) formation, an active compound that upregulates prostaglandin E2, which is directly correlated with an increased risk of CRC development. Furthermore, 4-HNE can dysregulate cell proliferation, cell survival, differentiation, autophagy, senescence, apoptosis, and necrosis via MAPK, PI3K/AKT, and protein kinase C signaling pathways [46].

Moreover, the adipose tissue of obese patients present M1 macrophage, which will secrete tumor-promoting molecules, such as TNF-α, IL-1β, IL-6, IL-8, IL-18, IL-32, interferon (IFN)-γ, vascular endothelial growth factor (VEGF), osteopontin (OPN), tenascin C (TNC), and monocyte chemoattractant protein (MCP)-1 [47]. During cancer development, TNF-α is involved in cellular transformation, promotion, survival, proliferation, invasion, angiogenesis, and metastasis [48].

Soltani G et al. conducted a study that included 693 patients who were evaluated for adenoma/adenocarcinoma and underwent colonoscopy. The study concluded that obese and diabetic patients present an increased risk of developing adenoma versus the control group. The research group did not detect any association between obesity, diabetes, and adenocarcinoma [49]. Another important risk factor for CRC may be considered the gut microbiota disruption [50]. Diet can influence the gut microbiota through production of metabolites. Butyric acid, an important source for colonocytes, protects the colonic epithelial cells from tumorigenesis, having anti-inflammatory and antineoplastic properties. Instead, protein fermentation and bile acid deconjugation will damage the colonic cells in proinflammatory and pro-neoplastic ways, leading to increased risk of developing CRC [51]. Moreover, the initial microflora plays a key role in maintaining the survival and health of the host organism, because it can activate antitumor cytokines and reduce the production of oxygen free radicals. In CRC patients, a significant intestinal decrease in intestinal microbiota diversity versus healthy people has been observed. Moreover, intestinal microbiome dysregulation can stimulate intestinal epithelial cells to activate the nuclear factor kappa-light-chain-enhancer of activated B cells (NF-κB) pathway that will trigger an inflammation stage [52]. Dysbiosis or imbalance of gut microbiota may cause chronic inflammation, which is recognized as one of the prime causes of CRC. Therefore, gut microbiota-derived phytometabolites can eliminate gut pathogenic organisms and reduce DNA oxidative damage and pro-inflammatory mediators, regulating normal cell division and apoptosis [53].

Patients diagnosed with long-standing ulcerative colitis and Crohn's disease have an elevated risk of developing CRC [54]. Furthermore, gut microbiota has effects on the immune cells in the lamina propria, which further influence the inflammation process and subsequently CRC [55]. The fermented fibers produce butyrate, which further induces tumor cell and T-cell apoptosis, which represents the source of colonic inflammation [56]. Saturated fats or the Western diet negatively alter the gut microbiota. Instead, a diet rich in n-3 PUFA has a positive effect on gut microbiome, increasing the production of good probiotics such as *Lactobacillus* and *Bifidobacteria* and reducing *Helicobacter and Fusobacteria nucleatum* [56]. Smoking is another risk factor for CRC, especially in the case of individuals that have smoked for over 30 years [57]. Moreover, bile acid synthesis, such as cholic acid, may be strongly associated with colon cancer development [58].

3. CRC Pathogenesis

According to the Cancer Genome Atlas, three molecular types of CRC tumors, hypermutated (13%), ultra-mutated (3%), and with chromosomal instability (CIN) (84%), have been identified [59]. CIN-CRC type is associated with inactivation or loss of *Adenomatous polyposis Coli (APC)* tumor suppressor gene as an early event in neoplasia development. The hypermutated-CRC type is characterized by DNA mismatch repair (MMR) and microsatellite instability (MSI) and is often associated with wild-type *TP53* gene mutation [60]. In total, 70% of CRC adenomas are correlated with early *APC* gene mutation, which usually progress to carcinoma by acquiring *KRAS* as well as *TP53* and *SMAD4* inactivated mutations. Moreover, a small subset of sporadic CRC cases has active *BRAF* mutations [61]. Approximately 15% of CRC have MSI due to either epigenetic silencing of *MLH1* or a germline mutation in one of the mismatch repair genes *MLH1, MSH2, MSH6*, or *PMS2* [62].

More than 80% of sporadic CRC cases manifest CIN and are characterized by chromosome changes such as gains, deletions, and translocations [59]. In sporadic CRC adenomas and adenocarcinomas, *APC* gene mutations are frequently reported as being nonsense or frame shift mutations that encode for truncated APC proteins [63]. CRC adenoma–carcinomas that are observed in most human CRC cases are 84% CIN tumors with DNA somatic alteration and mutations in *APC, TP53, KRAS, SMAD4*, and *PIK3CA* genes [59]. The *KRAS* gene, also known as *Kirsten Rat Sarcoma Viral* oncogene homologue, is located on human chromosome 12, which encodes for the KRAS protein [64]. Wang D and colleagues detected in 6364 CRC tumors that *KRAS* mutation is abundant among Chinese patients [65]. Familial adenomatous polyposis (FAP), which has an increased risk of CRC progression, is mainly caused by *APC* gene mutation [66], which may account for 87% and which are causative point mutations, while 10%–15% of them are intragenic deletions and duplications [67].

Currently, *APC* gene has become one of the most frequent mutations in CRC patients with a family history of polyposis [68]. Ye ZL and co-workers published the results of a study conducted over a period of 20 years (May 1998–December 2018), which included 1190 Chinese CRC patients who had undergone clinical genetic testing. The study reported that 582 CRC patients (48.9%) had gene mutations, among whom 19.7% presented two concurrent mutations and 1.0% with three concurrent mutations. The most common gene mutations were *KRAS* (36,1%) followed by *PIK3CA* (10.2%), *NRAS* (3.9%), *BRAF* (2.9%), *HRAS* (0.9%), and epidermal growth factor receptor *(EGFR)* (0.9%). Regarding the relationship between mutation and prognosis, the study did not find any association between *KRAS/NRAS/PIK3CA/BRAF* mutations and CRC prognosis. Instead, *BRAF* mutation was associated with poor prognosis in CRC patients who received anti-EGFR therapy [69].

The study conducted by Yaeger R and colleagues reported oncogenic alterations in *KRAS, BRAF, PIK3CA, AKT1, RNF43,* and *SMAD4* in 1,134 mCRC patients with right-sided primary tumor compared with left-sided primary tumors [70].

Paerlman R and his research team investigated the rate of gene mutations in 450 patients with early-onset CRC (younger than 50 years). The study reported that 16% of patients have 75 gene mutations. In addition, from 48 patients (10.7%) with MMR-deficient tumors, 40 patients (83.3%) had at least 1 gene mutation, and, from 402 patients (89.3%) with MMR-proficient tumors, 32 patients (8%) with at least 1 gene mutation [71]. Wang Q and co-workers reported that the *PIK3CA* gene mutation was 9.55% in 440 CRC patients. They also showed worse response to first-line chemotherapy, because this mutation causes PI3K/AKT signaling pathway activation, which increases LGR5$^+$ CRC stem cells survival and proliferation [72].

In addition, activation of *KRAS* mutations has been reported in various malignancies involved in cell proliferation, anti-apoptosis, and angiogenesis. More than 40% of *KRAS* mutations have been detected in CRC [64]. The *RAS* genes family, which includes *KRAS*, *NRAS*, and *HRAS*, plays crucial roles in EGFR-activated signaling pathways [46]. *APC* gene mutations promote β-catenin dysregulation, which further activates the wingless-type (Wnt) pathway; therefore, the mechanism of polyps' formation is activated leading to cancer progression [73]. In CRC, *NRAS* mutations are shown in about 3–5% of cases, while *HRAS* mutations are negligible events [74].

Recently, in eukaryotic cells, circRNAs have been detected, which are a class of ubiquitous and abundant RNA molecules, characterized by the absence of both 5′caps and 3′tails. CircRNAs play key roles in cancer growth, metastasis, stemness, and resistance to therapy, including CRC [75]. Therefore, during CRC progression, genetic alterations occur in the initiation, transformation, and progression steps of normal colonic stem cells into neoplastic, malignant, and metastatic cells [76].

An increased risk for CRC and polyposis formation is the germ-line mutations in the exonuclease domain of DNA polymerase Pol δ and Pol ε. Moreover, two recurrent pathogenic variants, *POLE* p.L424V and *POLD1* p.S478N, have been identified in CRC family cases [77]. On the other hand, stool DNA testing is more sensitive than the fecal occult blood test. Stool DNA testing is a noninvasive procedure based on colonocytes exfoliation from malignant lesions, which are higher compared with normal tissue [78].

From the molecular point of view, CRC has been classified in 4 consensus molecular subtypes (CMS). CMS1 presents MSI status, *BRAF* mutation, increased immune cell infiltration, and upregulation of checkpoint inhibitors, while CMS2 is characterized by CIN, Wnt/MYC pathway activation, and decreased immune cells infiltration. The CMS3 subtype has *KRAS* mutation, whereas CMS4 has a mesenchymal phenotype with transforming growth factor-β (TGF-β) activation and a high rate of stromal and immune cell infiltration [79]. The TGF-β family of cytokines inhibits normal growth of epithelial cells and may promote tumorigenesis when they lose their sensitivity. After the binding of TGF-β ligands to TGF-β type I and type II receptors (TGFBR1, TGFBR2), TGF-β signaling is activated, which further phosphorylates the receptor-activated SMADs (R-SMADs), SMAD2, and SMAD3, involved in transcriptional regulation. TGF-β pathway members' mutations are common in multiple human types of malignancies including CRC. Approximately, 10% and 15% of patients with sporadic CRC cases have *SMAD4* and *TGFBR2* mutations, respectively. Moreover, *TGFBR2* mutation is particularly present in MSI tumors [80].

Calcium-activated chloride channels (CLCA) are proteins involved in chloride transport across the plasma. The CLCA family proteins have 3 subtypes (CLCA1, CLCA2, and CLCA4) that have a high degree of homology regarding sequence and functions, but with differences in tissue distributions. CLCA4 expression is downregulated in human cancers including CRC. In CRC and many other cancers, CLCA4 mutation has a decreased prevalence (0.44% of CRC) [81].

The complex cooperation between tumor cells with stroma, immune cells, and endothelial cells will constitute the tumor microenvironment (TME) [82]. TME orchestrates the tumor proliferation, immune evasion, metastasis, and chemoresistance. The stromal cells, extracellular matrix (ECM) components, and exosomes are the main TME components. Endothelial cells, cancer-associated fibroblasts, pericytes, immune cells, lymphocytes, natural killer cells, regulatory T cells (Treg), tumor-associated macrophages (TAMs), myeloid-derived suppressor cell chemokines, matrix metalloproteinases-MMPS, and integrins can be detected in CRC-TME [83]. These cells suffer dynamic changes that will sustain the progression and metastasis of CRC tumors [83]. The myeloid cells sustain the survival and proliferation of neoplastic cells by the inflammatory cytokines (IL-6, IL-1, IL-23, and IL-17A) release or may induce an adaptive anti-tumoral immunity (IL-12, interferon gamma-IFN-γ) [84]. Zhang R and co-workers conducted experimental studies (in vivo and in vitro) and reported that cancer-associated fibroblasts attract monocytes by secreting IL-8 and subsequently promote M2 polarization of macrophages correlated with the suppression of the function of natural killer cells. In addition, IL-6 secretion promotes the adhesion between monocytes in CRC cells [85].

Therefore, TME has pro-tumorigenic effects through cytokines and growth factors (GFs) that will support cancer cell proliferation, survival, motility, and invasion [82]. The presence of inflammatory cells and inflammatory mediators such as chemokines and cytokines will facilitate CRC progression. Moreover, a single cytokine can activate signaling pathways, leading to tumor progression and development [86]. IL-6 level is increased in CRC patients' serum versus those of healthy subjects. Moreover, studies performed in vitro and in vivo revealed that IL-6 stimulates the invasiveness of human CRC cells, promoting colonic tumor growth [87]. Chemokines are small proteins that can bind to G-protein-coupled receptors that are involved in tumorigenesis, metastasis, and angiogenesis. However, there are chemokines with a positive impact. For example, CC ligand 19 (CCL19), also named as macrophage inflammatory protein 3-beta (MIP-3b), inhibits tumorigenesis, metastasis, and angiogenesis, and is associated with a good prognosis of CRC patients. Studies in vivo and in vitro performed by Xu Z et al. revealed that, in CRC cases, CCL19 may block angiogenesis by inhibiting tyrosine-protein kinase Met (Met)/extracellular signal regulated kinase (ERK)/Elk-1/hypoxia-inductible factor-1 alpha (HIF-1α)/VEGF-A pathway in a CCR7-dependent pattern [88].

De la Fuente López M and his research team have evaluated the levels of chemokines (CCL2, CCL3, CCL4, CCL5, and CX3CL1), TNF-α, and VEGF, in both plasma and tissue lysates of 48 CRC Chilean patients. Chemokines, TNF-α, and VEGF levels from tissue lysate of CRC patients statistically increased compared with healthy tissue. The plasma levels of CCL2, CCL3, CCL4, TNF- α, and VEGF were detected in 32 patients with CRC and 15 heathy subjects. From all chemokines measured, only CCL3 had a statistically higher level in CRC patients' plasma. The research team observed positive correlations between the plasmatic level of CCL4 with TNF -α and VEGF, correlations that reflect poor prognosis of CRC patients. Therefore, plasmatic levels of chemokines together with TNF-α and VEGF can be used as biomarkers for CRC prognosis [89].

Macrophages support neoplastic transformation and malignant progression by ROS release, which will be conducive to DNA damage and mutation in neighboring epithelial cells. Moreover, via NF-kB pathway, commensal bacteria, and microbial products induce the secretion of inflammatory cytokines, including IL-1β, IL-6, and IL-23, which further promote the proliferation and survival of neoplastic cells and pro-tumorigenic Th-17 T cells differentiation [90]. In CRC microenvironment, TAMs shift from M1 to M2 macrophages, which are induced by Th2 cytokines [91]. M1 macrophages possess anti-tumor properties, while M2 macrophages lead to immunosuppression and tumorigenesis [91]. CRC patients with M1 macrophages infiltration at the tumor site have been observed to be correlated with a better prognosis. Unfortunately, most CRC TAMs display the M2 phenotype [92].

The epithelial–mesenchymal transition (EMT) plays an important role in the metastasis process, being involved in the interaction between the tumor cells and TME. Moreover, EMT-programmed tumor cells release inflammatory mediators that change the cellular and noncellular components of TME [93]. Moreover, cytokines released from infiltrated inflammatory cells contribute to tumor initiation by ROS and reactive nitrogen species (RNS) increased levels production, because they change the epigenetic of tumor suppressor genes. On the other hand, cytokines and chemokines sustain tumor growth in the later stage of tumorigenesis by promoting angiogenesis and suppressing the anti-tumor immune response [94].

During EMT transition, the epithelial cells lose the epithelial phenotype and acquire the mesenchymal phenotype [95]. Tumor cells, including CRC, undergo metabolic reprogramming, including glycolysis, mitochondrial energy production, lactate, and fatty acid metabolism important for the maintenance of malignant features, which will lead to a rapid proliferation rate [96]. microRNAs (miRNAs) are important regulators of CRC metabolic reprogramming, which sustain the metabolic processes after interactions with enzymes, transporters, suppressors, and oncogenes. Moreover, due to its localization in CRC epithelial cells, MiR-181a detection can be a valorous prognostic biomarker for mCRC patients, which is correlated with distant metastasis and poor overall survival [96].

Under various stimuli, EGFR signaling regulates macrophage activation. EGFR phosphorylation occurs in macrophages and will have major effects on the expression of both M1 and M2 macrophages [97]. EGFR signaling has been mostly studied within the context of epithelial cell function and has been correlated with CRC initiation and progression [97]. Besides EGFR, VEGF receptor (VEGFR) is mostly expressed in endothelial cells including CRC [98].

In addition, obesity characterized by chronic inflammation contributes to CRC progression by several mechanisms, including insulin, IGF, leptin, adiponectin, microbiome, and cytokines [99]. The most important environmental and genetic factors involved in CRC pathogenesis are presented in Figure 1.

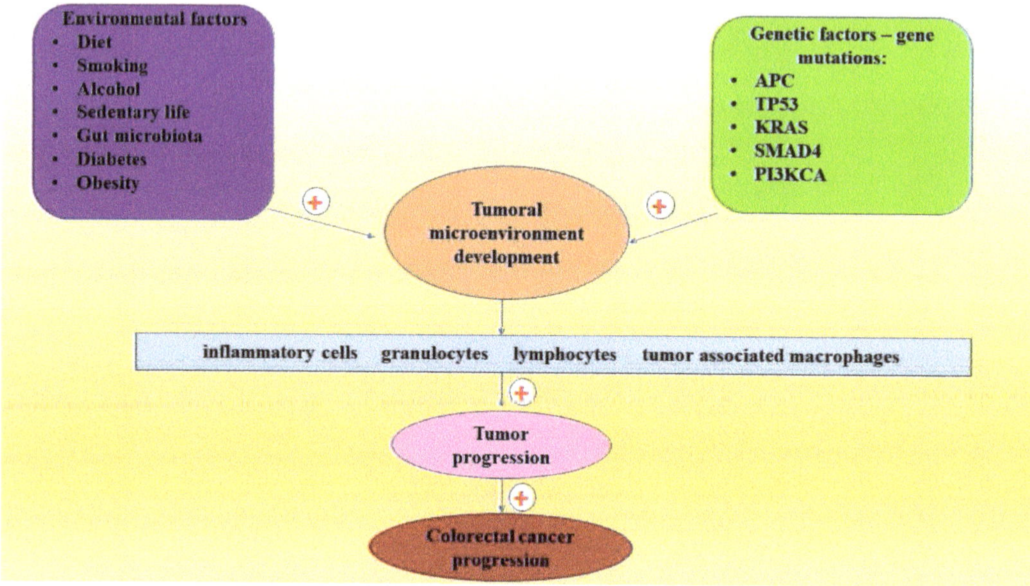

Figure 1. The environmental and genetic factors involved in colorectal cancer progression.

3.1. CRC and Insulin-Like Growth Factor Family (IGF)

The IGF family of proteins have three ligands, IGF1, IGF2, and insulin, which will bind to the following surface transmembrane receptors: IGF1R, IGF2R, and insulin receptor (IR) [100]. IGF-1 receptor (IGF-1R) is a transmembrane glycoprotein that acts as a tyrosine kinase receptor and presents two extracellular units and two cytoplasmic subunits [101]. Moreover, it is involved in many human cancers, favoring cell growth, proliferation, differentiation, apoptosis, and angiogenesis [101]. IGF-1R overexpression has been detected in CRC, pancreatic, gastric, and esophageal cancer [102]. In addition, IGF-1 and IGFBP-3 favor angiogenesis by increasing the VEGF gene transcription. An elevated serum ratio for IGF-1/insulin-like growth factor binding protein-3 (IGFBP-3) was associated with increased risk of CRC [103]. Both IGF-1 and STAT3 can induce CRC development and progression via cell-autonomous and microenvironmental effects [82]. During cancer progression and metastasis, insulin and IGF-1 have a functional role, especially in patients with hyperinsulinemia. Furthermore, insulin is able to induce mRNA expression of the matrix metalloproteinase-2 (MMP-2) by activating the signaling pathways insulin receptor substrate-1 (IRS1)/PI3K/ AKT and MAPK signaling in HCT-116 human colorectal cells [104]. Both MMP-2 and MMP-9 are involved in the regulation of the activity of cell receptors and growth factors. Moreover, MMP-2 is overexpressed in tumor tissues, including CRC [105]. The expression of IGF-1R is found in mild, longstanding inflamed colon, which will further lead to elevated levels of both mARN and protein. In these inflammatory conditions, epithelial cells may suffer pathological changes [106]. Moreover, in murine acute colitis, IGF-1-primed macrophages will suppress intestinal immune inflammation by producing IL-10 [106]. In inflammatory conditions, immune and epithelial cells release ROS and nitrogen species (RNS), which will induce DNA lesions [59]. Currently, IGF-1R has been recognized as a major determinant of cancers, while its biological roles and exact tumorigenesis mechanisms remain elusive [64]. IGF-1R plays crucial roles in mitochondrial respiratory chain regulation, which is a key element between colitis and CRC development [107]. IGF-1R, together with mesenchymal-epithelial transition (MET), is frequently overexpressed by various tumor types, including CRC [108]. Additionally, IRS-1 may present a certain association with colon cancer incidence [108].

Jiang B et al. reported that serum levels of leptin, insulin, IGF-1, and IGF-1/IGFBP3 in CRC patients were significantly elevated compared with healthy ones, while the IGFBP-3 level decreased compared with controls. These aspects suggest that serum detection of IGF-1 may be an early warning indicator [109]. To test the implication of hyperinsulinemia in CRC progression, various epidemiologic observations and experimental studies were performed [109,110]. Dietary-induced hyperinsulinemia and hypertriglyceridemia may affect the colon by producing aberrant crypt foci, a putative precursor of colon cancer [110]. Moreover, it has been observed that insulin influences the growth of the colon epithelial and carcinoma cells in vitro [110]. Hu J and his research team measured the expressions of IGF-1, ERK, GLUT4, and IRS-1 in CRC patients with metabolic patients compared with healthy controls [111]. The study concluded that the expression levels of IGF-1 and ERK were elevated in patients with metabolic syndrome with/without CRC versus the healthy controls [111]. The expression of GLUT4 was decreased in CRC patients with metabolic syndrome, compared with patients without metabolic syndrome and controls [111]. Moreover, the study observed that expression levels of ERK, IGF-1, and GLUT4 were correlated with CRC clinical characteristics, such as tumor size, distant metastasis, and advanced stages (III/IV) [111]. Peters G and co-workers detected the expression of IGF-1, IGF-2, and IGF-1R in CRC patients [112]. The expression of IGF-1 was observed in 7.5%, IGF-2 in 12.6%, while IGF-1R in 99.6% of the cases [112]. Moreover, the study detected few associations between IGF-1 and Ki-67, IGF-2, and tumor stage. In addition, IGF-2 was positively correlated with worse clinical outcomes [112]. Alagaratnam S et al. detected IGF-1Ec, an isoform of IGF-1, in 16 patients with CRC and 11 patients with colonic polyp. IGF-1EC has been identified to be overexpressed in cancers, such as prostate and neuroendocrine tumors [113]. The study revealed a significantly increased expression of IGF-1Ec in CRC patients ($p < 0.001$) and colorectal polyps ($p < 0.05$) compared with normal colonic tissues [113]. Furthermore, it has been postulated that markers of hyperinsulinemia such as IGF-1 and C-peptide may be correlated with an increased risk of CRC [114]. In addition, phosphorylated nuclear IGF-1R (nIGF-1R) is expressed in approximately 20% of mCRC and 50% of patients harboring mutations within the *BRAF* gene [115]. The inhibition of IGF-1/IGF-1R signaling will inactivate downstream AKT/mTOR signaling pathway [115].

3.2. CRC and Epidermal Growth Factor (EGF)

Dysregulation of the EGF receptor (EGFR) signaling pathway is frequently met in human cancers, including CRC [116]. The EGFR (ERB-1 or HER-1) is a member of the human EGFR (HER)-erbB family of tyrosine kinase receptors (RTKs), which includes three other members, such as HER2/C-neu (ErbB2), HER3 (ErbB3), and HER4 (ErbB4) [117]. EGFR is a glycoprotein that belongs to the ErbB family member of RTK, which presents an extracellular ligand-binding domain and an intracellular tyrosine kinase domain [118]. In the absence of specific ligands, such as EGF, TGF-α, epiregulin (EREG), betacellulin, heparin-binding EGF-like growth factor (HB-EGF), amphiregulin (AREG), epigen, heregulin, and neuregulins 1–4, EGFR is in a state of inhibition [118]. After the binding of one of the mentioned ligands to the extracellular domain, homo- or hetero-dimerization takes place, triggering the phosphorylation of the tyrosine kinase domain and activation of the RAS-RAF-MAPK signaling pathway, promoting tumor growth and progression [118]. EGFR can be found on the cell membrane surface, and its expression is elevated in cancer, moderate in adenoma, and very decreased in normal epithelia [119]. EGFR is an excellent candidate for targeted cancer therapy, being over-expressed in many types of cancers, including CRC [120]. Moreover, after binding to its receptor EGFR, EGF will activate the PI3K/AKT/mTOR signaling pathway, which is critical to cell survival, motility, and invasion [118,121]. Moreover, in CRC, EGFR mutation is rare [121]. The survival of patients with mCRC has been significantly improved with the introduction of the monoclonal antibodies that have as target EGFR [122]. The human epidermal growth factor receptor (HER-2) protein is involved in cancer cell proliferation, differentiation, and apoptosis [123,124]. HER-2 is a transmembrane tyrosine growth factor receptor that is considered to be a relevant therapeutic target in several human cancers, including CRC [123,124]. Moreover, HER-2 can be found on normal and malignant epithelial cells [123,124]. Lawan AI and co-workers explored the expression of EGFR in 54 patients with CRC carcinoma and reported that EGFR was expressed in 85.2% of the cancer cases [125]. Moreover, the study observed an association between EGFR status and depth of tumor invasion and tumor size. EGFR presence is correlated with a poor survival rate [125]. EGFR contributes to malignant behaviors of colon cancer cells in five ways—transformation of non-tumorigenic cells into tumorigenic cells, mitogenesis of polarizing colon cancer cells, cancer cells proliferation, cellular metastasis, and autophagy [126]. In addition, the tumorigenic effect of EGFR is attenuated in the presence of TGF-β signaling. Therefore, TGF-β may stimulate EGFR to create a beneficial microenvironment for metastasis [126]. Nemanqani DM and colleagues explored the expression of EGFR in 35 CRC specimens and observed its presence in 74% of the studied specimens [127]. The study also observed a higher EGFR expression mostly in grade II (85%) and stage T3 of tumors (69%) [127]. Thus, CRC EGFR over-expression could be a biomarker for an unfavorable prognosis [128].

3.3. Colorectal Cancer and Vascular Endothelial Growth Factor (VEGF)

Angiogenesis, the process of formation of new blood vessels, is fundamental for the growth of all tumor cells, including CRC [129]. VEGF is a member of the platelet-derived growth factor family that includes related glycoproteins, such as VEGF-A, VEGF-B, VEGF-C, and VEGF-D [130]. VEGF is one of the most important and specific factors that stimulate angiogenesis in both situations, physiological and pathological [131]. In addition, VEGF is excessively synthetized in epithelial, mesenchymal, and particularly in tumor cells. Additionally, elevated serum levels of IL-6, TNF-α, and VEGF are strongly associated with CRC and with the clinical stage of this disease [132]. VEGF has two receptors, VEGFR1 and VEGFR2, that act through tyrosine receptor kinases, which are implicated in angiogenesis, while VEGFR3 is involved in lymphangiogenesis [133]. However, VEGFRs are not only expressed in vascular endothelial cells, but also by the macrophages and monocytes [133].

VEGF regulates angiogenesis and vascular function. Thus, VEGF can promote angiogenesis in various pathologic conditions, including cancer, mediating endothelial cell proliferation and survival [134]. Mohamed SY et al. evaluated the expression of VEGF in 50 patients diagnosed with CRC [135]. VEGF was expressed in 70% of the cases, and presented a significant correlation with tumor size, grade, and advanced tumor stage [135]. Unfortunately, VEGF-A is correlated in CRC patients with poor clinical outcome, mainly in stages II and III [136]. Moreover, VEGF-A may be a prognostic factor in mCRC patients [137]. Jannuzzi AT et al. evaluated the VEGF single-nucleotide polymorphisms (VEGF −2578A > C, +936C > T, and −460C > T) in patients diagnosed with CRC [138]. The study illustrated that VEGF-2578A > C was significantly associated with CRC risk, while +936C > T and −460C > T genotypes did not present significant differences between CRC patients and controls [138]. Therefore, VEGF polymorphisms might play a role in CRC development [138]. VEGFA knockdown could inhibit CRC cell growth [139]. Moreover, EGFR and VEGF can be detected in CRC patients using fluorescence-Raman endoscopy [140]. In addition, VEGF-A expression in CRC tissue is associated with worse survival rate in male compared with females [141].

4. PI3K/AKT/mTOR and MAPK Signaling Pathways in Colorectal Cancer

PI3Ks are intracellular lipid kinases that are implicated in regulation of cellular proliferation, differentiation, and survival [142,143]. PI3K/AKT/mTOR signaling pathway overexpression has been reported in various cancers types, including CRC [142,143]. It is well known that PI3Ks are kinases promoting cellular proliferation [144]. Mutations that occur in *PIK3CA* gene encoding p110α catalytic subunit of PI3K have been detected in different human solid tumors, including CRC [144]. PI3K/AKT/mTOR signaling pathway plays a crucial role in cancer development including proliferation, metastasis, survival, and angiogenesis [144]. Moreover, AKT and mTOR are both downstream targets of VEGF-A [144].

Beside PI3K/AKT/mTOR signaling pathway, all the three major subfamilies of MAPK—ERK, the c-Jun N-terminal kinase or stress-activated protein kinases (JNK or SAPK), and MAPK14—are involved in CRC pathogenesis [144]. The ERK/MAPK plays a key role in cell proliferation. Moreover, the MAPK pathways are situated downstream of many GFs receptors, including EGF [144]. Therefore, the MAPK pathways are activated by various stimuli, such as peptide growth factors, cytokines, hormones, oxidative stress (OS) and endoplasmic reticulum stress, regulating cells' proliferation, differentiation, survival, and death [145]. The ERK signaling pathway plays a crucial role in tumorigenesis, including cancer cell proliferation, migration, and invasion, including in CRC [145]. In CRC tumors, the *PIK3CA* gene mutation has been identified in 10–20% of cases [146]. EGFR is a valuable therapeutic target in mCRC [147]. EGFR influences the tumorigenic cells' proliferation by activation of ERK1/ERK2, which is stimulated by Src, which further mediates a cross talk between EGFR and aryl hydrocarbons [3]. Therefore, MAPK and PI3K/AKT signaling pathways are responsible for cancer cell survival and invasion [148]. The Raf/mitogen-activated protein kinase (MEK)/ERK signaling pathways transmit signals from GFs receptors and further regulate gene expression and may prevent apoptosis [149]. After VEGF binds to VEGFR-2, phosphorylation at specific tyrosine residue occurs, and further activation of ERK1/ERK2 rapidly accelerates fibrosarcoma Raf/MEK1-MAPK, triggering increased cell proliferation [150]. mTOR pathway inhibition may induce suppression of invasion and migration of tumoral cells [151]. EGFR activates PI3K, which further catalyzes the phosphorylation of PIP_2 (phosphatidylinositol 4,5-bisphosphate) to PIP3 (phosphatidylinositol 3,4,5-triphosphate), an important second messenger involved in AKT recruitment, which activates mTOR, involved in the activation of growth, proliferation, and survival signaling responses [152]. The negative regulator of PI3K/AKT signaling cascade, Phosphatase and Tensin Homolog (PTEN), dephosphorylates PIP3 to PIP2 and is over-expressed in human colon cancer in around 60–70% patients [153,154]. In addition, increased levels of EGF trigger synthesis of hydrogen peroxide (H_2O_2), which stimulates Ribosomal protein S6 kinase beta-1 (S6K1) or p70S6K1 via the PI3K/AKT/mTOR signaling pathway, leading further to VEGF activation [155].

5. PI3K/AKT/mTOR and MAPK Signaling Pathways Inhibitors

Surgery, radiotherapy, and chemotherapy are the primary methods for treating CRC [154]. These medical techniques are accompanied by side effects, including reduced gastrointestinal function, reduced immunity, and increased pain after radio- or chemotherapy [154]. The most commonly used target drugs for CRC therapy are those that target EGFR and VEGFR [154]. Important drugs for CRC treatment include monoclonal antibodies and tyrosine kinase inhibitors that have been developed to inhibit EGFR, VEGF, and VEGFR [156]. Being involved in CRC progression, EGFR is an attractive target for therapy acting on monoclonal antibodies and the tyrosine kinase inhibitors [157]. The monoclonal antibodies used to target EGFR have been applied in mCRC treatment, with good results for patients [158]. In patients with mCRC, the anti-EGFR antibodies, cetuximab (an IgG1 recombinant human/mouse chimeric anti-EGFR mAb) and panitumumab (an IgG2κ recombinant, only human anti-EGFR mAb), have been used in several phase III clinical trials [159]. These antibodies present efficacy in terms of progression-free survival (PFS) and overall survival (OR) and are able to prolong patients' survival when are used as monotherapy or in combination with other drugs [159]. Cetuximab and panitumumab are target drugs against EGFR, while bevacizumab, ramucirumab, zivaflibercept, and regorafenib act against VEGF [160]. Among all, bevacizumab is the only VEGF-targeted agent approved by the US Food and Drug Administration (FDA) for mCRC patients [160] (Figure 2). In the case of CRC patients with extended *RAS* wild-type profiles, and with left-sided tumors, the EGFR antibodies therapy should be restricted [161]. The molecular alterations of the oncogenes such as *RAS*, *BRAF*, *PI3KCA*, and PTEN in the downstream pathway of EGFR, which activates MAPK/ERK signaling pathway, represent the novel mechanisms of resistance to anti-EGFR therapies [162]. Studies reported that among patients with CRC tumors carrying wild-type *KRAS*, *EGFR* gene copy number, mutations of *BRAF*, *PIK3CA*, or loss of PTEN expression develop resistance to anti-EGFR therapy [117]. The meta-analysis conducted by Therkildsen C and his research team demonstrated that mutations in *KRAS*, *NRAS*, *BRAF*, *PIK3CA*, and loss of PTEN will predict resistance to anti-EGFR therapies in the case of mCRC patients [163]. Canavese M et al. reported that EGFR therapy with monoclonal antibodies (cetuximab and panitumumab) improves outcomes in mCRC patients with wild-type *RAS* oncogene [164]. The treatment with the anti-EGFR moAb cetuximab activates the RAS-RAF-MEK-MAPK pathway, which is the main EGFR downstream effector [164]. Napolitano S et al. evaluated the cetuximab resistance in various human CRC models in combination with MEK inhibitors (MEKi) [165]. The in vivo and in vitro results performed on a CRC model demonstrated that the combined treatment between cetuximab and MEKi has synergic anti-proliferative and pro-apoptotic properties, combined with MAPK and PI3K/AKT/mTOR inhibition [165]. The anti-VEGF-A monoclonal antibody (Mab), bevacizumab, or Avastin was approved by the FDA for the treatment of mCRC [166]. Therefore, bevacizumab is used for solid tumor types and currently is the most widely used cancer therapeutic drug. Studies have shown that bevacizumab has a significant survival rate in patients with previously untreated mCRC when it is combined with fluoropyrimidine [166]. Furthermore, bevacizumab is the first therapy line against mCRC, demonstrating the fact that VEGF is a key mediator of tumor angiogenesis, and blocking angiogenesis is an important strategy to treat human cancer [167]. Currently, in clinical practice, EGFR is targeted by cetuximab, and VEGF by bevacizumab [140]. The detection of plasma or serum concentration of VEGF-A have been analyzed in relation to drug efficacy. The results were contradictory—after the bevacizumab treatment, the levels of serum VEGF-A may be decreased. But an elevated serum level of VEGF-A after an initial decrease has been associated with a poor response and a reactive resistance to chemotherapy with bevacizumab [168]. Bevacizumab-VEGF inhibitor, in combination with other anti-angiogenic agents (murine inhibitor) and ONC201 in both CRC xenograft and patient-derived xenograft (PDX) models, may lead to significant tumor regression or even complete tumor ablation [169]. Fruquintinib may be a promising oral drug in the CRC fight, being an active inhibitor of VEGFR-1, -2, -3 tyrosine kinases, inhibiting VEGFR-2 phosphorylation, endothelial cell proliferation, and tubule formation. Presently, it is used in China for mCRC treatment in patients that have failed at least two prior systemic antineoplastic therapies [170]. The resistance that appears in VEGFR inhibitors seems to be attributed to receptor mutations that appear in PIK3CA/AKT, ERK, HER-2, or even EGFR [171]. The effectiveness of two monoclonal antibodies, cetuximab and panitumumab, increases in combination with fluorouracil (5-FU) plus irinotecan (FOLFIRI) and 5-FU plus oxaliplatin (FOLFOX) by acting on EGFR, leading to RAS-RAF-MEK-ERK signaling pathway inhibition in mCRC patients [172]. 5-FU has been used in the medical practice for the management of CRC for decades and is now utilized in combination with other chemotherapeutic agents that may activate MAPK [173]. CRC patients may develop resistance to chemotherapeutic drugs, including cisplatin, irinotecan, and 5-FU, due to MAPK, p38α MAPK being a mediator of resistance [174]. p38 MAPKs have a dual role—they may mediate cell survival or promote cell death through different mechanisms [174]. Based on these aspects, the CRC growth in vitro and in preclinical models is significantly reduced by the combination of the following drugs, such as p38α inhibitors (SB202190, SB203580, and BIRB796), autophagy inhibitors (3MA and bafilomycin), MEK inhibitors (PD98059, UO126, and CI-1040), *HER2* inhibitors (lapatinib), multi-kinase inhibitors (sorafenib), or chemotherapeutic agents (5-FU, irinotecan, and cisplatin), which promote a higher rate of apoptosis versus the single treatment [174]. Cheng H and co-workers tested on CRC cell line the inhibitory effect of Naringin. The research team illustrated that Naringin stops the proliferation of CRC cells, promoting apoptosis by inhibiting the PI3K/AKT/mTOR signaling pathway in a dose-dependent manner [154]. Wang J and his research team tested on seven different colorectal cell lines the effect of W922, a novel PI3K/AKT/mTOR pathway inhibitor, as an efficient anti-tumoral. Between all cell lines used, the HCT116 line was the most sensitive to W922 treatment [175]. W922 was able to inhibit HCT116 cell viability and cell proliferation in vitro, in a concentration and time-dependent manner [143]. Under W922 treatment, the suppression of tumor growth was observed, as well as dephosphorylation of PI3K/AKT/mTOR proteins and mTOR inhibition [175]. Moreover, co-treatment of W922 and chloroquine leads to cells apoptosis, thus providing a promising therapeutic strategy for patients diagnosed with CRC [175]. Kallikrein-related peptidase 10 (KLK10) was identified in 1996 as normal epithelial cell-specific 1, involved in cancer development by regulation of cell growth, invasion, and apoptosis [176]. Moreover, using CRC cell lines, a negative correlation has been detected between KLK10 high expression and OR rate. Therefore, knockdown of KLK10 dramatically suppresses cell viability and induces apoptosis in CRC cell lines [176]. KLK10 acts by blocking the PI3K/AKT/mTOR signaling pathway, inhibiting cell growth and glucose metabolism [176]. Helmy MW et al. explored the effects of diosmin (DIO, a natural NF-κB inhibitor) and BEZ-235 (dactolisib, dual PI3K-mTOR inhibitor) in HCT-116 CRC cells [177]. The research team reported that co-administration of both drugs in two combinations inhibited the PI3K/AKT/mTOR/NF-κB signaling cascades, leading to apoptosis and cell proliferation inhibition, and altered the angiogenesis process [177]. Future preclinical and clinical studies must be carried out [162]. The study conducted by Li S et al. evaluated the effect of the extract *Selaginella doederleinii Hieron* ethyl acetate (SDEA) in vitro and in vivo [178]. Using HT29 and HCT116 cell lines, the anti-tumoral effect of SDEA was manifested by cell morphological changes, cell cycle arrest, autophagy, and apoptosis [178]. Moreover, the SDEA extract may induce the loss of the mitochondrial membrane potential, increases the autophagic flux, and will inhibit the PI3K/AKT/mTOR signaling pathways [178]. Therefore, in xenograft tumors, SDEA inhibits the growth in a dose-dependent manner [178]. The experimental studies conducted by Han YH and co-workers, performed on cell lines, explored the inhibitory effect of betulin in mCRC [179]. Studies performed in vitro illustrated that betulin can induce apoptosis, autophagy, and cell cycle arrest by PI3K/AKT/mTOR and MAPK signaling pathways inactivation [179]. In addition, oral administration of betulin significantly inhibits CT26 cell lung metastasis [179]. Li N and his research team evaluated in vitro if nobiletin may enhance the inhibitory effect of oxaliplatin on CRC cell lines [180]. The study reported that nobiletin increases CRC sensibility to oxaliplatin to induce CRC cells' apoptosis, as evidenced by the increased expression of pro-apoptotic proteins and the downregulation of anti-apoptotic protein Bcl-2 [180]. Moreover, this combination will downregulate the PI3K/AKT/mTOR pathway [180].

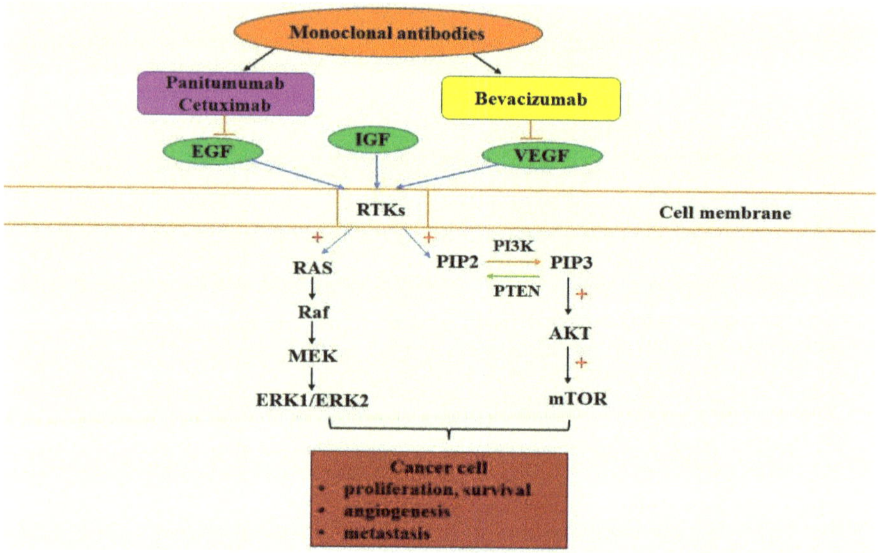

Figure 2. Growth factors, growth factor inhibitors, and PI3K/AKT/mTOR-MAPK signaling pathways in colorectal cancer development.

Studies performed in vivo observed that MEK and RAF inhibitors suppress colorectal tumor growth, but these cells develop resistance to these inhibitors by activating the PI3K/AKT/mTOR or JAK/STAT signaling pathways that mediate resistance. Understanding the mechanisms of CRC drug resistance will enhance the patients' survival rate [181]. The FRESCO Randomized Clinical Trial conducted by Li J et al. explored the efficacy and safety of oral fruquintinib, a VEGFR inhibitor, as third-line or later therapy in 519 patients (aged 18 to 75 years) diagnosed with mCRC. The study observed a median OR significantly improved by fruquintinib versus placebo (9.3 months compared with 6.6 months). Moreover, the median PFS was also significantly increased with fruquintinib (3.7 months versus 1.8 months) [182]. PI3K or AKT inhibitors may be used in CRC clinical trial with promising results, but drug resistance frequently appears, driven by β-catenin, which blocks FOXO 3A to induce apoptosis. Therefore, using Wnt/β -catenin signaling pathway inhibitors will reduce PI3K or AKT drug resistance in CRC patients [183]. Arques and his research team explored in clinical trials, which included CRC patients, if Wnt mediates resistance in patients treated with PI3K or AKT different inhibitors. The study used NVP-TNKS656—a Wnt/tankyrase inhibitor, to overcome PI3K or AKT resistance [183]. The study reported good results because Wnt/tankyrase inhibitor promotes apoptosis in PI3K or AKT inhibitor-resistant cells. For CRC patients, PI3K/AKT/mTOR and Wnt/β-catenin inhibitors represent an excellent strategy [183].

Everolimus—an mTOR inhibitor was administered in 12 patients with primary resectable rectal cancer 14 days prior to the start of chemoradiotherapy and continued throughout the four-week course with 5-FU and radiotherapy. The study detected no increase in toxicity at any of the doses with 5-FU and radiotherapy. Moreover, no significant increase in complete pathological response (pCR) was observed and the everolimus maximum tolerated dose was 10 mg. The study concluded that the combination of chemoradiotherapy and everolimus has feasible results over long time. Another mTOR inhibitor, rapamycin, was used in Phase I (13 patients) and II clinical trial (31 patients) with primary resectable rectal cancer, where patients received rapamycin one week before and during radiotherapy. The study illustrated a higher rate of post-operative complications in phase I. Regarding the patients included in phase II, it was observed that rapamycin was feasible correlated with a significant reduction of tumor metabolic activity [184]. Ganesan P and co-workers evaluated, in early-phase clinical trials, 191 CRC patients with diverse mutations, especially *KRAS* and the PI3K/AKT/mTOR inhibitors. Depending on the mutation, the patients received different drug inhibitors, such as for mTOR, PI3K, and AKT. The study concluded that the median PFS for patients with *PIK3CA* mutations and PI3K/AKT/mTOR inhibitors was 1.9 months, while there was no difference in median PFS in patients with *KRAS* mutations compared with patients with wild type *KRAS* [185]. Garrido-Laguna I and colleagues explored, in early-phase trials, the impact of PI3K/AKT/mTOR inhibitors on 238 patients with mCRC, with 51% *KRAS* mutations and 15% *PIK3CA* mutations. The treatment with different PI3K/AKT/mTOR inhibitors presented a limited activity in these patients, because of the MAPK activating mutations [186]. Kyriakopoulos CE et al. conducted a phrase I trial that evaluated the effect of tivantinib and temsirolimus in patients with advanced solid tumors including CRC. The doses administered in this study were overall well tolerated and demonstrated that this combination has an enhanced anti-tumoral activity [187]. In patients diagnosed with mCRC, everolimus—an oral mTOR inhibitor—presents efficacy. Ng K and his research team evaluated, in a sequential phase II study, the effect of everolimus in 100 patients with mCRC, which were refractory to bevacizumab-, fluoropyrimidine-, oxaliplatin-, and irinotecan. The patients received a high weekly dose of everolimus, while the daily dose has a lower concentration. Median PFS and OS were 1.8 and 4.9 months, and 1.8 and 5.9 months, respectively, for the weekly and daily administrated doses. Among the patients who received a daily dose of everolimus, those with *KRAS* mutations had a significantly shorter median OR versus those with wild-type *KRAS* mutations. The daily or weekly dose of everolimus was well tolerated but did not confer a significant efficacy in mCRC [188]. A phase I study included 27 CRC patients with *KRAS* mutations who received the pan-HER inhibitor dacomitinib in combination with MEK1/2 inhibitor PD-0325901. The patients received various drug doses by oral administration in cycles of 28 days, but the results revealed that the mentioned combination was not tolerated by most of the patients. These results may be explained by the activation of PI3K/AKT and MAPK signaling pathways, by the *KRAS* and *PIK3CA* mutations [189]. A phase I trial was initiated with the pan-HER inhibitor afatinib plus the MEK inhibitor selumetinib in 19 CRC patients with *KRAS* and *PIK3CA* wild-type mutations. In peripheral blood mononuclear cells, inhibition of phosphorylated ERK needs a specified concentration for both selumetinib and afatinib. Although the study reported limited clinical efficacy for the two drugs, several side effects have been reported after oral administration [190]. Folprecht G et al. conducted a clinical study that included 47 patients with mCRC who received EKB-569, an EGFR tyrosine kinase inhibitor, in combination with irinotecan, 5-FU, and leucovorin (FOLFIRI). At the recommended dose of EKB-569 (EKB-569/full dose FOLFIRI), the complete inhibition of phosphorylated EGFR occurs. Instead, FOLFIRI alone did not affect EGFR phosphorylation, but may inhibit epidermal proliferation and MAPK [191]. Tabernero J and his research team evaluated the cetuximab efficacy in tissue samples collected during a phase I as first-line therapy in 62 patients with mCRC. The patients received cetuximab monotherapy for 6 weeks, followed by the administration of cetuximab in combination with 5-FU, leucovorin, and irinotecan until disease progression. In 35 mCRC patients, cetuximab treatment was correlated with substantial downregulation of EGFR, MAPK, and STAT3 phosphorylation. The study reported that PFS was longer for patients with KRAS wild-type compared with *KRAS* mutant tumors [192].

Currently, 3 clinical trials are in progress. A phase II, comparative trial, AtezoTRIBE, includes unresectable and previously untreated mCRC patients that have received FOLFOXIRI treatment (fluorouracil, leucovorin, oxaliplatin, and irinotecan) plus bevacizumab up to 8 cycles (the standard treatment) or a combination with atezolizumab (the experimental treatment), followed by the treatment with 5-FU/leucovorin plus bevacizumab with or without atezolizumab according to disease progression. Until now, a few patients reported severe adverse reactions [193]. The second one, conducted by Damato A and his research team, is a prospective, open-label, multicentric phase II trial, which includes patients with mCRC and RAS/BRAF mutations who received, in the first line of treatment, nivolumab in combination with FOLFOXIRI/bevacizumab every 2 weeks for 8 cycles. After that period, the patients intravenously received bevacizumab plus nivolumab for another 2 weeks in a dose that depends on the patient's weight. The main aim of the study is to enhance the overall response rate from 66 to 80% [194]. The third, conducted by Meric-Bernstam F et al., is a phase 2a, multiple basket study, called MyPathway, which included patients with HER2-amplified mCRC. Initially, the patients intravenously received an increased dose of pertuzumab, and, every 3 weeks, the dose was reduced by half. For trastuzumab, the loading dose was in an increased concentration, followed by every 3 weeks by a reduced dose administrated intravenously. Although some patients reported several adverse reactions, the preliminary results reported that the dual therapy is well tolerated and could represent a favorable therapy for HER2-amplified mCRC patients [195].

Overall, experimental studies and clinical evidence revealed that polyphenols have an important role in CRC chemoprevention and exhibit cytotoxic effects on CRC cells [196].

6. Conclusions

Unfortunately, CRC has an increasing incidence among the young population, and adopting a healthy diet correlated with regular medical analysis may decrease the incidence of this malignancy. The CRC pathogenesis is very complex and assumes the presence of many genetic mutations that will be involved in cancer progression. Moreover, CRC progression is influenced by the presence of inflammatory cells and their inflammatory mediators, such as cytokines that can activate signaling pathways, leading to tumoral development. IGF-1R, EGFR, and VEGF can bind to RTKs, which will activate RAS-RAF-MAPK and PI3K/AKT/mTOR signaling pathways, promoting tumor growth, progression, cell survival, motility, and invasion.

Therefore, important drugs that have already been used in CRC clinical trials are EGFR and VEGFR monoclonal antibodies. The anti-EGFR antibodies, cetuximab and panitumumab, are used in phase III trials especially for mCRC patients with good results regarding PFS and OR. FDA approved bevacizumab—a VEGFR inhibitor for mCRC patients that can inhibit angiogenesis, as a key step in cancer therapy. Gene mutations, including *RAS, BRAF, PI3KCA, PTEN*, and *HER-2* activate MAPK or PI3K/AKT/mTOR leading to resistance to EGFR or VEGF therapy. Thus, a combination of EGFR/VEGF with RAS-RAF-MEK-MAPK and PI3K/AKT/mTOR inhibitors have anti-proliferative and anti-apoptotic properties. Currently, the experimental in vitro studies focus on blocking the PI3K/AKT/mTOR/NF-kB and MAPK signaling pathways, which are able to inhibit CRC cells growth, leading to apoptosis. These promising results may enhance the CRC patients' survival rate. Moreover, PI3K, AKT, or mTOR inhibitors alone are not very efficient in CRC treatment, because drug resistance appears to be driven by Wnt/β-catenin or by MAPK signaling pathways' components.

In this context, we conclude that a promising therapeutic strategy for CRC patients may be based on genetic mutation detections and targeting either EGFR/VEGFR in association with PI3K/AKT/mTOR, Wnt/β-catenin, or MAPK inhibitors. This approach could provide new perspectives and new hopes for CRC patients.

Author Contributions: Conceptualization, C.S., D.M., M.G., I.-I.S.-S., A.R.T., R.I.N., and M.J.; methodology, M.J., D.M., M.G., and I.-I.S.-S.; software, C.S., A.R.T., D.M., and M.G.; validation, M.J., A.R.T., R.I.N., and M.G.; formal analysis, C.S. and R.I.N.; investigation, M.J., D.M., M.G., I.-I.S.-S., C.S., A.R.T., and R.I.N.; resources, D.M., M.G., and C.S.; data curation, C.S. and A.R.T.; writing—original draft preparation, C.S., D.M., I.-I.S.-S., and M.G.; writing—review and editing, M.G., M.J., and C.S.; visualization, M.J., I.-I.S.-S., C.S., and R.I.N.; supervision, M.J., M.G., and A.R.T.; project administration, M.G., M.J., R.I.N., I.-I.S.-S., C.S., and A.R.T. All authors have read and agreed to the published version of the manuscript.

Funding: This research received no external funding.

Conflicts of Interest: The authors declare no conflict of interest.

References

1. Kuipers, E.J.; Grady, W.M.; Lieberman, D.; Seufferlein, T.; Sung, J.J.; Boelens, P.G.; van de Velde, C.J.; Watanabe, T. Colorectal cancer. *Nat. Rev. Dis. Primers* **2015**, *5*, 15065. [CrossRef] [PubMed]
2. Balhareth, A.; Aldossary, M.Y.; McNamara, D. Impact of physical activity and diet on colorectal cancer survivors' quality of life: A systematic review. *World J. Surg. Oncol.* **2019**, *31*, 153. [CrossRef] [PubMed]
3. Xu, W.; Yu, M.; Qin, J.; Luo, Y.; Zhong, M. LACTB regulates PIK3R3 to promote autophagy and inhibit EMT and proliferation through the PI3K/AKT/mTOR signaling pathway in colorectal cancer. *Cancer Manag. Res.* **2020**, *30*, 5181–5200. [CrossRef]
4. Brenner, H.; Kloor, M.; Pox, C.P. Colorectal cancer. *Lancet* **2014**, *26*, 1490–1502. [CrossRef]
5. Haraldsdottir, S.; Einarsdottir, H.M.; Smaradottir, A.; Gunnlaugsson, A.; Halfdanarson, T.R. Colorectal cancer—Review. *Laeknabladid* **2014**, *100*, 75–82.
6. Navarro, M.; Nicolas, A.; Ferrandez, A.; Lanas, A. Colorectal cancer population screening programs worldwide in 2016: An update. *World J. Gastroenterol.* **2017**, *28*, 3632–3642. [CrossRef]
7. Onyoh, E.F.; Hsu, W.F.; Chang, L.C.; Lee, Y.C.; Wu, M.S.; Chiu, H.M. The rise of colorectal cancer in Asia: Epidemiology, screening, and management. *Curr. Gastroenterol. Rep.* **2019**, *10*, 36. [CrossRef]
8. Bardou, M.; Barkun, A.N.; Martel, M. Obesity and colorectal cancer. *Gut* **2013**, *62*, 933–947. [CrossRef]
9. Schulpen, M.; van den Brandt, P.A. Mediterranean diet adherence and risk of colorectal cancer: The prospective Netherlands Cohort Study. *Eur. J. Epidemiol.* **2020**, *35*, 25–35. [CrossRef]
10. Mauri, G.; Sartore-Bianchi, A.; Russo, A.G.; Marsoni, S.; Bardelli, A.; Siena, S. Early-onset colorectal cancer in young individuals. *Mol. Oncol.* **2019**, *13*, 109–131. [CrossRef] [PubMed]
11. Connell, L.C.; Mota, J.M.; Braghiroli, M.I.; Hoff, P.M.; Connell, L.C.; Mota, J.M.; Braghiroli, M.I.; Hoff, P.M. The rising incidence of younger patients with colorectal cancer: Questions about screening, biology, and treatment. *Curr. Treat. Options Oncol.* **2017**, *18*, 23. [CrossRef]
12. Weinberg, B.A.; Marshall, J.L.; Salem, M.E. The Growing challenge of young adults with colorectal cancer. *Oncology* **2017**, *15*, 381–389.
13. O'Sullivan, D.E.; Hilsden, R.J.; Ruan, Y.; Forbes, N.; Heitman, S.J.; Brenner, D.R. The incidence of young-onset colorectal cancer in Canada continues to increase. *Cancer Epidemiol.* **2020**, *69*, 101828. [CrossRef]
14. Siegel, R.; Desantis, C.; Jemal, A. Colorectal cancer statistics—2014. *CA Cancer J. Clin.* **2014**, *64*, 104–117. [CrossRef]
15. Jeong, M.A.; Kang, H.W. Early-onset colorectal cancer. *Korean J. Gastroenterol.* **2019**, *25*, 4–10. [CrossRef]
16. Abualkhair, W.H.; Zhou, M.; Ahnen, D.; Yu, Q.; Wu, X.C.; Karlitz, J.J. Trends in incidence of early-onset colorectal cancer in the United States among those approaching screening age. *JAMA Netw. Open* **2020**, *3*, e1920407. [CrossRef]
17. Kim, S.E.; Paik, H.Y.; Yoon, H.; Lee, J.E.; Kim, N.; Sung, M.K. Sex- and gender-specific disparities in colorectal cancer risk. *World J. Gastroenterol.* **2015**, *7*, 5167–5175. [CrossRef]
18. Modest, D.P.; Pant, S.; Sartore-Bianchi, A. Treatment sequencing in metastatic colorectal cancer. *Eur. J. Cancer* **2019**, *109*, 70–83. [CrossRef]
19. Fornasier, G.; Francescon, S.; Baldo, P. An update of efficacy and safety of cetuximab in metastatic colorectal cancer: A narrative review. *Adv. Ther.* **2018**, *35*, 1497–1509. [CrossRef]
20. Dzunic, M.; Petkovic, I.; Cvetanovic, A.; Vrbic, S.; Pejcic, I. Current and future targets and therapies in metastatic colorectal cancer. *J. BUON* **2019**, *24*, 1785–1792.
21. Dienstmann, R.; Salazar, R.; Tabernero, J. Molecular subtypes and the evolution of treatment decisions in metastatic colorectal cancer. *Am. Soc. Clin. Oncol. Educ. Book* **2018**, *23*, 231–238. [CrossRef]
22. Meng, S.; Jian, Z.; Yan, X.; Li, J.; Zhang, R. LncRNA SNHG6 inhibits cell proliferation and metastasis by targeting ETS1 via the PI3K/AKT/mTOR pathway in colorectal cancer. *Mol. Med. Rep.* **2019**, *20*, 2541–2548. [CrossRef]
23. Cornish, A.J.; Tomlinson, I.P.M.; Houlston, R.S. Mendelian randomisation: A powerful and inexpensive method for identifying and excluding non-genetic risk factors for colorectal cancer. *Mol. Asp. Med.* **2019**, *69*, 41–47. [CrossRef]
24. Shaw, E.; Farris, M.S.; Stone, C.R.; Derksen, J.W.G.; Johnson, R.; Hilsden, R.J.; Friedenreich, C.M.; Brenner, D.R. Effects of physical activity on colorectal cancer risk among family history and body mass index subgroups: A systematic review and meta-analysis. *BMC Cancer* **2018**, *11*, 71. [CrossRef]
25. Simon, K. Colorectal cancer development and advances in screening. *Clin. Interv. Aging* **2016**, *19*, 967–976.
26. Song, M.; Garrett, W.S.; Chan, A.T. Nutrients, foods, and colorectal cancer prevention. *Gastroenterology* **2015**, *148*, 1244–1260. [CrossRef]
27. Azeem, S.; Gillani, S.W.; Siddiqui, A.; Jandrajupalli, S.B.; Poh, V.; Syed, S.S.A. Diet and colorectal cancer risk in Asia—A systematic review. *Asian Pac. J. Cancer Prev.* **2015**, *16*, 5389–5396. [CrossRef]
28. Gibson, D.C.; Prochaska, J.D.; Yu, X.; Kaul, S. An examination between census tract unhealthy food availability and colorectal cancer incidence. *Cancer Epidemiol.* **2020**, *67*, 101761. [CrossRef]
29. Diallo, A.; Deschasaux, M.; Latino-Martel, P.; Hercberg, S.; Galan, P.; Fassier, P.; Allès, B.; Guéraud, F.; Pierre, F.H.; Touvier, M. Red and processed meat intake and cancer risk: Results from the prospective nutrinnet-sante cohort study. *Int. J. Cancer* **2018**, *15*, 230–237. [CrossRef]

30. Martin, O.C.B.; Olier, M.; Ellero-Simatos, S.; Naud, N.; Dupuy, J.; Huc, L.; Taché, S.; Graillot, V.; Levêque, M.; Bézirard, V.; et al. Haem iron reshapes colonic luminal environment: Impact on mucosal homeostasis and microbiome through aldehyde formation. *Microbiome* **2019**, *6*, 72. [CrossRef]
31. Turner, N.D.; Lloyd, S.K. Association between red meat consumption and colon cancer: A systematic review of experimental results. *Exp. Biol. Med.* **2017**, *242*, 813–839. [CrossRef]
32. Ekine-Afolabi, B.A.; Njan, A.A.; Rotimi SO, R.I.A.; Elbehi, A.M.; Cash, E.; Adeyeye, A. The impact of diet on the involvement of non-coding RNAs, extracellular vesicles, and gut microbiome-virome in colorectal cancer initiation and progression. *Front. Oncol.* **2020**, *14*, 583372. [CrossRef]
33. Sánchez-Alcoholado, L.; Ordóñez, R.; Otero, A.; Plaza-Andrade, I.; Laborda-Illanes, A.; Medina, J.A.; Ramos-Molina, B.; Gómez-Millán, J.; Queipo-Ortuño, M.I. Gut microbiota-mediated inflammation and gut permeability in patients with obesity and colorectal cancer. *Int. J. Mol. Sci.* **2020**, *16*, 6782. [CrossRef]
34. Farinetti, A.; Zurlo, V.; Manenti, A.; Coppi, F.; Mattioli, A.V. Mediterranean diet and colorectal cancer: A systematic review. *Nutrition* **2017**, *43*, 83–88. [CrossRef]
35. Cai, L.; Bennedsen, A.L.B.; Qvortrup, C.; Gögenur, I. Increasing incidence of colorectal cancer in young patients. *Ugeskr. Laeger.* **2019**, *30*, V09190524.
36. Gram, I.T.; Park, S.Y.; Wilkens, L.R.; Haiman, C.A.; Le Marchand, L. Smoking-related risks of colorectal cancer by anatomical subsite and sex. *Am. J. Epidemiol.* **2020**, *1*, 543–553. [CrossRef]
37. Jung, Y.S.; Kim, N.H.; Yang, H.J.; Park, S.K.; Park, J.H.; Park, D.I.; Sohn, C.I. The impact of passive smoking on the risk of colorectal neoplasia in never, former, and current smokers. *J. Gastroenterol. Hepatol.* **2018**, *3*, 1023–1030. [CrossRef]
38. Van Blarigan, E.L.; Meyerhardt, J.A. Role of physical activity and diet after colorectal cancer diagnosis. *J. Clin. Oncol.* **2015**, *1*, 1825–1834. [CrossRef]
39. Baena, R.; Salinas, P. Diet and colorectal cancer. *Maturitas* **2015**, *80*, 258–264. [CrossRef]
40. Ustundag, H.; Zengin, N.; Andsoy, I.I.; Gul, A. Awareness of health sciences students about colorectal cancer risk factors. *Eur. J. Cancer Care* **2019**, *28*, e13016. [CrossRef] [PubMed]
41. Bradbury, K.E.; Murphy, N.; Key, T.J. Diet and colorectal cancer in UK Biobank: A prospective study. *Int. J. Epidemiol.* **2020**, *1*, 246–258. [CrossRef]
42. Na, H.K.; Lee, J.Y. Molecular basis of alcohol-related gastric and colon cancer. *Int. J. Mol. Sci.* **2017**, *24*, 1116. [CrossRef]
43. Jain, A.; Jain, S. Rising incidence of colorectal cancer in patients younger than age 50 in Hawai'i. *Hawaii J. Med. Public Health* **2019**, *78*, 195–199.
44. Cirillo, F.; Catellani, C.; Sartori, C.; Lazzeroni, P.; Amarri, S.; Street, M.E. Obesity, insulin resistance, and colorectal cancer: Could miRNA dysregulation play a role? *Int. J. Mol. Sci.* **2019**, *14*, 2922. [CrossRef]
45. Pietrzyk, L.; Torres, A.; Maciejewski, R.; Torres, K. Obesity and obese-related chronic low-grade inflammation in promotion of colorectal cancer development. *Asian Pac. J. Cancer Prev.* **2015**, *6*, 4161–4168. [CrossRef]
46. Martinez-Useros, J.; Garcia-Foncillas, J. Obesity and colorectal cancer: Molecular features of adipose tissue. *J. Transl. Med.* **2016**, *22*, 21. [CrossRef]
47. Ahechu, P.; Zozaya, G.; Martí, P.; Hernández-Lizoáin, J.L.; Baixauli, J.; Unamuno, X.; Frühbeck, G.; Catalán, V. NLRP3 inflammasome: A possible link between obesity-associated low-grade chronic inflammation and colorectal cancer development. *Front. Immunol.* **2018**, *11*, 2918. [CrossRef]
48. Zhang, J.; Guo, S.; Li, J.; Bao, W.; Zhang, P.; Huang, Y.; Ling, P.; Wang, Y.; Zhao, Q. Effects of high-fat diet-induced adipokines and cytokines on colorectal cancer development. *FEBS Open Bio* **2019**, *9*, 2117–2125. [CrossRef] [PubMed]
49. Soltani, G.; Poursheikhani, A.; Yassi, M.; Hayatbakhsh, A.; Kerachian, M.; Kerachian, M.A. Obesity, diabetes and the risk of colorectal adenoma and cancer. *BMC Endocr. Disord.* **2019**, *29*, 113. [CrossRef] [PubMed]
50. Gao, R.; Gao, Z.; Huang, L.; Qin, H. Gut microbiota and colorectal cancer. *Eur. J. Clin. Microbiol. Infect. Dis.* **2017**, *36*, 757–769. [CrossRef] [PubMed]
51. Yang, J.; Yu, J. The association of diet, gut microbiota and colorectal cancer: What we eat may imply what we get. *Protein Cell* **2018**, *9*, 474–487. [CrossRef]
52. Han, S.; Zhuang, J.; Wu, Y.; Wu, W.; Yang, X. Progress in research on colorectal cancer-related microorganisms and metabolites. *Cancer Manag. Res.* **2020**, *21*, 8703–8720. [CrossRef]
53. Ganesan, K.; Jayachandran, M.; Xu, B. Diet-derived phytochemicals targeting colon cancer stem cells and microbiota in colorectal cancer. *Int. J. Mol. Sci.* **2020**, *1*, 3976. [CrossRef] [PubMed]
54. Triantafillidis, J.K.; Nasioulas, G.; Kosmidis, P.A. Colorectal cancer and inflammatory bowel disease: Epidemiology, risk factors, mechanisms of carcinogenesis and prevention strategies. *Anticancer Res.* **2009**, *29*, 2727–2737. [PubMed]
55. Bultman, S.J. Interplay between diet, gut microbiota, epigenetic events, and colorectal cancer. *Mol. Nutr. Food Res.* **2017**, *61*, 1500902. [CrossRef]
56. Chapkin, R.S.; Navarro, S.L.; Hullar, M.A.J.; Lampe, J.W. Diet and gut microbes act coordinately to enhance programmed cell death and reduce colorectal cancer risk. *Dig. Dis. Sci.* **2020**, *65*, 840–851. [CrossRef] [PubMed]
57. Botteri, E.; Iodice, S.; Bagnardi, V.; Raimondi, S.; Lowenfels, A.B.; Maisonneuve, P. Smoking and colorectal cancer: A meta-analysis. *JAMA* **2008**, *17*, 2765–2778. [CrossRef]

58. Li, S.; Ung, T.T.; Nguyen, T.T.; Sah, D.K.; Park, S.Y.; Jung, Y.D. Cholic Acid stimulates MMP-9 in human colon cancer cells via activation of MAPK, AP-1, and NF-kappaB Activity. *Int. J. Mol. Sci.* **2020**, *12*, 3420. [CrossRef]
59. Kasprzak, A.; Adamek, A. Insulin-Like Growth Factor 2 (IGF2) signaling in colorectal cancer-from basic research to potential clinical applications. *Int. J. Mol. Sci.* **2019**, *3*, 4915. [CrossRef]
60. Müller, M.F.; Ibrahim, A.E.; Arends, M.J. Molecular pathological classification of colorectal cancer. *Virchows Arch.* **2016**, *469*, 125–134. [CrossRef]
61. La Vecchia, S.; Sebastián, C. Metabolic pathways regulating colorectal cancer initiation and progression. *Semin. Cell Dev. Biol.* **2020**, *98*, 63–70. [CrossRef] [PubMed]
62. Vilar, E.; Gruber, S.B. Microsatellite instability in colorectal cancer-the stable evidence. *Nat. Rev. Clin. Oncol.* **2010**, *7*, 153–162. [CrossRef] [PubMed]
63. De Filippo, C.; Luceri, C.; Caderni, G.; Pacini, M.; Messerini, L.; Biggeri, A.; Mini, E.; Tonelli, F.; Cianchi, F.; Dolara, P. Mutations of the APC gene in human sporadic colorectal cancers. *Scand. J. Gastroenterol.* **2002**, *37*, 1048–1053. [CrossRef]
64. Yu, I.S.; Cheung, W.Y. Epidermal growth factor receptor immunohistochemistry: New opportunities in metastatic colorectal cancer. *J. Transl. Med.* **2015**, *7*, 217.
65. Wang, D.; Liang, W.; Duan, X.; Liu, L.; Shen, H.; Peng, Y.; Li, B. Detection of KRAS gene mutations in colorectal carcinoma: A study of 6 364 patients. *Zhonghua Bing Li Xue Za Zhi* **2014**, *43*, 583–587. [PubMed]
66. Wang, D.; Zhang, Z.; Li, Y.; Xu, C.; Yu, Y.; Li, M.; Chen, C.; Zhang, X. Adenomatous polyposis coli gene mutations in 22 Chinese pedigrees with familial adenomatous polyposis. *Med. Sci. Monit.* **2019**, *22*, 3796–3803. [CrossRef]
67. Nallamilli, B.R.R.; Hegde, M. Detecting APC Gene mutations in familial adenomatous polyposis (FAP). *Curr. Protoc. Hum. Genet.* **2017**, *11*, 10–18. [CrossRef]
68. Wachsmannova, L.; Mego, M.; Stevurkova, V.; Zajac, V.; Ciernikova, S. Novel strategies for comprehensive mutation screening of the APC gene. *Neoplasma* **2017**, *64*, 338–343. [CrossRef]
69. Ye, Z.L.; Qiu, M.Z.; Tang, T.; Wang, F.; Zhou, Y.X.; Lei, M.J.; Guan, W.L.; He, C.Y. Gene mutation profiling in Chinese colorectal cancer patients and its association with clinicopathological characteristics and prognosis. *Cancer Med.* **2020**, *9*, 745–756. [CrossRef]
70. Yaeger, R.; Chatila, W.K.; Lipsyc, M.D.; Hechtman, J.F.; Cercek, A.; Sanchez-Vega, F.; Jayakumaran, G.; Middha, S.; Zehir, A.; Donoghue, M.T.A. Clinical sequencing defines the genomic landscape of metastatic colorectal cancer. *Cancer Cell* **2018**, *8*, 125–136. [CrossRef]
71. Pearlman, R.; Frankel, W.L.; Swanson, B.; Zhao, W.; Yilmaz, A.; Miller, K.; Bacher, J.; Bigley, C.; Nelsen, L.; Goodfellow, P.J.; et al. Prevalence and spectrum of germline cancer susceptibility gene mutations among patients with early-onset colorectal cancer. *JAMA Oncol.* **2017**, *1*, 464–471. [CrossRef] [PubMed]
72. Wang, Q.; Shi, Y.L.; Zhou, K.; Wang, L.L.; Yan, Z.X.; Liu, Y.L.; Xu, L.L.; Zhao, S.W.; Chu, H.L.; Shi, T.T.; et al. PIK3CA mutations confer resistance to first-line chemotherapy in colorectal cancer. *Cell Death Dis.* **2018**, *3*, 739. [CrossRef]
73. Afrin, S.; Giampieri, F.; Gasparrini, M.; Forbes-Hernández, T.Y.; Cianciosi, D.; Reboredo-Rodriguez, P.; Zhang, J.; Manna, P.P.; Daglia, M.; Atanasov, A.G.; et al. Dietary phytochemicals in colorectal cancer prevention and treatment: A focus on the molecular mechanisms involved. *Biotechnol. Adv.* **2020**, *38*, 107322. [CrossRef] [PubMed]
74. Zhao, B.; Wang, L.; Qiu, H.; Zhang, M.; Sun, L.; Peng, P.; Yu, Q.; Yuan, X. Mechanisms of resistance to anti-EGFR therapy in colorectal cancer. *Oncotarget* **2017**, *17*, 3980–4000. [CrossRef]
75. Chen, L.Y.; Wang, L.; Ren, Y.X.; Pang, Z.; Liu, Y.; Sun, X.D.; Tu, J.; Zhi, Z.; Qin, Y.; Sun, L.N.; et al. The circular RNA circ-ERBIN promotes growth and metastasis of colorectal cancer by miR-125a-5p and miR-138-5p/4EBP-1 mediated cap-independent HIF-1α translation. *Mol. Cancer* **2020**, *23*, 164. [CrossRef] [PubMed]
76. Carethers, J.M.; Jung, B.H. Genetics and genetic biomarkers in sporadic colorectal cancer. *Gastroenterology* **2015**, *149*, 1177–1190.e3. [CrossRef] [PubMed]
77. Bellido, F.; Pineda, M.; Aiza, G.; Valdés-Mas, R.; Navarro, M.; Puente, D.A.; Pons, T.; González, S.; Iglesias, S.; Darder, E.; et al. POLE and POLD1 mutations in 529 kindred with familial colorectal cancer and/or polyposis: Review of reported cases and recommendations for genetic testing and surveillance. *Genet. Med.* **2016**, *18*, 325–332. [CrossRef] [PubMed]
78. Armengol, G.; Sarhadi, V.K.; Ghanbari, R.; Doghaei-Moghaddam, M.; Ansari, R.; Sotoudeh, M.; Puolakkainen, P.; Kokkola, A.; Malekzadeh, R.; Knuutila, S. Driver gene mutations in stools of colorectal carcinoma patients detected by targeted next-generation sequencing. *J. Mol. Diagn.* **2016**, *18*, 471–479. [CrossRef]
79. Timar, J.; Kashofer, K. Molecular epidemiology and diagnostics of KRAS mutations in human cancer. *Cancer Metastasis Rev.* **2020**, *39*, 1029–1038. [CrossRef]
80. Fleming, N.I.; Jorissen, R.N.; Mouradov, D.; Christie, M.; Sakthianandeswaren, A.; Palmieri, M.; Day, F.; Li, S.; Tsui, C.; Lipton, L. SMAD2, SMAD3 and SMAD4 mutations in colorectal cancer. *Cancer Res.* **2013**, *15*, 725–735. [CrossRef]
81. Mo, H.Y.; Lee, J.H.; Kim, M.S.; Yoo, N.J.; Lee, S.H. Frameshift mutations and loss of expression of CLCA4 gene are frequent in colorectal cancers with microsatellite instability. *Appl. Immunohistochem. Mol. Morphol.* **2020**, *28*, 489–494. [CrossRef] [PubMed]
82. Sanchez-Lopez, E.; Flashner-Abramson, E.; Shalapour, S.; Zhong, Z.; Taniguchi, K.; Levitzki, A.; Karin, M. Targeting colorectal cancer via its microenvironment by inhibiting IGF-1 receptor-insulin receptor substrate and STAT3 signaling. *Oncogene* **2016**, *19*, 2634–2644. [CrossRef] [PubMed]
83. Zhang, Q.; Wang, W.; Zhou, Q.; Chen, C.; Yuan, W.; Liu, J.; Li, X.; Sun, Z. Roles of circRNAs in the tumour microenvironment. *Mol. Cancer* **2020**, *23*, 14. [CrossRef] [PubMed]

84. Yin, Z.; Li, C.; Wang, J.; Xue, L. Myeloid-derived suppressor cells: Roles in the tumor microenvironment and tumor radiotherapy. *Int. J. Cancer* **2019**, *144*, 933–946. [CrossRef] [PubMed]
85. Zhang, R.; Qi, F.; Zhao, F.; Li, G.; Shao, S.; Zhang, X.; Yuan, L.; Feng, Y. Cancer-associated fibroblasts enhance tumor-associated macrophages enrichment and suppress NK cells function in colorectal cancer. *Cell Death Dis.* **2019**, *20*, 273. [CrossRef] [PubMed]
86. Manzat-Saplacan, R.M.; Balacescu, L.; Gherman, C.; Chira, R.I.; Craiu, A.; Mircea, P.A.; Lisencu, C.; Balacescu, O. The role of PDGFs and PDGFRs in colorectal cancer. *Mediators Inflamm.* **2017**, *2017*, 4708076. [CrossRef]
87. Lucas, C.; Barnich, N.; Nguyen, H.T.T. Microbiota, inflammation and colorectal cancer. *Int. J. Mol. Sci.* **2017**, *20*, 1310. [CrossRef]
88. Xu, Z.; Zhu, C.; Chen, C.; Zong, Y.; Feng, H.; Liu, D.; Feng, W.; Zhao, J.; Lu, A. CCL19 suppresses angiogenesis through promoting miR-206 and inhibiting Met/ERK/Elk-1/HIF-1alpha/VEGF-A pathway in colorectal cancer. *Cell Death Dis.* **2018**, *24*, 974. [CrossRef]
89. De la Fuente López, M.; Landskron, G.; Parada, D.; Dubois-Camacho, K.; Simian, D.; Martinez, M.; Romero, D.; Roa, J.C.; Chahuán, I.; Gutiérrez, R.; et al. The relationship between chemokines CCL2, CCL3, and CCL4 with the tumor microenvironment and tumor-associated macrophage markers in colorectal cancer. *Tumour Biol.* **2018**, *40*, 1010428318810059. [CrossRef]
90. Mola, S.; Pandolfo, C.; Sica, A.; Porta, C. The Macrophages-microbiota interplay in colorectal cancer (CRC)-related inflammation: Prognostic and therapeutic significance. *Int. J. Mol. Sci.* **2020**, *18*, 6866. [CrossRef] [PubMed]
91. Zhang, D.; Qiu, X.; Li, J.; Zheng, S.; Li, L.; Zhao, H. TGF-beta secreted by tumor-associated macrophages promotes proliferation and invasion of colorectal cancer via miR-34a-VEGF axis. *Cell Cycle* **2018**, *17*, 2766–2778. [CrossRef] [PubMed]
92. Pedrosa, L.; Esposito, F.; Thomson, T.M.; Maurel, J. The tumor microenvironment in colorectal cancer therapy. *Cancers* **2019**, *14*, 1172. [CrossRef]
93. Lin, X.; Wang, S.; Sun, M.; Zhang, C.; Wei, C.; Yang, C.; Dou, R.; Liu, Q.; Xiong, B. MiR-195-5p/NOTCH2-mediated EMT modulates IL-4 secretion in colorectal cancer to affect M2-like TAM polarization. *J. Hematol. Oncol.* **2019**, *26*, 20. [CrossRef]
94. Mizuno, R.; Kawada, K.; Sakai, Y. Prostaglandin E2/EP signaling in the tumor microenvironment of colorectal cancer. *Int. J. Mol. Sci.* **2019**, *11*, 6254. [CrossRef] [PubMed]
95. Song, J.J.; Li, W. MiR-10b suppresses the growth and metastasis of colorectal cancer cell by targeting FGF13. *Eur. Rev. Med. Pharmacol. Sci.* **2019**, *23*, 576–587.
96. Wai Hon, K.; Zainal Abidin, S.A.; Othman, I.; Naidu, R. Insights into the role of microRNAs in colorectal cancer (CRC) metabolism. *Cancers* **2020**, *31*, 2462. [CrossRef]
97. Hardbower, D.M.; Coburn, L.A.; Asim, M.; Singh, K.; Sierra, J.C.; Barry, D.P.; Gobert, A.P.; Piazuelo, M.B.; Washington, M.K.; Wilson, K.T. EGFR-mediated macrophage activation promotes colitis-associated tumorigenesis. *Oncogene* **2017**, *6*, 3807–3819. [CrossRef]
98. D'Haene, N.; Koopmansch, C.; Van Eycke, Y.R.; Hulet, F.; Allard, J.; Bouri, S.; Rorive, S.; Remmelink, M.; Decaestecker, C.; Maris, C.; et al. The prognostic value of the combination of low VEGFR-1 and high VEGFR-2 expression in endothelial cells of colorectal cancer. *Int. J. Mol. Sci.* **2018**, *9*, 3536. [CrossRef] [PubMed]
99. Na, S.Y.; Myung, S.J. Obesity and colorectal cancer. *Korean J. Gastroenterol.* **2012**, *59*, 16–26. [CrossRef]
100. Fettig, L.M.; Yee, D. Advances in insulin-like growth factor biology and -directed cancer therapeutics. *Adv. Cancer Res.* **2020**, *147*, 229–257. [PubMed]
101. Codony-Servat, J.; Cuatrecasas, M.; Asensio, E.; Montironi, C.; Martínez-Cardús, A.; Marín-Aguilera, M.; Horndler, C.; Martínez-Balibrea, E.; Rubini, M.; Jares, P.; et al. Nuclear IGF-1R predicts chemotherapy and targeted therapy resistance in metastatic colorectal cancer. *Br. J. Cancer* **2017**, *5*, 1777–1786. [CrossRef] [PubMed]
102. Shiratsuchi, I.; Akagi, Y.; Kawahara, A.; Kinugasa, T.; Romeo, K.; Yoshida, T.; Ryu, Y.; Gotanda, Y.; Kage, M.; Shirouzu, K. Expression of IGF-1 and IGF-1R and their relation to clinicopathological factors in colorectal cancer. *Anticancer Res.* **2011**, *31*, 2541–2545. [PubMed]
103. Ciulei, G.; Orasan, O.H.; Coste, S.C.; Cozma, A.; Negrean, V.; Procopciuc, L.M. Vitamin D and the insulin-like growth factor system: Implications for colorectal neoplasia. *Eur. J. Clin. Invest.* **2020**, *50*, 13265. [CrossRef] [PubMed]
104. Hosseini, S.A.; Zand, H.; Cheraghpour, M. The influence of curcumin on the downregulation of MYC, Insulin and IGF-1 receptors: A possible mechanism underlying the anti-growth and anti-migration in chemoresistant colorectal cancer cells. *Medicina* **2019**, *3*, 90. [CrossRef]
105. Ramezani, P.; Abnous, K.; Taghdisi, S.M.; Zahiri, M.; Ramezani, M.; Alibolandi, M. Targeted MMP-2 responsive chimeric polymersomes for therapy against colorectal cancer *Colloids Surf. B Biointerfaces* **2020**, *193*, 111135. [CrossRef] [PubMed]
106. Sipos, F.; Székely, H.; Kis, I.D.; Tulassay, Z.; Műzes, G. Relation of the IGF/IGF1R system to autophagy in colitis and colorectal cancer. *World J. Gastroenterol.* **2017**, *14*, 8109–8119. [CrossRef]
107. Wang, S.Q.; Yang, X.Y.; Cui, S.X.; Gao, Z.H.; Qu, X.J. Heterozygous knockout insulin-like growth factor-1 receptor (IGF-1R) regulates mitochondrial functions and prevents colitis and colorectal cancer. *Free Radic. Biol. Med.* **2019**, *134*, 87–98. [CrossRef]
108. Shali, H.; Ahmadi, M.; Kafil, H.S.; Dorosti, A.; Yousefi, M. IGF1R and c-met as therapeutic targets for colorectal cancer. *Biomed. Pharmacother.* **2016**, *82*, 528–536. [CrossRef]
109. Jiang, B.; Zhang, X.; Du, L.L.; Wang, Y.; Liu, D.B.; Han, C.Z.; Jing, J.X.; Zhao, X.W.; Xu, X.Q. Possible roles of insulin, IGF-1 and IGFBPs in initiation and progression of colorectal cancer. *World J. Gastroenterol.* **2014**, *14*, 1608–1613. [CrossRef]
110. Sandhu, M.S.; Dunger, D.B.; Giovannucci, E.L. Insulin, insulin-like growth factor-I (IGF-I), IGF binding proteins, their biologic interactions, and colorectal cancer. *J. Natl. Cancer Inst.* **2002**, *3*, 972–980. [CrossRef]

111. Hu, J.; Liu, X.; Chi, J.; Che, K.; Feng, Y.; Zhao, S.; Wang, Z.; Wang, Y. Expressions of IGF-1, ERK, GLUT4, IRS-1 in metabolic syndrome complicated with colorectal cancer and their associations with the clinical characteristics of CRC. *Cancer Biomark.* **2018**, *21*, 883–891. [CrossRef] [PubMed]
112. Peters, G.; Gongoll, S.; Langner, C.; Mengel, M.; Piso, P.; Klempnauer, J.; Rüschoff, J.; Kreipe, H.; von Wasielewski, R. IGF-1R, IGF-1 and IGF-2 expression as potential prognostic and predictive markers in colorectal-cancer. *Virchows Arch.* **2003**, *443*, 139–145. [CrossRef] [PubMed]
113. Alagaratnam, S.; Loizidou, M.; Yang, S.Y.; Fuller, B.; Ramesh, B. Increased expression of IGF-1Ec with increasing colonic polyp dysplasia and colorectal cancer. *J. Cancer Res. Clin. Oncol.* **2020**, *146*, 2861–2870. [CrossRef]
114. Pankaj, J.; Kumari, J.R.; Kim, W.; Lee, S.A. Insulin-like Growth Factor-1, IGF-binding Protein-3, C-peptide and colorectal cancer: A case-control study. *Asian Pac. J. Cancer Prev.* **2015**, *16*, 3735–3740. [PubMed]
115. Hu, X.; Zheng, W.; Luo, Y.; Ou, X.; Song, L.; Zhang, S.; He, T.; Guo, Z.; Zhu, J.; Shi, H.; et al. Arca subcrenata polypeptides inhibit human colorectal cancer HT-29 cells growth via suppression of IGF-1R/Akt/mTOR signaling and ATP production. *Nutr. Cancer* **2020**, *72*, 260–272. [CrossRef]
116. Daveri, E.; Adamo, A.M.; Alfine, E.; Zhu, W.; Oteiza, P.I. Hexameric procyanidins inhibit colorectal cancer cell growth through both redox and non-redox regulation of the epidermal growth factor signaling pathway. *Redox Biol.* **2021**, *38*, 101830. [CrossRef]
117. Custodio, A.; Feliu, J. Prognostic and predictive biomarkers for epidermal growth factor receptor-targeted therapy in colorectal cancer: Beyond KRAS mutations. *Crit. Rev. Oncol. Hematol.* **2013**, *85*, 45–81. [CrossRef]
118. Khan, K.; Valeri, N.; Dearman, C.; Rao, S.; Watkins, D.; Starling, N.; Chau, I.; Cunningham, D. Targeting EGFR pathway in metastatic colorectal cancer- tumour heterogeniety and convergent evolution. *Crit. Rev. Oncol. Hematol.* **2019**, *143*, 153–163. [CrossRef] [PubMed]
119. Miyamoto, Y.; Muguruma, N.; Fujimoto, S.; Okada, Y.; Kida, Y.; Nakamura, F.; Tanaka, K.; Nakagawa, T.; Kitamura, S.; Okamoto, K.; et al. Epidermal growth factor receptor-targeted molecular imaging of colorectal tumors: Detection and treatment evaluation of tumors in animal models. *Cancer Sci.* **2019**, *110*, 1921–1930. [CrossRef]
120. de Mello, R.A.; Marques, A.M.; Araújo, A. Epidermal growth factor receptor and metastatic colorectal cancer: Insights into target therapies. *World J. Gastroenterol.* **2013**, *14*, 6315–6318. [CrossRef]
121. Grossmann, A.H.; Samowitz, W.S. Epidermal growth factor receptor pathway mutations and colorectal cancer therapy. *Arch. Pathol. Lab. Med.* **2011**, *135*, 1278–1282. [CrossRef] [PubMed]
122. Nappi, A.; Berretta, M.; Romano, C.; Tafuto, S.; Cassata, A.; Casaretti, R.; Silvestro, L.; Divitiis, C.; Alessandrini, L.; Fiorica, F.; et al. Metastatic colorectal cancer: Role of target therapies and future perspectives. *Curr. Cancer Drug Targets* **2018**, *18*, 421–429. [CrossRef]
123. Yang, W.J.; Shen, X.J.; Ma, X.X.; Tan, Z.G.; Song, Y.; Guo, Y.T.; Yuan, M. Correlation of human epidermal growth factor receptor protein expression and colorectal cancer. *World J. Gastroenterol.* **2015**, *28*, 8687–8696. [CrossRef]
124. Greally, M.; Kelly, C.M.; Cercek, A. HER2: An emerging target in colorectal cancer. *Curr. Probl. Cancer* **2018**, *42*, 560–571. [CrossRef]
125. Lawan, A.I.; Ogunbiyi, J.O. Epidermal growth factor receptor expression of colorectal carcinoma in Nigerian patients. *West. Afr. J. Med.* **2020**, *37*, 100–105.
126. Xu, W.; Jing, H.; Zhang, F.; Xu, W.; Jing, H.; Zhang, F. Epidermal growth factor receptor-targeted therapy in colorectal cancer. *Front. Biosci.* **2016**, *1*, 410–418. [CrossRef]
127. Nemanqani, D.M.; Aftab, K.; Al-Malki, S.H.; Al-Sufyani, W.M. Expression of epidermal growth factor receptor in colorectal adenocarcinoma and its correlation with clinicopathological factors. *J. Coll. Phys. Surg. Pak.* **2018**, *28*, 527–531. [CrossRef]
128. Yun, S.; Kwak, Y.; Nam, S.K.; Seo, A.N.; Oh, H.K.; Kim, D.W.; Kang, S.B.; Lee, H.S. Ligand-independent epidermal growth factor receptor overexpression correlates with poor prognosis in colorectal cancer. *Cancer Res. Treat.* **2018**, *50*, 1351–1361. [CrossRef]
129. Wojtukiewicz, M.Z.; Mysliwiec, M.; Sierko, E.; Sobierska, M.; Kruszewska, J.; Lipska, A.; Radziwon, P.; Tucker, S.C.; Honn, K.V. Elevated microparticles, thrombin-antithrombin and VEGF Levels in colorectal cancer patients undergoing chemotherapy. *Pathol. Oncol. Res.* **2020**, *26*, 2499–2507. [CrossRef]
130. Wu, Q.B.; Chen, J.; Zhu, J.W.; Yin, X.; You, H.Y.; Lin, Y.R.; Zhu, H.Q. MicroRNA-125 inhibits RKO colorectal cancer cell growth by targeting VEGF. *Int. J. Mol. Med.* **2018**, *42*, 665–673. [CrossRef]
131. Lan, J.; Li, H.; Luo, X.; Hu, J.; Wang, G. BRG1 promotes VEGF-A expression and angiogenesis in human colorectal cancer cells. *Exp. Cell Res.* **2017**, *15*, 236–242. [CrossRef] [PubMed]
132. Coşkun, Ö.; Öztopuz, Ö.; Özkan, Ö.F. Determination of IL-6, TNF-alpha and VEGF levels in the serums of patients with colorectal cancer. *Cell Mol. Biol.* **2017**, *20*, 97–101. [CrossRef]
133. Mathonnet, M.; Perraud, A.; Christou, N.; Akil, H.; Melin, C.; Battu, S.; Jauberteau, M.O.; Denizot, Y. Hallmarks in colorectal cancer: Angiogenesis and cancer stem-like cells. *World J. Gastroenterol.* **2014**, *21*, 4189–4196. [CrossRef]
134. Bhattacharya, R.; Fan, F.; Wang, R.; Ye, X.; Xia, L.; Boulbes, D.; Ellis, L.M. Intracrine VEGF signalling mediates colorectal cancer cell migration and invasion. *Br. J. Cancer* **2017**, *5*, 848–855. [CrossRef]
135. Mohamed, S.Y.; Mohammed, H.L.; Ibrahim, H.M.; Mohamed, E.M.; Salah, M. Role of VEGF, CD105, and CD31 in the prognosis of colorectal cancer cases. *J. Gastrointest. Cancer* **2019**, *50*, 23–34. [CrossRef]

136. Dinami, R.; Porru, M.; Amoreo, C.A.; Sperduti, I.; Mottolese, M.; Buglioni, S.; Marinelli, D.; Maugeri-Saccà, M.; Sacconi, A.; Blandino, G.; et al. TRF2 and VEGF-A: An unknown relationship with prognostic impact on survival of colorectal cancer patients. *J. Exp. Clin. Cancer Res.* **2020**, *15*, 111. [CrossRef]
137. Karpuz, T.; Araz, M.; Korkmaz, L.; Kılınc, I.; Findik, S.; Karaagaç, M.; Eryilmaz, M.K.; Artac, M. The prognostic value of serum semaphorin3A and VEGF Levels in patients with metastatic colorectal cancer. *J. Gastrointest. Cancer* **2020**, *51*, 491–497. [CrossRef]
138. Jannuzzi, A.T.; Özhan, G.; Yanar, H.T.; Alpertunga, B. VEGF gene polymorphisms and susceptibility to colorectal cancer. *Genet. Test. Mol. Biomark.* **2015**, *19*, 133–137. [CrossRef]
139. Chen, X.; Xu, X.; Pan, B.; Zeng, K.; Xu, M.; Liu, X.; He, B.; Pan, Y.; Sun, H.; Wang, S. miR-150-5p suppresses tumor progression by targeting VEGFA in colorectal cancer. *Aging* **2018**, *26*, 3421–3437. [CrossRef]
140. Kim, Y.I.; Jeong, S.; Jung, K.O.; Song, M.G.; Lee, C.H.; Chung, S.J.; Park, J.Y.; Cha, M.G.; Lee, S.G.; Jun, B.H.; et al. Simultaneous detection of EGFR and VEGF in colorectal cancer using fluorescence-Raman Endoscopy. *Sci. Rep.* **2017**, *21*, 1035. [CrossRef]
141. Herichova, I.; Reis, R.; Hasakova, K.; Vician, M.; Zeman, M. Sex-dependent regulation of estrogen receptor beta in human colorectal cancer tissue and its relationship with clock genes and VEGF-A expression. *Physiol. Res.* **2019**, *20*, S297–S305. [CrossRef]
142. Narayanankutty, A. PI3K/ Akt/ mTOR pathway as a therapeutic target for colorectal cancer: A review of preclinical and clinical evidence. *Curr. Drug Targets* **2019**, *20*, 217–1226. [CrossRef] [PubMed]
143. Moafian, Z.; Maghrouni, A.; Soltani, A.; Hashemy, S.I. Cross-talk between non-coding RNAs and PI3K/AKT/mTOR pathway in colorectal cancer. *Mol. Biol. Rep.* **2021**, *48*, 4797–4811. [CrossRef]
144. Fang, J.Y.; Richardson, B.C. The MAPK signalling pathways and colorectal cancer. *Lancet Oncol.* **2005**, *6*, 322–327. [CrossRef]
145. Kim, E.K.; Choi, E.J. Pathological roles of MAPK signaling pathways in human diseases. *Biochim. Biophys. Acta* **2010**, *1802*, 396–405. [CrossRef]
146. Lech, G.; Słotwiński, R.; Słodkowski, M.; Krasnodębski, I.W. Colorectal cancer tumour markers and biomarkers: Recent therapeutic advances. *World J. Gastroenterol.* **2016**, *7*, 1745–1755. [CrossRef]
147. Martinelli, E.; Ciardiello, D.; Martini, G.; Troiani, T.; Cardone, C.; Vitiello, P.P.; Normanno, N.; Rachiglio, A.M.; Maiello, E.; Latiano, T.; et al. Implementing anti-epidermal growth factor receptor (EGFR) therapy in metastatic colorectal cancer: Challenges and future perspectives. *Ann. Oncol.* **2020**, *31*, 30–40. [CrossRef]
148. Pape, J.; Magdeldin, T.; Stamati, K.; Nyga, A.; Loizidou, M.; Emberton, M.; Cheema, U. Cancer-associated fibroblasts mediate cancer progression and remodel the tumouroid stroma. *Br. J. Cancer* **2020**, *123*, 1178–1190. [CrossRef]
149. Yao, W.; Lin, Z.; Shi, P.; Chen, B.; Wang, G.; Huang, J.; Sui, Y.; Liu, Q.; Li, S.; Lin, X.; et al. Delicaflavone induces ROS-mediated apoptosis and inhibits PI3K/AKT/mTOR and Ras/MEK/Erk signaling pathways in colorectal cancer cells. *Biochem. Pharmacol.* **2020**, *171*, 113680. [CrossRef]
150. Hybel, T.E.; Dietrichs, D.; Sahana, J.; Corydon, T.J.; Nassef, M.Z.; Wehland, M.; Krüger, M.; Magnusson, N.E.; Bauer, J.; Utpatel, K.; et al. Simulated microgravity influences VEGF, MAPK, and PAM signaling in prostate cancer cells. *Int. J. Mol. Sci.* **2020**, *13*, 1263. [CrossRef]
151. Duan, S.; Huang, W.; Liu, X.; Liu, X.; Chen, N.; Xu, Q.; Hu, Y.; Song, W.; Zhou, J. IMPDH2 promotes colorectal cancer progression through activation of the PI3K/AKT/mTOR and PI3K/AKT/FOXO1 signaling pathways. *J. Exp. Clin. Cancer Res.* **2018**, *5*, 304. [CrossRef]
152. Wei, R.; Xiao, Y.; Song, Y.; Yuan, H.; Luo, J.; Xu, W. FAT4 regulates the EMT and autophagy in colorectal cancer cells in part via the PI3K-AKT signaling axis. *J. Exp. Clin. Cancer Res.* **2019**, *4*, 112. [CrossRef]
153. Pandurangan, A.K. Potential targets for prevention of colorectal cancer: A focus on PI3K/Akt/mTOR and Wnt pathways. *Asian Pac. J. Cancer Prev.* **2013**, *14*, 2201–2205. [CrossRef]
154. Cheng, H.; Jiang, X.; Zhang, Q.; Ma, J.; Cheng, R.; Yong, H.; Shi, H.; Zhou, X.; Ge, L.; Gao, G. Naringin inhibits colorectal cancer cell growth by repressing the PI3K/AKT/mTOR signaling pathway. *Exp. Ther. Med.* **2020**, *19*, 3798–3804. [CrossRef]
155. Aggarwal, V.; Tuli, H.S.; Varol, A.; Thakral, F.; Yerer, M.B.; Sak, K.; Varol, M.; Jain, A.; Khan, M.A.; Sethi, G. Role of reactive oxygen species in cancer progression: Molecular mechanisms and recent advancements. *Biomolecules* **2019**, *13*, 735. [CrossRef]
156. Winder, T.; Lenz, H.J. Vascular endothelial growth factor and epidermal growth factor signaling pathways as therapeutic targets for colorectal cancer. *Gastroenterology* **2010**, *138*, 2163–2176. [CrossRef]
157. Yarom, N.; Jonker, D.J. The role of the epidermal growth factor receptor in the mechanism and treatment of colorectal cancer. *Discov. Med.* **2011**, *11*, 95–105.
158. Foroughi, S.; Tie, J.; Gibbs, P.; Burgess, A.W. Epidermal growth factor receptor ligands: Targets for optimizing treatment of metastatic colorectal cancer. *Growth Factors* **2019**, *37*, 209–225. [CrossRef]
159. Martini, G.; Troiani, T.; Cardone, C.; Vitiello, P.; Sforza, V.; Ciardiello, D.; Napolitano, S.; Della Corte, C.M.; Morgillo, F.; Raucci, A.; et al. Present and future of metastatic colorectal cancer treatment: A review of new candidate targets. *World J. Gastroenterol.* **2017**, *14*, 4675–4688. [CrossRef]
160. Weinberg, B.A.; Hartley, M.L.; Salem, M.E. Precision medicine in metastatic colorectal cancer: Relevant carcinogenic pathways and targets-part 1: Biologic therapies targeting the epidermal growth factor receptor and vascular endothelial growth factor. *Oncology* **2017**, *15*, 539–548.
161. Price, T.J.; Tang, M.; Gibbs, P.; Haller, D.G.; Peeters, M.; Arnold, D.; Segelov, E.; Roy, A.; Tebbutt, N.; Pavlakis, N.; et al. Targeted therapy for metastatic colorectal cancer. *Expert Rev. Anticancer Ther.* **2018**, *18*, 991–1006. [CrossRef] [PubMed]

162. Martini, G.; Ciardiello, D.; Vitiello, P.P.; Napolitano, S.; Cardone, C.; Cuomo, A.; Troiani, T.; Ciardiello, F.; Martinelli, E. Resistance to anti-epidermal growth factor receptor in metastatic colorectal cancer: What does still need to be addressed? *Cancer Treat. Rev.* **2020**, *86*, 102023. [CrossRef] [PubMed]
163. Therkildsen, C.; Bergmann, T.K.; Henrichsen-Schnack, T.; Ladelund, S.; Nilbert, M. The predictive value of KRAS, NRAS, BRAF, PIK3CA and PTEN for anti-EGFR treatment in metastatic colorectal cancer: A systematic review and meta-analysis. *Acta Oncol.* **2014**, *53*, 852–864. [CrossRef] [PubMed]
164. Canavese, M.; Ngo, D.T.; Maddern, G.J.; Hardingham, J.E.; Price, T.J.; Hauben, E. Biology and therapeutic implications of VEGF-A splice isoforms and single-nucleotide polymorphisms in colorectal cancer. *Int. J. Cancer* **2017**, *5*, 2183–2191. [CrossRef] [PubMed]
165. Napolitano, S.; Matrone, N.; Muddassir, A.L.; Martini, G.; Sorokin, A.; De Falco, V.; Giunta, E.F.; Ciardiello, D. Triple blockade of EGFR, MEK and PD-L1 has antitumor activity in colorectal cancer models with constitutive activation of MAPK signaling and PD-L1 overexpression. *J. Exp. Clin. Cancer Res.* **2019**, *16*, 492. [CrossRef]
166. Itatani, Y.; Yamamoto, T.; Zhong, C.; Molinolo, A.A.; Ruppel, J.; Hegde, P.; Taketo, M.M.; Ferrara, N. Suppressing neutrophil-dependent angiogenesis abrogates resistance to anti-VEGF antibody in a genetic model of colorectal cancer. *Proc. Natl. Acad. Sci. USA* **2020**, *1*, 21598–21608. [CrossRef]
167. Ferrara, N.; Hillan, K.J.; Gerber, H.P.; Novotny, W. Discovery and development of bevacizumab, an anti-VEGF antibody for treating cancer. *Nat. Rev. Drug Discov.* **2004**, *3*, 391–400. [CrossRef]
168. Marisi, G.; Scarpi, E.; Passardi, A.; Nanni, O.; Ragazzini, A.; Valgiusti, M.; Casadei-Gardini, A.; Neri, L.M.; Frassineti, G.L.; Amadori, D.; et al. Circulating VEGF and eNOS variations as predictors of outcome in metastatic colorectal cancer patients receiving bevacizumab. *Sci. Rep.* **2017**, *2*, 1293. [CrossRef]
169. Wagner, J.; Kline, C.L.; Zhou, L.; Khazak, V.; El-Deiry, W.S. Anti-tumor effects of ONC201 in combination with VEGF-inhibitors significantly impacts colorectal cancer growth and survival in vivo through complementary non-overlapping mechanisms. *J. Exp. Clin. Cancer Res.* **2018**, *22*, 11. [CrossRef]
170. Deng, Y.; Li, X. Fruquintinib and its use in the treatment of metastatic colorectal cancer. *Future Oncol.* **2019**, *15*, 2571–2576. [CrossRef]
171. Abdalla, A.N.; Malki, W.H.; Qattan, A.; Shahid, I.; Hossain, M.A.; Ahmed, M. Chemosensitization of HT29 and HT29-5FU cell lines by a combination of a multi-tyrosine kinase inhibitor and 5FU downregulates ABCC1 and inhibits PIK3CA in light of their importance in Saudi colorectal cancer. *Molecules* **2021**, *11*, 334. [CrossRef]
172. Miyamoto, Y.; Suyama, K.; Baba, H. Recent advances in targeting the EGFR signaling pathway for the treatment of metastatic colorectal cancer. *Int. J. Mol. Sci.* **2017**, *2*, 752. [CrossRef]
173. Pranteda, A.; Piastra, V.; Stramucci, L.; Fratantonio, D.; Bossi, G. The p38 MAPK signaling activation in colorectal cancer upon therapeutic treatments. *Int. J. Mol. Sci.* **2020**, *16*, 2773. [CrossRef]
174. Grossi, V.; Peserico, A.; Tezil, T.; Simone, C. p38alpha MAPK pathway: A key factor in colorectal cancer therapy and chemoresistance. *World J. Gastroenterol.* **2014**, *7*, 9744–9758. [CrossRef]
175. Wang, J.; Liang, D.; Zhang, X.P.; He, C.F.; Cao, L.; Zhang, S.Q.; Xiao, X.; Li, S.J.; Cao, Y.X. Novel PI3K/Akt/mTOR signaling inhibitor, W922, prevents colorectal cancer growth via the regulation of autophagy. *Int. J. Oncol.* **2021**, *58*, 70–82. [CrossRef]
176. Wei, H.; Dong, C.; Shen, Z. Kallikrein-related peptidase (KLK10) cessation blunts colorectal cancer cell growth and glucose metabolism by regulating the PI3K/Akt/mTOR pathway. *Neoplasma* **2020**, *67*, 889–897. [CrossRef]
177. Helmy, M.W.; Ghoneim, A.I.; Katary, M.A.; Elmahdy, R.K. The synergistic anti-proliferative effect of the combination of diosmin and BEZ-235 (dactolisib) on the HCT-116 colorectal cancer cell line occurs through inhibition of the PI3K/Akt/mTOR/NF-kappaB axis. *Mol. Biol. Rep.* **2020**, *47*, 2217–2230. [CrossRef]
178. Li, S.; Wang, X.; Wang, G.; Shi, P.; Lin, S.; Xu, D.; Chen, B.; Liu, A.; Huang, L.; Lin, X.; et al. Ethyl acetate extract of selaginella doederleinii hieron induces cell autophagic death and apoptosis in colorectal cancer via PI3K-Akt-mTOR and AMPKalpha-signaling pathways. *Front. Pharmacol.* **2020**, *19*, 565090. [CrossRef] [PubMed]
179. Han, Y.H.; Mun, J.G.; Jeon, H.D.; Kee, J.Y.; Hong, S.H. Betulin inhibits lung metastasis by inducing cell cycle arrest, autophagy, and apoptosis of metastatic colorectal cancer cells. *Nutrients* **2019**, *26*, 66. [CrossRef]
180. Li, N.; Zhang, Z.; Jiang, G.; Sun, H.; Yu, D. Nobiletin sensitizes colorectal cancer cells to oxaliplatin by PI3K/Akt/MTOR pathway. *Front. Biosci.* **2019**, *1*, 303–312.
181. Shahi Thakuri, P.; Luker, G.D.; Tavana, H. Cyclical treatment of colorectal tumor spheroids induces resistance to MEK inhibitors. *Transl. Oncol.* **2019**, *12*, 404–416. [CrossRef]
182. Li, J.; Qin, S.; Xu, R.H.; Shen, L.; Xu, J.; Bai, Y.; Yang, L.; Deng, Y.; Chen, Z.D.; Zhong, H.; et al. Effect of fruquintinib vs placebo on overall survival in patients with previously treated metastatic colorectal cancer: The FRESCO randomized clinical trial. *JAMA* **2018**, *26*, 486–2496. [CrossRef]
183. Arqués, O.; Chicote, I.; Puig, I.; Tenbaum, S.P.; Argilés, G.; Dienstmann, R.; Fernández, N.; Caratù, G.; Matito, J.; Silberschmidt, D.; et al. Tankyrase inhibition blocks Wnt/beta-catenin pathway and reverts resistance to PI3K and AKT inhibitors in the treatment of colorectal cancer. *Clin. Cancer Res.* **2016**, *1*, 644–656. [CrossRef]
184. Wanigasooriya, K.; Tyler, R.; Barros-Silva, J.D.; Sinha, Y.; Ismail, T.; Beggs, A.D. Radiosensitising cancer using phosphatidylinositol-3-Kinase (PI3K), protein kinase B (AKT) or mammalian target of rapamycin (mTOR) inhibitors. *Cancers* **2020**, *18*, 1278. [CrossRef]

185. Ganesan, P.; Janku, F.; Naing, A.; Hong, D.S.; Tsimberidou, A.M.; Falchook, G.S.; Wheler, J.J.; Piha-Paul, S.A.; Fu, S.; Stepanek, V.M.; et al. Target-based therapeutic matching in early-phase clinical trials in patients with advanced colorectal cancer and PIK3CA mutations. *Mol. Cancer Ther.* **2013**, *12*, 2857–2863. [CrossRef] [PubMed]
186. Garrido-Laguna, I.; Hong, D.S.; Janku, F.; Nguyen, L.M.; Falchook, G.S.; Fu, S.; Wheler, J.J.; Luthra, R.; Naing, A.; Wang, X.; et al. KRASness and PIK3CAness in patients with advanced colorectal cancer: Outcome after treatment with early-phase trials with targeted pathway inhibitors. *PLoS ONE* **2012**, *7*, e38033. [CrossRef]
187. Kyriakopoulos, C.E.; Braden, A.M.; Kolesar, J.M.; Eickhoff, J.C.; Bailey, H.H.; Heideman, J.; Liu, G.; Wisinski, K.B. A phase I study of tivantinib in combination with temsirolimus in patients with advanced solid tumors. *Invest. New Drugs* **2017**, *35*, 290–297. [CrossRef]
188. Ng, K.; Tabernero, J.; Hwang, J.; Bajetta, E.; Sharma, S.; Del Prete, S.A.; Arrowsmith, E.R.; Ryan, D.P.; Sedova, M.; Jin, J.; et al. Phase II study of everolimus in patients with metastatic colorectal adenocarcinoma previously treated with bevacizumab-, fluoropyrimidine-, oxaliplatin-, and irinotecan-based regimens. *Clin. Cancer Res.* **2013**, *15*, 3987–3995. [CrossRef]
189. Van Geel, R.M.J.M.; Tabernero, J.; Elez, E.; Bendell, J.C.; Spreafico, A.; Schuler, M.; Yoshino, T.; Delord, J.P.; Yamada, Y.; Lolkema, M.P.; et al. A Phase Ib dose-escalation study of encorafenib and cetuximab with or without alpelisib in metastatic BRAF-mutant colorectal cancer. *Cancer Discov.* **2017**, *7*, 610–619. [CrossRef] [PubMed]
190. Van Brummelen, E.M.J.; Huijberts, S.; van Herpen, C.; Desar, I.; Opdam, F.; van Geel, R.; Marchetti, S.; Steeghs, N.; Monkhorst, K.; Thijssen, B.; et al. Phase I study of afatinib and selumetinib in patients with KRAS mutated colorectal, non-small cell lung, and pancreatic cancer. *Oncologist* **2021**, *26*, 290–e545. [CrossRef]
191. Folprecht, G.; Tabernero, J.; Köhne, C.H.; Zacharchuk, C.; Paz-Ares, L.; Rojo, F.; Quinn, S.; Casado, E.; Salazar, R.; Abbas, R.; et al. Phase I pharmacokinetic/pharmacodynamic study of EKB-569, an irreversible inhibitor of the epidermal growth factor receptor tyrosine kinase, in combination with irinotecan, 5-fluorouracil, and leucovorin (FOLFIRI) in first-line treatment of patients with metastatic colorectal cancer. *Clin. Cancer Res.* **2008**, *1*, 215–223.
192. Tabernero, J.; Cervantes, A.; Rivera, F.; Martinelli, E.; Rojo, F.; von Heydebreck, A.; Macarulla, T.; Rodriguez-Braun, E.; Eugenia Vega-Villegas, M.; Senger, S.; et al. Pharmacogenomic and pharmacoproteomic studies of cetuximab in metastatic colorectal cancer: Biomarker analysis of a phase I dose-escalation study. *J. Clin. Oncol.* **2010**, *1*, 1181–1189. [CrossRef]
193. Antoniotti, C.; Borelli, B.; Rossini, D.; Pietrantonio, F.; Morano, F.; Salvatore, L.; Lonardi, S.; Marmorino, F.; Tamberi, S.; Corallo, S.; et al. AtezoTRIBE: A randomised phase II study of FOLFOXIRI plus bevacizumab alone or in combination with atezolizumab as initial therapy for patients with unresectable metastatic colorectal cancer. *BMC Cancer* **2020**, *22*, 683. [CrossRef] [PubMed]
194. Damato, A.; Iachetta, F.; Antonuzzo, L.; Nasti, G.; Bergamo, F.; Bordonaro, R.; Maiello, E.; Zaniboni, A.; Tonini, G.; Romagnani, A.; et al. Phase II study on first-line treatment of NIVolumab in combination with folfoxiri/bevacizumab in patients with advanced colorectal cancer RAS or BRAF mutated—NIVACOR trial (GOIRC-03-2018). *BMC Cancer* **2020**, *31*, 822. [CrossRef] [PubMed]
195. Meric-Bernstam, F.; Hurwitz, H.; Raghav, K.P.S.; McWilliams, R.R.; Fakih, M.; Vander Walde, A.; Swanton, C.; Kurzrock, R.; Burris, H.; Sweeney, C.; et al. Pertuzumab plus trastuzumab for HER2-amplified metastatic colorectal cancer (MyPathway): An updated report from a multicentre, open-label, phase 2a, multiple basket study. *Lancet Oncol.* **2019**, *20*, 518–530. [CrossRef]
196. Mileo, A.M.; Nisticò, P.; Miccadei, S. Polyphenols: Immunomodulatory and Therapeutic Implication in colorectal cancer. *Front. Immunol.* **2019**, *11*, 729. [CrossRef]

International Journal of *Molecular Sciences*

MDPI

Review

ELTD1—An Emerging Silent Actor in Cancer Drama Play

Ani-Simona Sevastre [1,†], Iuliana M. Buzatu [2,†], Carina Baloi [2,†], Alexandru Oprita [2], Alexandra Dragoi [2], Ligia G. Tataranu [3], Oana Alexandru [4], Stefania Tudorache [5,*] and Anica Dricu [2]

1 Department of Pharmaceutical Technology, Faculty of Pharmacy, University of Medicine and Pharmacy, 200349 Craiova, Romania; anifetea_umf@yahoo.com
2 Department of Biochemistry, Faculty of Medicine, University of Medicine and Pharmacy, 200349 Craiova, Romania; buzatu_iuliana@yahoo.com (I.M.B.); carina_baloi@yahoo.com (C.B.); opr.alexandru@gmail.com (A.O.); alexa.dragoi@yahoo.com (A.D.); anica.dricu@live.co.uk (A.D.)
3 Department of Neurosurgery, Bagdasar-Arseni Hospital, 041915 Bucharest, Romania; ttranu@gmail.com
4 Department of Neurology, Faculty of Medicine, University of Medicine and Pharmacy, 200349 Craiova, Romania; oanale@hotmail.com
5 Department of Obstetrics and Gynecology, Faculty of Medicine, University of Medicine and Pharmacy, 200349 Craiova, Romania
* Correspondence: stefania.tudorache@gmail.com
† These authors contributed equally.

Abstract: The epidermal growth factor, latrophilin, and seven transmembrane domain–containing protein 1 (ELTD1), is a member of the G–protein coupled receptors (GPCRs) superfamily. Although discovered in 2001, ELTD1 has been investigated only by a few research groups, and important data about its role in normal and tumor cells is still missing. Even though its functions and structure are not yet fully understood, recent studies show that ELTD1 has a role in both physiological and pathological angiogenesis, and it appears to be a very important biomarker and a molecular target in cancer diseases. Upregulation of ELTD1 in malignant cells has been reported, and correlated with poor cancer prognosis. This review article aims to compile the existing data and to discuss the current knowledge on ELTD1 structure and signaling, and its role in physiological and neoplastic conditions.

Keywords: ELTD1; biomarker; angiogenesis; cancer

Citation: Sevastre, A.-S.; Buzatu, I.M.; Baloi, C.; Oprita, A.; Dragoi, A.; Tataranu, L.G.; Alexandru, O.; Tudorache, S.; Dricu, A. ELTD1—An Emerging Silent Actor in Cancer Drama Play. *Int. J. Mol. Sci.* **2021**, 22, 5151. https://doi.org/10.3390/ijms22105151

Academic Editor: Jose Carlos Garcia-Borron

Received: 26 March 2021
Accepted: 10 May 2021
Published: 13 May 2021

1. Introduction

With major implications in cancer, the tyrosine kinases receptors (RTKs) are one of the most analyzed and reviewed research topics [1–7]. New ways of activating and transmitting the intracellular signal involved in cancer are constantly being discovered. Data indicates that one route of RTKs activation may be through agonists of G protein–coupled receptors (GPCRs) [8–10]. This phenomenon is called transactivation, and is considered to be an important pathway, involved in growth–promoting activity of GPCR ligands [11].

Comprising over 900 members and with more than 2% of the genes encoded by the human genome, the GPCRs family is by far the largest family of cell–surface signaling molecules. Of ancient origin, the adhesion GPCRs seem to have had a role in allowing cells to adhere and intercommunicate during the metazoan multicellularity evolution [12]. The GPCRs are involved in the control of the most important physiological functions, such as: neurotransmission, immune response, hormone and enzyme release, and contraction of smooth and cardiac muscles, with at least 15 of their receptors being dysregulated in a wide range of human chronic diseases, especially in tumors [13,14].

Discovered in developing cardiomyocytes, the epidermal growth factor, latrophilin, and seven transmembrane domain–containing protein on chromosome 1 (ELTD1), also known as the adhesion G protein–coupled receptor L4 (ADGRL4), is a member of the GPCR superfamily [15] and one of the 33 members of the "adhesion family" [16], characterized by specific large extracellular domains with adhesion items, absent in other GPCR families [17].

It has been shown that, amongst other functions, ELTD1 regulates brain angiogenesis and promotes tumor growth and metastasis [18]. Furthermore, its expression in normal vasculature was found to be regulated by two angiogenic pathways: increased by vascular endothelial growth factor (VEGF) pathway and repressed by Delta–like ligand 4 (DLL4) from NOTCH intercellular signaling pathway [18–20]. Since its discovery, ELTD1 was associated with cardiac and renal function, glioblastoma, and colorectal cancer [18,21–25].

Clinical application of the therapies involving VEGF and NOTCH signaling pathways proved to be unsuccessful. Despite the broad antitumoral spectrum of VEGF inhibition and the preclinical optimistic results, implementing these strategies into clinical studies did not improve overall survival(OS), possibly due to tumor resistance [26]. It can be hypothesized that in time, tumors subjected to selective antiangiogenic treatments may be able to activate alternative parallel angiogenic pathways [27]. Therefore, such pathways represent important topics for current research and ELTD1 is a promising drug target. Several research groups published data from experiments using polyclonal (pAb) and monoclonal (mAb) antibody against ELTD1, showing that these novel treatment strategies may have high potential in glioma preclinical mouse models [28–30]. Furthermore, an increased level of cytotoxicity in glioblastoma cell lines could be achieved by silencing ELTD1 via siRNA [31,32]. Yet, little is known about ELTD1 functions and mechanisms of action. This review aims to summarize the current knowledge regarding ELTD1 and to highlight its importance as a possible candidate to be a part of an innovative therapeutic strategy, either alone or in combination with other conventional approaches already in use.

2. Roles

The ELTD1 receptor was discovered by Nechiporuk and colleagues using murine models, and it was found to be highly expressed in cardiomyocytes, blood vessels, and bronchi's smooth muscle cells [15].

Recently, Olaniru and colleagues studied the distribution of adhesion G–protein coupled receptors in human tissues [33]. The ELTD1 is highly distributed in some tissues, such as: adipose tissue, brain, liver, skeletal muscle, gastrointestinal tract, and pancreas [34,35] The distribution in different types of tissues may be observed in Figure 1, generated by accessing the Biogps portal database [35].

The GPCR superfamily of ELTD1 influences different processes, like: smell, taste, vision, chemotaxis, hormone secretion, and inflammation [36,37]. The orphan ELTD1 receptor was reported to be involved in angiogenesis [18,22,24,38] and cardiac hypertrophyin rat [21], sensitivity of anesthetics [39], subcutaneous fat thickness in pig [40], and tick burden in cattle [41].

ELTD1 involvement in various pathologies was also investigated. Studies revealed its upregulation in malignancies, such as: renal, colorectal, head, neck, and ovarian cancers [18].

Wallgard and colleagues suggested that ELTD1 is an important marker of microvascular endothelium in malignant diseases [38] and Towner et al. found that it is a valuable tumor marker in cases of glioma [42]. Moreover, Dai and colleagues proved that tumor progression may be inhibited by miR–139–5p, via targeting ELTD1 [43]. Then, in 2017 Ziegler and colleagues targeted ELTD1 for its antiangiogenic effect in glioma xenograft models [25]. In the same year, Serban F. and colleagues showed that by silencing of ELTD1, cell death may occur in glioblastoma cell lines [31].

Several studies positively associated the upregulation of ELTD1 on chromosome 1 in malignant cells with poor cancer prognosis. In 2019, a study reported that ELTD1 facilitates proliferation, migration, and invasion in glioma by activating the signaling axis JAK/STAT3/HIF–1α. Finally, high levels of ELTD1 were correlated with poor prognosis in human glioma in another study [44]. This data suggests that ELTD1 may be a potential target for prevention and treatment of glioma. However, the implication of ELTD1 in cancer behavior still remains mostly unknown.

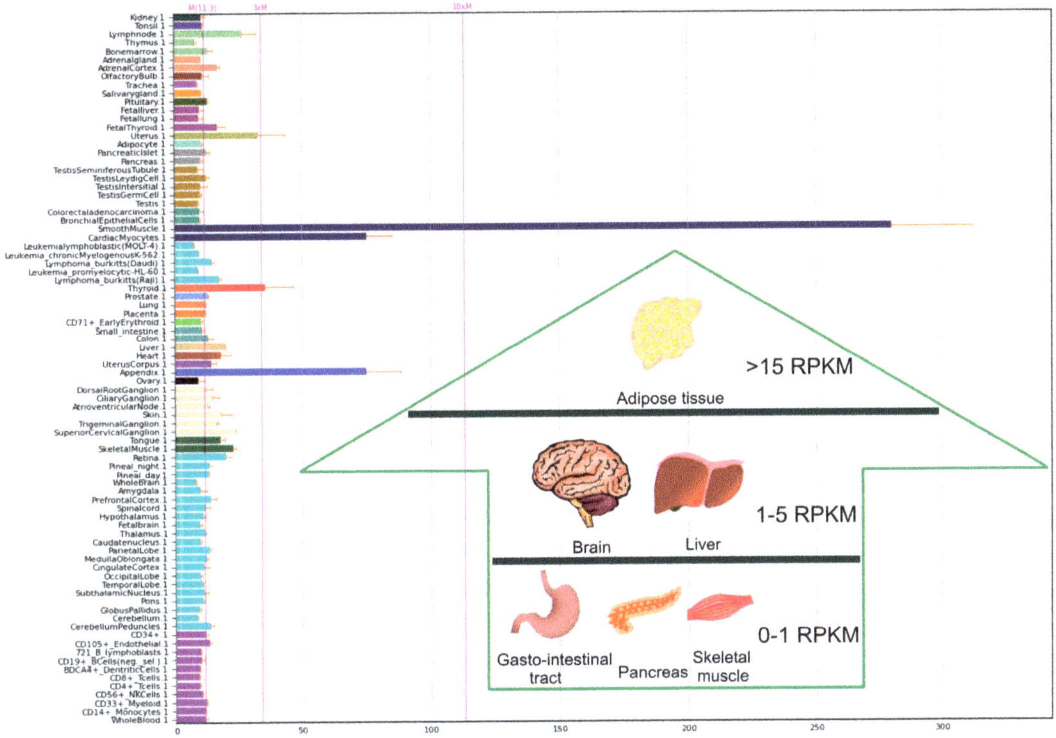

Figure 1. ELTD1 distribution in tissues. The results were retrieved from Genomics Biogps portal database [35]. Selective data regarding the distribution in the main organs was obtained from published RNAseq analysis [34]. Abbreviations: RPKM—reads per kilobase million.

3. Digging into Knowing the ELTD1 Mechanism

First discovered in 2001, ELTD1 expression has been shown in cardiomyocytes, vascular and bronchiolar smooth muscle cells (SMCs) in rat heart and lungs, and it has been suspected to be involved in cardiomyocyte differentiation and coronary angiogenesis [15].

Later, in 2008, Wallgard et al. found that ELTD1 mRNA was a broad marker for vascular endothelial cells in mouse [38]. Porto and colleagues proved that ELTD1 gene DNA variation was associated with tick burden in cattle [41]. In the same year, Lee at al suggested that ELTD1 gene is one of the eight neuronal genes influencing subcutaneous fat thickness in humans and pigs [40].

In 2013, another research group led by Towner et al. introduced the idea of "ELTD1 as novel biomarker for glioma" [42], and their studies were later expanded by others, then Xiao and colleagues linked the cardiac hypertrophy to the low ELTD1 expression in mice [21]. By microsatellite scanning of the immunogenome in transplantation of hematopoietic stem cell, Harkensee associated the ELTD1 and MAPK14 with graft–versus–host disease [45].

Simultaneously, by the assiduous work of Masiero et al., the endothelial orphan receptor ELTD1 was identified as an important regulator of angiogenesis [18]. Based on this discovery, other studies followed and generated valuable data and interesting hypothesis [24].

In 2015, Carty et al. conducted a meta–analysis of genome–wide association and identified ELTD1 as one of the genetic risk factors for stroke in the population of African Americans [46].

In the same year, Ziegler and colleagues begun the adventure in developing new therapies based on ELTD1 against mouse glioma models [47] followed by other studies that independently proved the role of this receptor, as molecular target, in glioma therapy [20,25].

In 2017, Favara et al. reported that ELTD1 is upregulated in breast cancer endothelium that in turn induced lipid metabolism downregulation [48] and later, in 2019, the same group demonstrated that ELTD1 silencing alters the cell metabolic profile in endothelial cells [32].

A year later, Kan and colleagues proved that the tumor microenvironment is regulating ELTD1 function in hepatocellular carcinoma [49].

Treatments targeting ELTD1 started to be investigated by several research groups. Zalles et al., provided some options for glioblastoma treatment using monoclonal antibodies and scFvantibody fragment, in a G55 xenograft mouse model [29,30]. One of the most recent research conducted by Niivirta et al. provided data regarding ELTD1, as a predictive marker for the treatment of renal cancer patients. Their results identified ELTD1 expression in tumor vessels as a positive predictive marker for sunitinib–treatment in metastatic renal cell cancer patients [22].

In the same time, the research area regarding this receptor expanded to other pathologies. For example, based on the fact that ELTD1 has been found to be associated to cannabis use disorder [50], Zhang and colleagues suggested a strong association of this receptor with schizophrenia [51].

By using cell lines and orthotopic xenograft mouse model, Santiago and colleagues recently demonstrated that ELTD1 is a potential target in retinoblastoma. They found that, without affecting normal cell viability, cell migration, and metastasis were reduced by ELTD1 disruption [52].

The most important studies involving ELTD1 are organized in chronological order in the Figure 2.

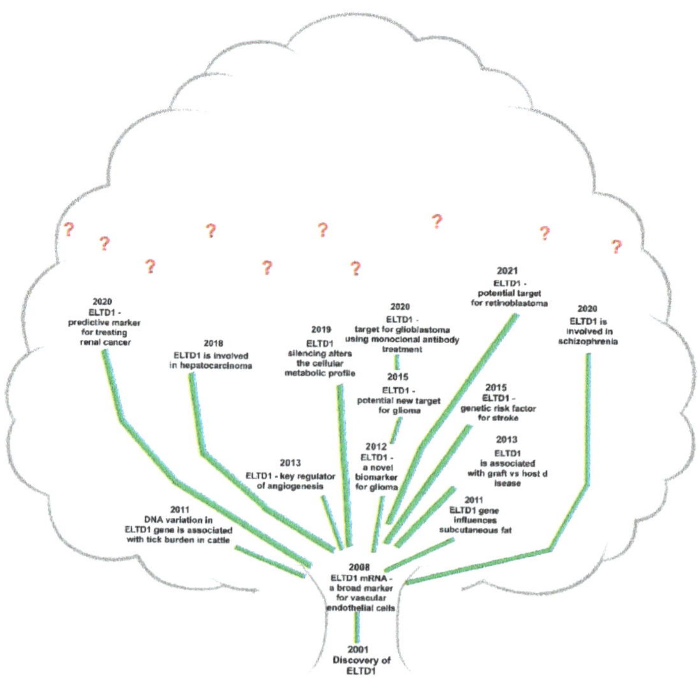

Figure 2. Branching the ELTD1 knowledge.

3.1. Structure and Signaling

ELTD1 (ADGRL4) is a member of the GPCR big family of receptors, which contains more than 900 members divided into five families: glutamate family, rhodopsin family, adhesion family, frizzled family, and secretin family [53].

Initially, the "adhesion family" was a part of the secretin family, but later, due to distinct characteristics, such as unusually elongated N–terminal ectodomain with adhesion–linked motifs [54], it was created as a distinct family. The orphan ELTD1 receptor and the 1, 2, and 3 latrophilin receptors are the fourmembers of the adhesion family, grouped in the latrophilin–like subfamily [17].

Topographically, the ELTD1 receptor consists of an intracellular domain (ICD), a 7–trans–membrane domain (7TMD) and an extracellular domain (ECD) composed of an epidermal growth factor(EGF) domain, an EGF Ca2+ binding domain and a GPCR autoproteolysis site [15,27]. The adhesion motifs are represented by the EGF domain and the EGF Ca2+ binding domain. The ICD contains a tyrosine kinase phosphorylation region, possibly involved in signaling pathway of ELTD1, for which scant data is available. Besides this, based on the cleavage compartmentation criteria, the ELTD1 receptor consists of an N–terminal Fragment and a C–terminal Fragment [24,32]. The structure is represented in Figure 3.

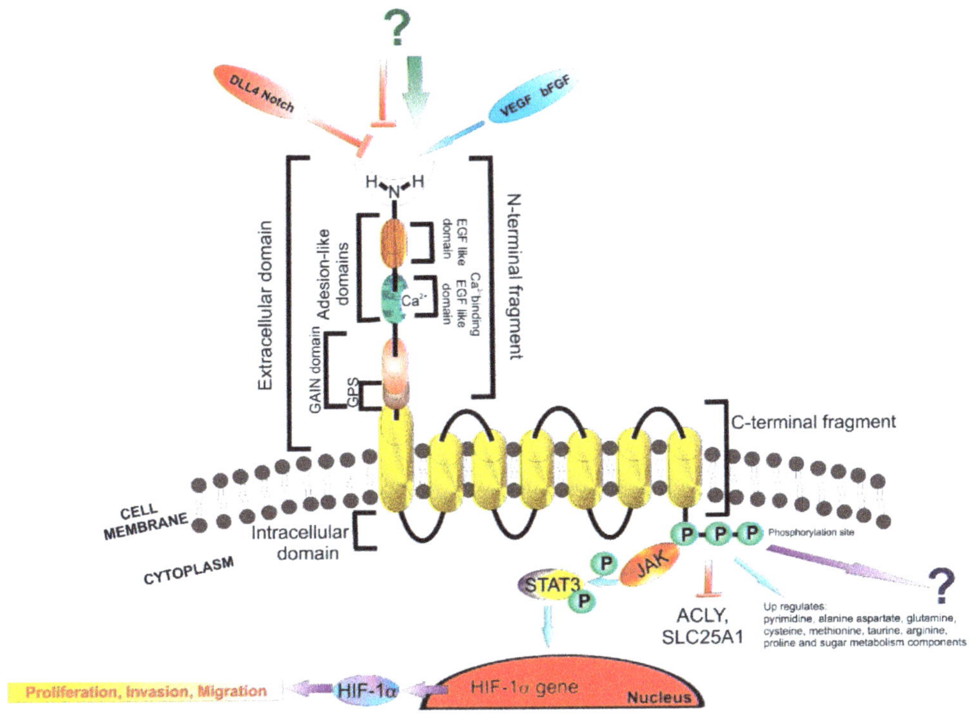

Figure 3. Structure of ELTD1. Abbreviations: VEGF—vascular epithelial growth factor, bFGF—fibroblast growth factor, DLL4—delta like ligand 4, Notch–GPS—G–protein coupled protein hormone receptor proteolysis site, EGF—epidermal growth factor, GAIN domain—G–protein–coupled receptor (GPCR) autoproteolysis–inducing domain. P—phosphate, JAK—Janus kinases, STAT3—signal transducer and activator of transcription protein 3, HIF–1α—Hypoxia–inducible factor 1–alpha, point arrow—activation, block arrow—inhibition.

There are many aspects regarding ELTD1 signaling pathways that have not been clarified, although few important steps forwards have been recently made.

A recent study on cell lines showed that silencing ELTD1 can regulate the endothelial metabolism by suppressing the mitochondrial gene of solute carrier family 25 member 1 (SLC25A1) and the ATP citrate lyase gene (ACLY) [32]. Furthermore, silencing the ELTD1 induced the expression of the hematopoietic stem cell regulator (KIT) [32]. These findings suggest that there is a relationshipbetween ELTD1 and the Notch signaling pathway. The Notch signaling pathway was influenced by suppressing Hes Family BHLH Transcription Factor 2 (HES2) and Jagged Canonical Notch Ligand 1 (JAG1), and by upregulating the Delta Like Canonical Notch Ligand 4 (DLL4) [32]. Additionally, by silencing the ELTD1 in endothelial cells, some components involved in metabolism of pyrimidine, alanine aspartate, glutamine, cysteine, methionine, taurine, arginine, proline, and sugar were found to be upregulated [32]. This study showed non–significant increase in components involved in glycolysis. It is still unclear why ELTD1 silencing leads to ACLY upregulation. This data suggests that regulation of ACLY and SLC25A1 expression by ELTD1 may help to maintain an equilibrium in endothelial metabolism and homeostasis as shown in Figure 3.

More recently, a study conducted by Li et al. showed that JAK/STAT3 signaling pathway (Janus kinases/signal transducer and activator of transcription protein 3) is involved in ELTD1 regulation of proliferation, migration, and invasion of glioma cells. By performing knockdown of ELTD1 in U–87MG and U–138MG cells, they found that the JAK/STAT signaling pathway was inhibited, without effect on other signaling pathways [44]. By using a nude mice orthotopic tumor model, they also found that ELTD1 upregulates the protein expression of HIF–1α (hypoxia–inducible factor 1–alpha), a regulator of tumor formation (cell proliferation, colony formation, migration, and invasion). This data suggests that by ELTD1 silencing, tumor growth could be inhibited and its effect in vivo could be suppressed by HIF–1α overexpression [44] (Figure 3).

Moreover, considering that ELTD1 regulation by the VEGF ligand has been established [24], VEGFR2 association with ELTD1 in glioma was studied using targeted antibody inhibition, proving again that ELTD1 has a key role in angiogenesis, both in vitro and in vivo [18,24].

By treating G55 glioma–bearing mice with either anti–ELTD1 or anti–VEGFR2 antibodies, it was observed that VEGFR2 levels were decreased after anti–ELTD1 antibody treatment, and vice versa, ELTD1 levels were decreased after anti–VEGFR2 antibody treatment, compared to untreated tumors [20]. ELTD1 and VEGFR2 colocalization was also demonstrated by immunohistochemistrystudies. The treatment using anti–ELTD1 antibody significantly increased animal survival, and decreased tumor volumes, compared to IgG–treated or untreated tumor bearing mice [20].

3.2. Ligands

Not muchis known about how aGPCRs (in general) and ELTD1 (in particular) are functioning, because of the lack of data on known ligands, receptor activation or its signaling pathways.

One study published by Favara and colleagues specifically investigated ELTD1's evolution, concluding that its gene appeared cca. 435 million years ago in bony fish and is a highly conserved early core angiogenic gene, with three evolutionary variants [55].

aGPCR signaling is initiated when a tethered agonist binds to a specific extracellular portion of the seven transmembrane helices [56]. Based on that, the conservation mapping of ELTD1 across orthologues was used to hypothesize that its highly conserved external 7TM regions (external loops 2–3, and 4–5) could potentially represent important sites to bind ligands that will initiate the signal transduction of ELTD1. It has been hypothesized that, because ELTD1's exons are probably an ancestral characteristic, the conservation mapping across vertebrate orthologues could be used to clarify its activation. Furthermore, a functional overlap was detected between ELTD1 and at least one member from a different family, with ability to stimulate angiogenesis based on integrin [57]. Sincethe extracellular matrix is known to have a key role in angiogenesis, it was hypothesized that the extracellular matrix ligands that bind to other aGPCR family members could also bind to ELTD1 [58].

ELTD1 was demonstrated to be regulated by two angiogenic ligands: upregulated by VEGF (vascular endothelial growth factor) and downregulated by DLL4 (Notch ligand delta–like ligand 4) [25,27,31,42], a startingpoint for further investigation.Recently, the Stachelhypothesis has been used to solve the activation mechanism of some orphan aGPCRs [59–61]. This hypothesis suggests that, except for GPR123, all aGPCRs express a short 10–20 amino acid tethered agonist called the Stachel peptide, situated C–terminally to the GPS cleavage site, essential for the activation via striking the seven transmembrane receptor loops that initiates signaling [62]. The G–protein–coupled receptor (GPCR) autoproteolysis–inducing domain (GAIN domain) is important for the activity of the GPS cleavage site [63].

Despite all the relevant and productive research work described, the mechanism of ELTD1 activation remains unclarified.

4. ELTD1 an Effective Target in a Wide Range of Diseases

Alterations of ELTD1 have been found in several non malignant diseases, but it is also considered to be a potential treatment target in different types of cancers.

For example, the involvement of the ELTD1 receptor has been studied in the following non malignant diseases: multiplesclerosis, schizophrenia, and stroke.

Perturbation in central nervous system (CNS) vasculature is a distinguishing feature in many diseases. ELTD1 antibody therapy was found to affect molecular pathways involved in multiplesclerosis (MS). Towner and colleagues showed that ELTD1 is highly detectable in the brain of mice with experimental autoimmune encephalomyelitis (EAE), as MS model, showing that ELTD1 may represent a promising biomarker for CNS inflammation [64]. In the past, ELTD1 was linked to the cannabis use disorders [50]. Recently, in 2020, based on the symptomatology and psychopharmacology of some CNS disorders, similarities between psychiatric disorders were suggested [65] regarding ELTD1 involvement in schizophrenia development [51].

The identification of some genomic regions, and genes associated with social genetic effects, could represent the basis to better understand the genetic implication for social average daily gain (ADG). By using the genome–wide association strategy in pigs, ELTD1 gene was linked with social genetic effects, suggesting that this receptor could be used as a marker for ADG. Three single nucleotide polymorphisms (SNP) were located upstream the ELTD1 gene, between 161 and 191 kb. Furthermore, it would be of interest to study the association between ELTD1, prostaglandin F2α receptor (PTGFR), and interferon–induced protein 44 (IFI44) genes [66].

A study performed in 2015 focused on African American patients diagnosed with stroke and genome–wide single nucleotide polymorphism (SNP). The study called COMPASS collaboration was the first large–scale GWAS meta–analysis in African Americans individuals with stroke. The data reported showed that the 15q21.3 locus, related with hypertension and high lipid levels, was associated with total stroke. Amongst other, variants of the ELTD1 gene showed nominal associations with various degree of stroke in African Americans individuals [46].

Malignant Diseases

Between the public health issues worldwide, cancer is one of the most relevant [67]. Studies involving antibodies and small molecules that target specific types of cancer are continuously growing in number, suggesting the importance of this therapeutic approach [68]. These targeted cancer therapies are being studied as single strategies, or in combination with others [69–71]. Although it could represent a major step forward in personalized medicine [72], the molecularly targeted therapy has substantial limitations [73], which are motivating the researchers to develop novel approaches based on emerging technologies [74,75].

By analyzing the alterations of ELTD1 genetic sequence in different cancers usingcBioPortal for cancer genomic database [76], in Figure 4 it can be observed that ELTD1 alterations have a very low frequency in several malignancies such as glioblastoma, ocular melanoma, renal cell carcinoma, colorectal cancer, thyroid cancer, and ovarian cancer, which may suggest that mutation levels are not influencing the tumor evolution, even though high levels of its expression may indicate the presence of the respective malignancy [18,76,77].

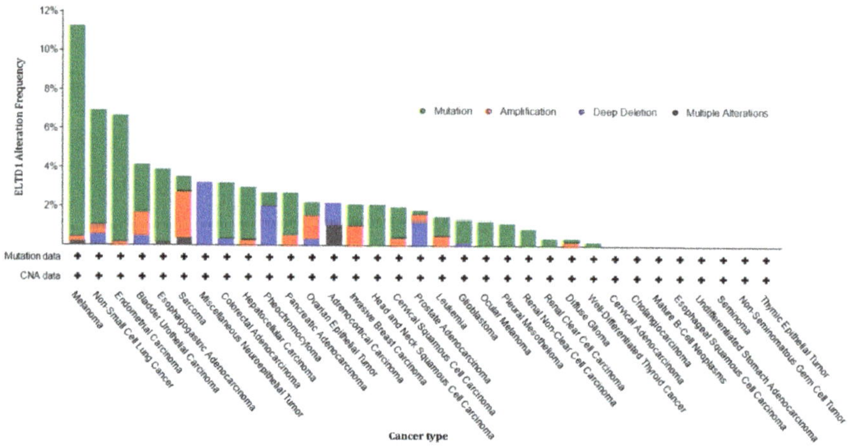

Figure 4. Alteration frequency of ELTD1/ADGRL4 in different types of human cancers. The results were retrieved from the database of BioPortal for Cancer Genomics [76].

Hepatocarcinoma: In a recent study, Kann et al. showed that silencing of ELTD1 drastically reduced hepatocellular carcinoma cells invasiveness [49], confirming previous studies that linked ELTD1 to mechanismsinvolved in the metastatic process [25].

Retinoblastoma: In January 2021, Guihurt Santiago and colleagues reported differences regarding the expression and functional roles of ELTD1 and G–protein receptor 125 (GPR125/ADRGRA3)—two adhesion–GPCRs in retinoblastoma (Rb) [52]. The investigation demonstrated for the first time, that ELTD1, and not GPR125, was overexpressed in Rb compared to fetal retinas. By disrupting ELTD1, in vitro cell migration and in vivo metastasis were reduced without affecting cell viability. This data suggests that ELTD1 may be a potential target for prevention of extraocular Rb and for treatment ofmetastatic Rb [52].

Renal and colorectal cancer: It is well known that ELTD1 and GPR116 are two members of the adhesion G–protein–coupled receptor family expressed in endothelial cells [27]. A study performed by Lu and colleagues focused on their functions using mice lacking ELTD1 and G–protein receptor 116 (GPR116) [78]. The renal and cardiovascular functions were not influenced by the loss of either ELTD1 or GPR116, while the loss of both receptors led to perinatal lethality in half of the mutants, due to cardiovascular malformations (aortic arch arteries and cardiac outflow tract). In addition, the surviving mice showed hemolysis, splenomegaly, and renal thrombotic microangiopathy, with a significant mortality. Meanwhile, the loss of ELTD1 and GPR116 in neural crest–derived cells and endothelial cells did not lead to repetition of any of the phenotypes detected in ELTD1–GPR116 deficient mice, suggesting that loss of these two receptors materialized in cardiovascular and renal defects [78].Common treatment strategies in metastatic renal cell cancer (mRCC) consists in inhibiting the development of new blood vessels by using sunitinib, a tyrosine kinase inhibitor of the vascular endothelial growth factor (VEGF) and platelet–derived growth factor (PDGF) receptors signaling [79]. In 2020, Niinivirta et al. linked the ELTD1 expression level to the progression free survival (PFS) after sunitinib treatment. The expression of ELTD1 in tumor vessels was a positive predictive marker for the sunitinib treatment in patients with renal cancer [22]. A significantly higher PFS after sunitinib treatment was observed in patients with high ELTD1 expression compared to low ELTD1 expression (8 months vs. 5.5 months). On the contrary, the expression level of VEGFR2 had no correlation with sunitinib response. Moreover, this study showed that for sunitinib therapy, ELTD1 may be considered a predictive and not a prognostic marker [22].

Head and neckcancer: In 2013, Massiero et al. analyzed the genes that could potentially be involved in angiogenesis, by profiling the in vivo expression and characterized of the most important candidates using in vitro and in vivo models. By comparing head and neck tumors with normal tissues, a significant increase was observed in ELTD1 expression in tumor–associated cultured endothelial cells (ECs) [18]. Furthermore, increased ELTD1 levels in endothelial cells were also correlated with high microvascular density in head and neck cancers, suggesting its involvement in tumor angiogenesis [18]. Additionally, by profiling the ELTD1 expression in head and neck squamous cell carcinoma, a significant inverse correlation between *CA9ELTD1* (a hypoxia–inducible gene) and ELTD1 mRNA levels was observed [80], probably because of a better perfusion in high ELTD1 tumors [18].

Ovarian cancer: In the same study performed by Massiero and colleagues, primary ovarian human tumor samples were used to study the ELTD1 protein expression. Upregulation of EC ELTD1 expression was observed in neoplastic ovarian tissue, compared to normal tissue. Additionally, higher EC ELTD1 was significantly correlated with increased OS in ovarian tumors [18]. A study performed by Favara et al. showed that ELTD1 was upregulated in several types of cancer cells, including ovarian vascular smooth muscle cells and in tumor–associated endothelial cells, both in zebrafish and humans [55]. By silencing Eltd1 gene in ovarian tumor xenografts in mice, the tumor growth was substantially limited by inhibiting tumor–vessel angiogenesis. In human ovarian cancer patients, increased tumor–vessel endothelial ELTD1 expression was linked to improved OS in patients treated with anticancer therapy. These results show that ELTD1 is a prognostic marker of favorable outcome inhead, neck, and ovarian cancer patients, may be because increased ELTD1 expression could correlate with higher microvessel density, allowing an improved anticancer targeted drug delivery [55].

Glioblastoma:In the beginning of 2013, Towner and colleagues conducted a study proving that ELTD1 expression is a marker for high grade glioma [42]. Few years later, Ziegler et al. used ELTD1 as an antiangiogenic target for treating glioma in mouse and human xenograft glioma models [25]. Furthermore, Serban et al. found that ELTD1 silencing induced cell death in glioblastoma [24,31]. Supporting the above data, Dai S et al. proved that miR–139–5p inhibited tumor progression by targeting ELTD1 [43]. Recently, in vitro and in vivo experiments demonstrated that ELTD1 plays a very important role in proliferation, migration, and invasion of glioma cells, and its overexpression was correlated with poor overall survival (OS) and disease–free survival (DFS) rates in glioma patients. The same study offered evidence that the JAK/STAT3/HIF–1α signaling could control this process [44].

In the last years, ELTD1 was found to be highly expressed in human gliomas, a very aggressive type of brain cancer, and treatments have been started using anti–ELTD1 polyclonal antibodies in glioma preclinical models, but with promiscuous pAb binding [25]. A further study was performed using monoclonal anti–ELTD1 in G55 xenograft glioma mice models, with promising results [29]. The treatment using monoclonal anti–ELTD1 antibody showed high binding specificity, increasing the lifespan, normalizing the vasculature and reducing the tumor volume, compared with the untreated or polyclonal–treated mice. Additionally, a very important result was that ELTD1 interacted and interrupted Notch1 signaling pathway [29]. These data support the idea that ELTD1 may represent a drug target in glioblastoma therapy.

Oligodendroglioma:A network–based strategy was developed to identify novel cancer gene candidates in the region of the 1p/19q codeletion, responsible for some primary human brain tumors, such as oligodendrogliomas. Yet, there is scant evidence regarding the pathology of the above chromosomal mutation. ELTD1, a glioblastoma validated oncogene located on 1p, was predicted to have strong pushing impact on signaling and metabolic pathways involved in oligodendroglioma development [81].

Table 1 summarizes the above information regarding ELTD1 role in malignant diseases.

Table 1. ELDT1 role in cancer.

Type of Cancer	Presumed Role	Observations
Hepatocarcinoma	ELTD1 supports the tumor invasiveness	By silencing of ELTD1, the hepatocellular carcinoma cells invasiveness was drastically reduced [49]
Retinoblastoma	ELTD1 is overexpressed in Rb	ELTD1, was found to be overexpressed in Rb compared to fetal retinas. By disrupting ELTD1, in vitro cell migration and in vivo metastasis were reduced [52]
Renal and Colorectal Cancer	ELTD1 is involved in renal thrombotic microangiopathy and may represent a positive predictive marker after sunitinib treatment	The mice lacking ELTD1 and G–protein receptor 116 (GPR116) showed hemolysis, splenomegaly and renal thrombotic microangiopathy, [78]. A significantly higher PFS after sunitinib treatment was observed in patients with high ELTD1 expression compared to low ELTD1 expression. ELTD1 may be considered a predictive and not a prognostic marker [22]
Head and Neck Cancer	ELTD1 is involved in angiogenesis	Increased ELTD1 levels in endothelial cells were correlated with high microvascular density in head and neck cancers, suggesting its involvement in tumor angiogenesis [18], and also a significant inverse correlation between a the CA9ELTD1 and the ELTD1 mRNA levels was observed [18,80].
Ovarian Cancer	ELTD1 is overexpressed in ovarian cancer	Upregulation of EC ELTD1 expression was observed in neoplastic ovarian tissue, compared to normal tissue. [18,55]. ELTD1 may be a putative prognostic marker with favorable outcome in head, neck and ovarian cancer patients
Glioblastoma	ELTD1 expression is a marker for high grade glioma and a suitable antiangiogenic target	ELTD1 was used as an anti–angiogenic target for treating glioma in mouse and human xenograft glioma models [25,42]. Furthermore, by silencing ELTD1, cellular death was induced in glioblastoma [24,31]. Supporting the above data, miR–139–5p inhibited tumor progression by targeting ELTD1 [43]. In vitro and in vivo experiments showed that ELTD1 has an important role in proliferation, migration, and invasion of glioma cells. There are evidences that the JAK/STAT3/HIF–1α signaling could control this process [44].
Oligodendroglioma	ELTD1 has a strong pushing impact on oligodendroma signaling and metabolic pathways	ELTD1, a glioblastoma validated oncogene located on 1p, was predicted to have strong pushing impact on signaling and metabolic pathways involved in oligodendroglioma development [81].

The preclinical trials targeting ELTD1 are organized in the Table 2.

At present, the use of ELTD1 as therapeutic target in clinical practice has not been studied in large. Even if ELTD1 is of clinical and therapeutic interest, it is the most poorly studied of the GPCR protein families. Relatively littleis known about the receptor intracellular signaling or its activating ligand. Available studies about X–ray crystal structures of ELTD1/ligand complex or ELTD1/intracellular proteins complexes, to validate the binding site on the protein–protein interface, does not practically exist in the literature. In particular, small–molecule inhibitors for ELTD1 have not yet been identified. In a recent study, published in 2021 by MarjutNiinivirta et al., was reported that high expression of ELTD1 in the tumor vasculature predicts a favorable response to sunitinib treatment, in patients with metastatic renal cell cancer [22]. However, no proposed, ongoing or completed clinical trials involving ELTD1 are currently reported in the literature.

Table 2. ELTD1 preclinical trials.

Preclinical Trial	Observations
ELTD1, an effective antiangiogenic target for gliomas: preclinical assessment in mouse GL261 and human G55 xenograft glioma models	Data regarding tumor volume and OS showed that by using antibodies against ELTD1, glioma growth could be inhibited even more if compared with other therapeutic targets (VEGFR). Untreated GL261 mouses had significantly higher ELTD1 levels compared with mouse normal brain. The therapy involving antibody against ELTD1 had an anti–angiogenic effect observed in microvessel density, magnetic resonance angiography and perfusion measurements, with decreased vascularization compared with controls [25]
ELTD1 as a biomarker for multiple sclerosis: Preclinical molecular–targeted studies in a mouse experimental autoimmune encephalomyelitis model	ELTD1 antibody therapy affected the molecular pathways involved in multiplesclerosis, with a high level of ELTD1 expressionis in the brain of mice experimentally induced with autoimmune encephalomyelitis [64]

5. Conclusions

There are many unclear aspects regarding ELTD1 structure, associated ligands, and-mechanisms of action. It is well known currently that ELTD1 is highly expressed in tumor endothelial cells in many cancers and recent evidence shows that. Being associated with angiogenesis, it may be a putative predictive biomarker.

All the cited studies support a specific role of ELTD1 in migration and invasion of cancer cells and shed a new light on a new path to better understand the tumor behavior with the hope of identifying and developing new therapeutic strategies in cancer therapy.

However, there are still many questions that need an answer, like:

- Are ELTD1 and other angiogenesis genes reciprocally affected?
- What other ligands may bind to ELTD1 receptor, apart from VEGFR and DLL4?
- What other molecules are involved in the signaling pathways of ELTD1?

Theanswers to these questions still need to be provided and new research fields need to be explored, in order to provide a more complete elucidation of ELTD1 role in malignant diseases, before its introduction into clinics.

Author Contributions: Conceptualization, A.-S.S., A.D. (Anica Dricu); methodology, I.M.B., C.B.; investigation, A.-S.S., I.M.B., C.B.; resources, A.O., A.D. (Alexandra Dragoi); writing—original draft preparation, L.G.T.; writing—review and editing, A.-S.S., O.A.; visualization, S.T., A.D. (Anica Dricu); supervision, A.-S.S., S.T., A.D. (Anica Dricu). All authors have read and agreed to the published version of the manuscript.

Funding: This research was funded by Grant PN–II–ID–PCE–3–1041, by the UEFISCDI Authority, Romania.

Institutional Review Board Statement: Not applicable.

Informed Consent Statement: Not applicable.

Data Availability Statement: Not applicable.

Conflicts of Interest: The authors declare no conflict of interest.

References

1. Cordover, E.; Minden, A. Signaling pathways downstream to receptor tyrosine kinases: Targets for cancer treatment. *J. Cancer Metastasis Treat.* **2020**, 6–45. [CrossRef]
2. Carapancea, M.; Alexandru, O.; Fetea, A.S.; Dragutescu, L.; Castro, J.; Georgescu, A.; Popa–Wagner, A.; Bäcklund, M.L.; Lewensohn, R.; Dricu, A. Growth factor receptors signaling in glioblastoma cells: Therapeutic implications. *J. Neurooncol.* **2009**, *92*, 137–147. [CrossRef] [PubMed]

3. Schmidt–Arras, D.; Böhmer, F.D. Mislocalisation of Activated Receptor Tyrosine Kinases—Challenges for Cancer Therapy. *Trends Mol. Med.* **2020**, *26*, 833–847. [CrossRef]
4. Alexandru, O.; Horescu, C.; Sevastre, A.S.; Cioc, C.E.; Baloi, C.; Oprita, A.; Dricu, A. Receptor tyrosine kinase targeting in glioblastoma: Performance, limitations and future approaches. *Contemp. Oncol. (Pozn)* **2020**, *24*, 55–66. [CrossRef] [PubMed]
5. Oprita, A.; Baloi, S.C.; Staicu, G.A.; Alexandru, O.; Tache, D.E.; Danoiu, S.; Micu, E.S.; Sevastre, A.S. Updated Insights on EGFR Signaling Pathways in Glioma. *Int. J. Mol. Sci.* **2021**, *22*, 587. [CrossRef] [PubMed]
6. Alexandru, O.; Sevastre, A.S.; Castro, J.; Artene, S.A.; Tache, D.E.; Purcaru, O.S.; Sfredel, V.; Tataranu, L.G.; Dricu, A. Platelet–Derived Growth Factor Receptor and Ionizing Radiation in High Grade Glioma Cell Lines. *Int. J. Mol. Sci.* **2019**, *20*, 4663. [CrossRef] [PubMed]
7. Singh, H.; Li, Y.Y.; Spurr, L.F.; Shinagare, A.B.; Abhyankar, R.; Reilly, E.; Brais, L.K.; Nag, A.; Ducar, M.D.; Thorner, A.R.; et al. Molecular Characterization and Therapeutic Targeting of Colorectal Cancers Harboring Receptor Tyrosine Kinase Fusions. *Clin. Cancer Res.* **2021**, *27*, 1695–1705. [CrossRef] [PubMed]
8. Delcourt, N.; Bockaert, J.; Marin, P. GPCR–jacking: From a new route in RTK signalling to a new concept in GPCR activation. *Trends Pharm. Sci.* **2007**, *28*, 602–607. [CrossRef]
9. Neves, M.; Perpiñá–Viciano, C.; Penela, P.; Hoffmann, C.; Mayor, F., Jr. Modulation of CXCR4–Mediated Gi1 Activation by EGF Receptor and GRK2. *ACS Pharm. Transl. Sci.* **2020**, *3*, 627–634. [CrossRef] [PubMed]
10. Wachira, J.; Hughes–Darden, C.; Nkwanta, A. Investigating Cell Signaling with Gene Expression Datasets. *CourseSource* **2019**, *6*, 10. [CrossRef]
11. Schafer, A.E.; Blaxall, B.C. G Protein Coupled Receptor–mediated Transactivation of Extracellular Proteases. *J. Cardiovasc. Pharm.* **2017**, *70*, 10–15. [CrossRef]
12. De Mendoza, A.; Sebé–Pedrós, A.; Ruiz–Trillo, I. The evolution of the GPCR signaling system in eukaryotes: Modularity, conservation, and the transition to metazoan multicellularity. *Genome Biol. Evol.* **2014**, *6*, 606–619. [CrossRef] [PubMed]
13. Pierce, K.L.; Premont, R.T.; Lefkowitz, R.J. Seventransmembrane receptors. *Nature Rev. Mol. Cell Biol.* **2002**, *3*, 639–650. [CrossRef] [PubMed]
14. Aust, G.; Zhu, D.; Van Meir, E.G.; Xu, L. Adhesion GPCRs in tumorigenesis. *Handb. Exp. Pharm.* **2016**, *234*, 369–396. [CrossRef]
15. Nechiporuk, T.; Urness, L.D.; Keating, M.T. ETL, a novel seven-transmembrane receptor that is developmentally regulated in the heart. ETL is a member of the secretin family and belongs to the epidermal growth factor-seven-transmembrane subfamily. *J. Biol. Chem.* **2001**, *276*, 4150–4157. [CrossRef] [PubMed]
16. Hamann, J.; Aust, G.; Araç, D.; Engel, F.B.; Formstone, C.; Fredriksson, R.; Hall, R.A.; Harty, B.L.; Kirchhoff, C.; Knapp, B.; et al. International Union of Basic and Clinical Pharmacology. XCIV. Adhesion G protein–coupled receptors. *Pharm. Rev.* **2015**, *67*, 338–367. [CrossRef] [PubMed]
17. Bjarnadottir, T.K.; Fredriksson, R.; Hoglund, P.J.; Gloriam, D.E.; Lagerstrom, M.C.; Schioth, H.B. The human and mouse repertoire of the adhesion family of G–protein–coupled receptors. *Genomics* **2004**, *84*, 23–33. [CrossRef]
18. Masiero, M.; Simoes, F.C.; Han, H.D.; Snell, C.; Peterkin, T.; Bridges, E.; Mangala, L.S.; Wu, S.Y.; Pradeep, S.; Li, D.; et al. A core human primary tumor angiogenesis signature identifies the endothelial orphan receptor ELTD1 as a key regulator of angiogenesis. *Cancer Cell* **2013**, *24*, 229–241. [CrossRef] [PubMed]
19. Dieterich, L.C.; Mellberg, S.; Langenkamp, E.; Zhang, L.; Zieba, A.; Salomäki, H.; Teichert, M.; Huang, H.; Edqvist, P.H.; Kraus, T.; et al. Transcriptional profiling of human glioblastoma vessels indicates a key role of VEGF-A and TGFbeta2 in vascular abnormalization. *J. Pathol.* **2012**, *228*, 378–390. [CrossRef]
20. Ziegler, J.; Zalles, M.; Smith, N.; Saunders, D.; Lerner, M.; Fung, K.M.; Patel, M.; Wren, J.D.; Lupu, F.; Battiste, J.; et al. Targeting ELTD1, an angiogenesis marker for glioblastoma (GBM), also affects VEGFR2: Molecular–targeted MRI assessment. *Am. J. Nucl. Med. Mol. Imaging* **2019**, *9*, 93–109.
21. Xiao, J.; Jiang, H.; Zhang, R.; Fan, G.; Zhang, Y.; Jiang, D.; Li, H. Augmented cardiac hypertrophy in response to pressure overload in mice lacking ELTD1. *PLoS ONE* **2012**, *7*, 1–9. [CrossRef] [PubMed]
22. Niinivirta, M.; Georganaki, M.; Enblad, G.; Lindskog, C.; Dimberg, A.; Ullenhag, G.J. Tumor endothelial ELTD1 as a predictive marker for treatment of renal cancer patients with sunitinib. *BMC Cancer* **2020**, *20*, 339. [CrossRef] [PubMed]
23. McNamara, M.G.; Sahebjam, S.; Mason, W.P. Emerging biomarkers in glioblastoma. *Cancers* **2013**, *5*, 1103–1119. [CrossRef]
24. Serban, F.; Artene, S.A.; Georgescu, A.M.; Purcaru, S.O.; Tache, D.E.; Alexandru, O.; Dricu, A. Epidermal growth factor, latrophilin, and seven transmembrane domain-containing protein I marker, a novel angiogenesis marker. *Onco Targets Ther.* **2015**, *8*, 3767–3774. [CrossRef]
25. Ziegler, J.; Pody, R.; De Souza, P.C.; Evans, B.; Saunders, D.; Smith, N.; Mallory, S.; Njoku, C.; Dong, Y.; Chen, H.; et al. ELTD1, an effective anti–angiogenic target for gliomas: Preclinical assessment in mouse GL261 and human G55 xenograft glioma models. *Neuro–Oncology* **2016**, *19*, 175–185. [CrossRef]
26. Bergers, G.; Hanahan, D. Modes of resistance to anti–angiogenic therapy. *Nat. Rev. Cancer* **2008**, *8*, 592–603. [CrossRef]
27. Favara, D.M.; Banham, A.H.; Harris, A.L. A review of ELTD1, a pro–angiogenic adhesion GPCR. *Biochem. Soc. Trans.* **2014**, *42*, 1658–1664. [CrossRef] [PubMed]
28. Smith, N.; Saunders, D.; Towner, R.; Zalles, M. EXTH-07. Optimization of targeting eltd1 in glioblastoma using a molecular targeting approach. *Neuro-Oncology* **2019**, *21*, vi83. [CrossRef]

29. Zalles, M.; Smith, N.; Ziegler, J.; Saunders, D.; Remerowski, S.; Thomas, L.; Gulej, R.; Mamedova, N.; Lerner, M.; Fung, K.-M.; et al. Optimized monoclonal antibody treatment against ELTD1 for GBM in a G55 xenograft mouse model. *J. Cell. Mol. Med.* **2020**, *24*, 1738–1749. [CrossRef]
30. Zalles, M.; Smith, N.; Saunders, D.; Saran, T.; Thomas, L.; Gulej, R.; Lerner, M.; Fung, K.M.; Chung, J.; Hwang, K.; et al. Assessment of an scFv Antibody Fragment Against ELTD1 in a G55 Glioblastoma Xenograft Model. *Transl. Oncol.* **2020**, *13*, 100737. [CrossRef]
31. Serban, F.; Daianu, O.; Tataranu, L.G.; Artene, S.A.; Emami, G.; Georgescu, A.M.; Alexandru, O.; Purcaru, S.O.; Tache, D.E.; Danciulescu, M.M.; et al. Silencing of epidermal growth factor, latrophilin and seven transmembrane domain-containing protein 1 (ELTD1) via siRNA–induced cell death in glioblastoma. *J. Immunoass. Immunochem.* **2017**, *38*, 21–33. [CrossRef]
32. Favara, D.M.; Zois, C.E.; Haider, S.; Pires, E.; Sheldon, H.; McCullagh, J.; Banham, A.H.; Harris, A.L. ADGRL4/ELTD1 Silencing in Endothelial Cells Induces ACLY and SLC25A1 and Alters the Cellular Metabolic Profile. *Metabolites* **2019**, *9*, 287. [CrossRef]
33. Olaniru, O.E.; Persaud, S.J. Adhesion G–protein coupled receptors: Implications for metabolic function. *Pharm. Ther.* **2019**, *198*, 123–134. [CrossRef]
34. Fagerberg, L.; Hallström, B.M.; Oksvold, P.; Kampf, C.; Djureinovic, D.; Odeberg, J.; Habuka, M.; Tahmasebpoor, S.; Danielsson, A.; Edlund, K.; et al. Analysis of the Human Tissue–specific Expression by Genome–wide Integration of Transcriptomics and Antibody–based Proteomics. *Mol. Cell. Proteom.* **2014**, *13*, 397–406. [CrossRef]
35. Biogps Portal. Available online: http://biogps.org/#goto=genereport&id=64123 (accessed on 25 February 2021).
36. Eo, H.S.; Choi, J.P.; Noh, S.J.; Hur, C.G.; Kim, W. A combined approach for the classification of G protein–coupled receptors and its application to detect GPCR splice variants. *Comput. Biol. Chem.* **2007**, *31*, 246–256. [CrossRef] [PubMed]
37. Lefkowitz, R.J. The superfamily of heptahelical receptors. *Nat. Cell Biol.* **2000**, *2*, E133–E136. [CrossRef]
38. Wallgard, E.; Larsson, E.; He, L.; Hellström, M.; Armulik, A.; Nisancioglu, M.H.; Genove, G.; Lindahl, P.; Betsholtz, C. Identification of a core set of 58 gene transcripts with broad and specific expression in the microvasculature. *Arterioscler. Thromb. Vasc. Biol.* **2008**, *28*, 1469–1476. [CrossRef] [PubMed]
39. Agrawal, A.; Pergadia, M.L.; Saccone, S.F.; Lynskey, M.T.; Wang, J.C.; Martin, N.G.; Statham, D.; Henders, A.; Campbell, M.; Garcia, R.; et al. An autosomal linkage scan for cannabis use disorders in the nicotine addiction genetics project. *Arch. Gen. Psychiatry* **2008**, *65*, 713–721. [CrossRef]
40. Lee, K.T.; Byun, M.J.; Kang, K.S.; Park, E.W.; Lee, S.H.; Cho, S.; Kim, H.; Kim, K.W.; Lee, T.; Park, J.E.; et al. Neuronal genes for subcutaneous fat thickness in human and pig are identified by local genomic sequencing and combined SNP association study. *PLoS ONE* **2011**, *6*, e16356. [CrossRef] [PubMed]
41. Porto Neto, L.R.; Bunch, R.J.; Harrison, B.E.; Barendse, W. DNA variation in the gene ELTD1 is associated with tick burden in cattle. *Anim. Genet.* **2011**, *42*, 50–55. [CrossRef]
42. Towner, R.A.; Jensen, R.L.; Colman, H.; Vaillant, B.; Smith, N.; Casteel, R.; Saunders, D.; Gillespie, D.L.; Silasi–Mansat, R.; Lupu, F.; et al. ELTD1, a potential new biomarker for gliomas. *Neurosurgery* **2013**, *72*, 77–90; discussion 91. [CrossRef]
43. Dai, S.; Wang, X.; Li, X.; Cao, Y. MicroRNA–139–5p acts as a tumor suppressor by targeting ELTD1 and regulating cell cycle in glioblastoma multiforme. *Biochem. Biophys. Res. Commun.* **2015**, *467*, 204–210. [CrossRef]
44. Li, J.; Shen, J.; Wang, Z.; Xu, H.; Wang, Q.; Chai, S.; Fu, P.; Huang, T.; Anas, O.; Zhao, H.; et al. ELTD1 facilitates glioma proliferation, migration and invasion by activating JAK/STAT3/HIF–1α signaling axis. *Sci. Rep.* **2019**, *9*, 13904. [CrossRef]
45. Harkensee, C.; Oka, A.; Onizuka, M.; Middleton, P.G.; Inoko, H.; Nakaoka, H.; Gennery, A.R.; Ando, K.; Morishima, Y. Japan Marrow Donor Programme (JMDP). Microsatellite scanning of the immunogenome associates MAPK14 and ELTD1 with graft–versus–host disease in hematopoietic stem cell transplantation. *Immunogenetics* **2013**, *65*, 417–427. [CrossRef]
46. Carty, C.L.; Keene, K.L.; Cheng, Y.C.; Meschia, J.F.; Chen, W.M.; Nalls, M.; Bis, J.C.; Kittner, S.J.; Rich, S.S.; Tajuddin, S.; et al. COMPASS and METASTROKE Consortia. Meta-Analysis of Genome-Wide Association Studies Identifies Genetic Risk Factors for Stroke in African Americans. *Stroke* **2015**, *46*, 2063–2068. [CrossRef]
47. Ziegler, J.; Pody, R.; Rodriguez, L.; Smith, N.; Saunders, D.; Souza, P.C.; Wren, J.; Towner, R. Abstract 205: ELTD1 and Plexin–B2 as novel antibody therapies against glioma biomarkers. *Cancer Res.* **2015**, *75*, 205.
48. Favara, D.M.; Nambiar, M.; Sheldon, H.; Masiero, M.; Li, D.; Jazayeri, A.; Banham, A.H.; Harris, A.L. ELTD1/ADGRL4, a novel adhesion GPCR regulator of tumour angiogenesis, suppresses lipid metabolism in endothelial cells, and is upregulated in breast cancer endothelium and epithelium. *Cancer Res.* **2017**, *77*, 777. [CrossRef]
49. Kan, A.; Le, Y.; Zhang, Y.F.; Duan, F.T.; Zhong, X.P.; Lu, L.H.; Ling, Y.H.; Guo, R.P. ELTD1 Function in Hepatocellular Carcinoma is Carcinoma–Associated Fibroblast–Dependent. *J. Cancer* **2018**, *9*, 2415–2427. [CrossRef] [PubMed]
50. Agrawal, A.; Lynskey, M.T. Candidate genes for cannabis use disorders: Findings, challenges and directions. *Addiction* **2009**, *104*, 518–532. [CrossRef] [PubMed]
51. Zhang, Z.; Chen, G. A logical relationship for schizophrenia, bipolar, and major depressive disorder. Part 1: Evidence from chromosome 1 high density association screen. *J. Comp. Neurol.* **2020**, *528*, 2620–2635. [CrossRef]
52. Guihurt Santiago, J.; Burgos-Tirado, N.; Lafontaine, D.D.; Mendoza Sierra, J.C.; Herrera Camacho, R.; Vecchini Rodríguez, C.M.; Morales-Tirado, V.; Flores-Otero, J. Adhesion G protein–coupled receptor, ELTD1, is a potential therapeutic target for retinoblastoma migration and invasion. *BMC Cancer* **2021**, *21*, 53. [CrossRef] [PubMed]
53. Schiöth, H.B.; Fredriksson, R. The GRAFS classification system of G-protein coupled receptors in comparative perspective. *Gen. Comp. Endocrinol.* **2005**, *142*, 94–101. [CrossRef] [PubMed]

54. Nordström, K.J.; Lagerström, M.C.; Wallér, L.M.; Fredriksson, R.; Schiöth, H.B. The Secretin GPCRs descended from the family of Adhesion GPCRs. *Mol. Biol. Evol.* **2009**, *26*, 71–84. [CrossRef] [PubMed]
55. Favara, D.M.; Banham, A.H.; Harris, A.L. ADGRL4/ELTD1 is a highly conserved angiogenesis–associated orphan adhesion GPCR that emerged with the first vertebrates and comprises 3 evolutionary variants. *BMC Evol. Biol.* **2019**, *19*, 143. [CrossRef]
56. Stoveken, H.M.; Hajduczok, A.G.; Xu, L.; Tall, G.G. Adhesion G protein–coupled receptors are activated by exposure of a cryptic tethered agonist. *Proc. Natl. Acad. Sci. USA* **2015**, *112*, 6194–6199. [CrossRef] [PubMed]
57. Wang, T.; Ward, Y.; Tian, L.; Lake, R.; Guedez, L.; Stetler-Stevenson, W.G.; Kelly, K. CD97, an adhesion receptor on inflammatory cells, stimulates angiogenesis through binding integrin counterreceptors on endothelial cells. *Blood* **2005**, *105*, 2836–2844. [CrossRef]
58. Sottile, J. Regulation of angiogenesis by extracellular matrix. *Biochim. Biophys. Acta* **2004**, *1654*, 13–22. [CrossRef] [PubMed]
59. Muller, A.; Winkler, J.; Fiedler, F.; Sastradihardja, T.; Binder, C.; Schnabel, R.; Kungel, J.; Rothemund, S.; Hennig, C.; Schoneberg, T.; et al. Oriented cell division in the C. elegans embryo is coordinated by G–protein signaling dependent on the adhesion GPCR LAT–1. *PLoS Genet.* **2015**, *11*, e1005624. [CrossRef]
60. Wilde, C.; Fischer, L.; Lede, V.; Kirchberger, J.; Rothemund, S.; Schoneberg, T.; Liebscher, I. The constitutive activity of the adhesion GPCR GPR114/ADGRG5 is mediated by its tethered agonist. *FASEB J.* **2016**, *30*, 666–673. [CrossRef]
61. Brown, K.; Filuta, A.; Ludwig, M.G.; Seuwen, K.; Jaros, J.; Vidal, S.; Arora, K.; Naren, A.P.; Kandasamy, K.; Parthasarathi, K.; et al. Epithelial Gpr116 regulates pulmonary alveolar homeostasis via Gq/11 signaling. *JCI Insight* **2017**, *2*, e93700. [CrossRef]
62. Liebscher, I.; Schon, J.; Petersen, S.C.; Fischer, L.; Auerbach, N.; Demberg, L.M.; Mogha, A.; Coster, M.; Simon, K.U.; Rothemund, S.; et al. A tethered agonist within the Ectodomain activates the adhesion G protein–coupled receptors GPR126 and GPR133. *Cell Rep.* **2014**, *9*, 2018–2026. [CrossRef] [PubMed]
63. Arac, D.; Aust, G.; Calebiro, D.; Engel, F.B.; Formstone, C.; Goffinet, A.; Hamann, J.; Kittel, R.J.; Liebscher, I.; Lin, H.H.; et al. Dissecting signaling and functions of adhesion G protein–coupled receptors. *Ann. N. Y. Acad. Sci.* **2012**, *1276*, 1–25. [CrossRef]
64. Towner, R.A.; Smith, N.; Zalles, M.; Morris, S.; Toliver, M.; Saunders, D.; Lerner, M.; Kumar, G.; Axtell, R.C. ELTD1 as a biomarker for multiple sclerosis: Pre–clinical molecular–targeted studies in a mouse experimental autoimmune encephalomyelitis model. *Mult. Scler. Relat. Disord.* **2021**, *49*, 102786. [CrossRef] [PubMed]
65. Aukes, M.F.; Laan, W.; Termorshuizen, F.; Buizer-Voskamp, J.E.; Hennekam, E.A.; Smeets, H.M.; Ophoff, R.A.; Boks, M.P.; Kahn, R.S. Familial clustering of schizophrenia, bipolar disorder, and major depressive disorder. *Genet. Med. Off. J. Am. Coll. Med. Genet.* **2012**, *14*, 338–341. [CrossRef]
66. Hong, J.K.; Jeong, Y.D.; Cho, E.S.; Choi, T.J.; Kim, Y.M.; Cho, K.H.; Lee, J.B.; Lim, H.T.; Lee, D.H. A genome–wide association study of social genetic effects in Landrace pigs. *Asian–Australas. J. Anim. Sci.* **2018**, *31*, 784–790. [CrossRef]
67. Siegel, R.L.; Miller, K.D.; Fuchs, H.E.; Jemal, A. Cancer Statistics, 2021. *CA Cancer J. Clin.* **2021**, *71*, 7–33. [CrossRef]
68. Miller, K.D.; Nogueira, L.; Mariotto, A.B.; Rowland, J.H.; Yabroff, K.R.; Alfano, C.M.; Jemal, A.; Kramer, J.L.; Siegel, R.L. Cancer treatment and survivorship statistics, 2019. *CA Cancer J. Clin.* **2019**, *69*, 363–385. [CrossRef]
69. Taniguchi, H.; Sen, T.; Rudin, C.M. Targeted Therapies and Biomarkers in Small Cell Lung Cancer. *Front. Oncol.* **2020**, *10*, 741. [CrossRef]
70. Horescu, C.; Cioc, C.E.; Tuta, C.; Sevastre, A.S.; Tache, D.E.; Alexandru, O.; Artene, S.A.; Danoiu, S.; Dricu, A.; Purcaru, S.O. The effect of temozolomide in combination with doxorubicin in glioblastoma cells in vitro. *J. Immunoass. Immunochem.* **2020**, *41*, 1033–1043. [CrossRef]
71. Alexandru, O.; Dragutescu, L.; Tataranu, L.; Ciubotaru, V.; Sevastre, A.; Georgescu, A.M.; Purcaru, O.; Danoiu, S.; Bäcklund, L.M.; Dricu, A. Helianthin induces antiproliferative effect on human glioblastoma cells in vitro. *J. Neurooncol.* **2011**, *102*, 9–18. [CrossRef]
72. Sevastre, A.-S.; Horescu, C.; Carina Baloi, S.; Cioc, C.E.; Vatu, B.I.; Tuta, C.; Artene, S.A.; Danciulescu, M.M.; Tudorache, S.; Dricu, A. Benefits of Nanomedicine for Therapeutic Intervention in Malignant Diseases. *Coatings* **2019**, *9*, 628. [CrossRef]
73. Balik, K.; Modrakowska, P.; Maj, M.; Kaźmierski, Ł.; Bajek, A. Limitations of Molecularly targeted therapy. *Med. Res. J.* **2019**, *4*, 99–105. [CrossRef]
74. Horesh Bergquist, S.; Lobelo, F. The Limits and Potential Future Applications of Personalized Medicine to Prevent Complex Chronic Disease. *Public Health Rep.* **2018**, *133*, 519–522. [CrossRef] [PubMed]
75. Oprita, A.; Sevastre, A.S. New pharmaceutical dosage forms used in the treatment of breast cancer. Polymeric micelles. *Med. Oncol.* **2020**, *1*, 38–52.
76. Cbioportal. Available online: http://www.cbioportal.org (accessed on 25 February 2021).
77. Cerami, E.; Gao, J.; Dogrusoz, U.; Gross, B.E.; Sumer, S.O.; Aksoy, B.A.; Jacobsen, A.; Byrne, C.J.; Heuer, M.L.; Larsson, E.; et al. The cBio cancer genomics portal: An open platform for exploring multidimensional cancer genomics data. *Cancer Discov.* **2012**, *2*, 401–404. [CrossRef] [PubMed]
78. Lu, S.; Liu, S.; Wietelmann, A.; Kojonazarov, B.; Atzberger, A.; Tang, C.; Schermuly, R.T.; Gröne, H.J.; Offermanns, S. Developmental vascular remodeling defects and postnatal kidney failure in mice lacking Gpr116 (Adgrf5) and Eltd1 (Adgrl4). *PLoS ONE* **2017**, *12*, e0183166. [CrossRef] [PubMed]
79. Schmid, T.A.; Gore, M.E. Sunitinib in the treatment of metastatic renal cell carcinoma. *Ther. Adv. Urol.* **2016**, *8*, 348–371. [CrossRef]

80. Buffa, F.M.; Harris, A.L.; West, C.M.; Miller, C.J. Large meta–analysis of multiple cancers reveals a common, compact and highly prognostic hypoxia metagene. *Br. J. Cancer* **2010**, *102*, 428–435. [CrossRef]
81. Gladitz, J.; Klink, B.; Seifert, M. Network–based analysis of oligodendrogliomas predicts novel cancer gene candidates within the region of the 1p/19q co–deletion. *Acta Neuropathol. Commun.* **2018**, *6*, 49. [CrossRef]

International Journal of
Molecular Sciences

Review

The Interference between SARS-CoV-2 and Tyrosine Kinase Receptor Signaling in Cancer

Oana-Stefana Purcaru [1,†], Stefan-Alexandru Artene [1,†], Edmond Barcan [1], Cristian Adrian Silosi [2], Ilona Stanciu [3], Suzana Danoiu [4], Stefania Tudorache [5], Ligia Gabriela Tataranu [6,*] and Anica Dricu [1]

1 Department of Biochemistry, Faculty of Medicine, University of Medicine and Pharmacy of Craiova, Str. Petru Rares nr. 2-4, 710204 Craiova, Romania; stoapo@yahoo.com (O.-S.P.); stefan.artene@yahoo.com (S.-A.A.); edmond.barcan@gmail.com (E.B.); anica.dricu@live.co.uk (A.D.)
2 Department of Surgery, Faculty of Medicine, University of Medicine and Pharmacy of Craiova, Str. Petru Rares nr. 2-4, 710204 Craiova, Romania; cristian_silosi@yahoo.fr
3 "Victor Babeş" Clinical Hospital of Infectious Diseases and Pneumophtisiology, Craiova, Str. Calea Bucuresti, nr. 126, 200525 Craiova, Romania; stanciuilona@yahoo.com
4 Department of Physiopathology, Faculty of Medicine, University of Medicine and Pharmacy of Craiova, Str. Petru Rares nr. 2-4, 710204 Craiova, Romania; suzanadanoiu@yahoo.com
5 Department of Obstetrics and Gynecology, University of Medicine and Pharmacy Craiova, 710204 Craiova, Romania; stefania.tudorache@gmail.com
6 Department of Neurosurgery, "Bagdasar-Arseni" Emergency Hospital, Soseaua Berceni 12, 041915 Bucharest, Romania
* Correspondence: ttranu@gmail.com
† These authors contributed equally to this work.

Citation: Purcaru, O.-S.; Artene, S.-A.; Barcan, E.; Silosi, C.A.; Stanciu, I.; Danoiu, S.; Tudorache, S.; Tataranu, L.G.; Dricu, A. The Interference between SARS-CoV-2 and Tyrosine Kinase Receptor Signaling in Cancer. *Int. J. Mol. Sci.* **2021**, *22*, 4830. https://doi.org/10.3390/ijms22094830

Academic Editor: Michael Welsh

Received: 2 April 2021
Accepted: 30 April 2021
Published: 2 May 2021

Abstract: Cancer and viruses have a long history that has evolved over many decades. Much information about the interplay between viruses and cell proliferation and metabolism has come from the history of clinical cases of patients infected with virus-induced cancer. In addition, information from viruses used to treat some types of cancer is valuable. Now, since the global coronavirus pandemic erupted almost a year ago, the scientific community has invested countless time and resources to slow down the infection rate and diminish the number of casualties produced by this highly infectious pathogen. A large percentage of cancer cases diagnosed are strongly related to dysregulations of the tyrosine kinase receptor (TKR) family and its downstream signaling pathways. As such, many therapeutic agents have been developed to strategically target these structures in order to hinder certain mechanisms pertaining to the phenotypic characteristics of cancer cells such as division, invasion or metastatic potential. Interestingly, several authors have pointed out that a correlation between coronaviruses such as the SARS-CoV-1 and -2 or MERS viruses and dysregulations of signaling pathways activated by TKRs can be established. This information may help to accelerate the repurposing of clinically developed anti-TKR cancer drugs in COVID-19 management. Because the need for treatment is critical, drug repurposing may be an advantageous choice in the search for new and efficient therapeutic compounds. This approach would be advantageous from a financial point of view as well, given that the resources used for research and development would no longer be required and can be potentially redirected towards other key projects. This review aims to provide an overview of how SARS-CoV-2 interacts with different TKRs and their respective downstream signaling pathway and how several therapeutic agents targeted against these receptors can interfere with the viral infection. Additionally, this review aims to identify if SARS-CoV-2 can be repurposed to be a potential viral vector against different cancer types.

Keywords: coronavirus; pandemic; tyrosine kinase; receptor; signaling pathway; EGFR

1. Introduction

Coronaviruses are RNA viruses that affect mammals, having an affinity for the respiratory apparatus in humans. Strains of coronavirus, namely severe acute respiratory syndrome coronavirus (SARS-CoV) and Middle East respiratory syndrome coronavirus (MERS-CoV), have previously caused a large number of cases before completely disappearing. SARS-CoV-2's origin is currently still unknown, but bats are a very likely source, as SARS-CoV and MERS-CoV, similar coronaviruses, have been associated with bats [1,2]. SARS-CoV-2 and bat-CoV RaTG13 share a 96.2% genome sequence identity, demonstrating a common ancestry between the two viruses [3]. The incidence of COVID-19, the infectious disease caused by SARS-CoV-2, is constantly increasing, with almost 62 million confirmed cases and almost 1.5 million deaths worldwide. SARS-CoV-2's human-to-human transmission is mainly sustained through direct contact or through coughing and sneezing droplets received from an infected individual [4].

SARS-CoV-2 is the newest strain of beta coronaviruses, known to have an incubation period of 5.2 days [5]. However, cases with longer incubation periods, up to 24 days, have been reported [6]. This long incubation period, through which the patients present no symptoms but are contagious, is considered one of the main reasons why SARS-CoV-2 has spread so fast around the world [5]. After this asymptomatic period, the symptoms that usually appear are the following: fever, fatigue, cough, headache, difficulty in breathing, hemoptysis, sputum production, sore throat and diarrhea [7,8].

The pathogenesis of the virus is mainly represented by the attachment of the spike (S)-glycoprotein located on the surface of the coronavirus to the angiotensin conversion enzyme 2 (ACE2) receptor from the human cells [9]. The S-glycoprotein is composed of two subunits, S1 and S2. S1's main purpose is determining the virus–host range and cellular tropism with the key function domain, the receptor-binding domain (RBD), while S2 mediates virus–cell membrane fusion through two tandem domains, heptane repeats (HR) 1 and 2 [10].

Furthermore, research has been conducted regarding the ability of the SARS-CoV-2 S1 RBD to bind heparin. Heparins are drugs used for their anticoagulant/thrombotic properties and are known for being safe, stable and highly effective. They also present antiviral activity, which was never fully explored in a clinical setting. Interestingly, coronaviruses are also targeted by heparin because of SARS-CoV's envelope proteins containing positively charged amino acids that are prone to interact with the negatively charged sulfate groups of heparin sulfate proteoglycans [11].

The innate immune system is activated, and pattern recognition receptors (PRRs) are used to recognize the pathogen-associated molecular patterns (PAMP). PRRs consist predominantly of toll-like receptor (TLR), RIG-I-like receptor (RLR) (also previously demonstrated in MERS-CoV [12]), NOD-like receptor (NLR), C-type lectin-like receptors (C_{Lmin}) [13], cytosolic receptor melanoma differentiation-associated gene 5 (MDA5) and nucleotidyl transferase cyclic GMP-AMP synthase (cGAS) [14].

The aforementioned complex factors catalyze the activation of the transcription factor nuclear factor-κB (NF-κB) and interferon regulatory factor 3 (IRF3), leading to the production of type I interferons (IFN-α/β) and a series of proinflammatory cytokines [15,16].

Oncolytic virotherapy is a novel therapy consisting of the use of replicating viruses, through the genetic modification that they produce in cells, as a means of treating cancer. The viruses' tropism is restricted in order to infect only certain cell types. Furthermore, exogenous genes can be added in order to make the virus more aggressive, hence inducing the host's immune response against the specifically targeted cancer cells [17,18].

2. Growth Factors, Tyrosine Kinase Receptors and SARS-CoV-2: A Complex Equation

Growth factor receptors (GFRs) possess the important role of binding extracellular polypeptide growth factors, which determines a cascade of signaling events with the final purpose of regulating cell growth [19]. GFRs are also relevant for the entry of multiple viruses, including coronaviruses, which makes them a central topic of discussion regarding the SARS-CoV-2 pandemic. Drugs inhibiting GFRs that are used for antitumoral purposes are presented in Figure 1 [20–22].

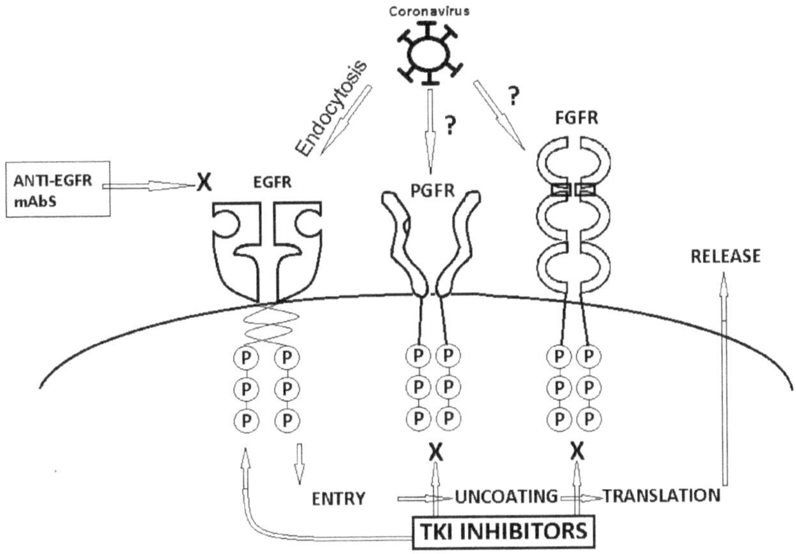

Figure 1. The potential role of several GFRs in coronavirus infection.

2.1. The Epithelial Growth Factor Receptor

The epidermal growth factor receptor (EGFR) is a member of the ErbB family of TKRs with important functions in epithelial cell physiology [23]. It is well known for presenting overexpressions and mutations in a multitude of human cancers, hence becoming the target for multiple cancer therapies [24]. EGFR tyrosine kinase inhibitors (TKIs) have been well documented in numerous clinical studies and are used in the treatment of several types of cancer, most notably non-small-cell lung cancer (NSCLC), for almost two decades now [25].

The EGFR can play a role in the internalization of coronaviruses through binding to the S protein. Transmissible gastroenteritis virus (TGEV) is an alpha-coronavirus that infects the epithelial cells of the intestine, causing severe, potentially lethal, diarrhea in piglets. In a study, the mechanism of infection of TGEV was analyzed, concentrating on the binding with the EGFR. The internalization of the virus was achieved through clathrin- and caveolin-mediated endocytosis. Afterwards, the virus was bound to the EGFR, promoting successive clathrin-mediated endocytosis [26].

After the TGEV spike protein binds with EGFR, the phosphoinositide 3 kinase (PI3K) pathway is activated, inducing the phosphorylation of cofilin and the polymerization of F-actin via Rac1/Cdc42 GTPases. EGFR activates the MAPK pathway, correlated with F-actin reorganization, thus proving again the important involvement of the EGFR in the coronavirus endocytosis [27]. TGEV infection can be also treated with A9, a TKI of the tyrphostin class. In a preclinical in vitro study, the A9 inhibitory activity of the TGEV was mediated by the p38 MAPK signaling pathway. A study by Dong et al. proved the potential of targeting p38 as a means of treating coronaviruses [28].

Researchers explored the possibility of how SARS-CoV infection can influence EGFR signaling and consequently amplify the effect of the receptor's activation. The authors tested the potential of EGFR to provoke fibrosis and how much it varies depending on the presence of viral infection. To their surprise, the overregulation of EGFR signaling followed by SARS-CoV infection determined higher levels of inflammation in the lungs than it would be normally expected alongside interstitial edema [29].

EGFR TKIs are known to have the side effect of promoting interstitial lung disease in the patients receiving these drugs [30]. An important similarity between this interstitial lung disease and the characteristics of COVID-19 has been observed, from the clinical symptoms (fever, cough, fatigue, sputum production, shortness of breath, myalgia, etc.) to radiological findings (ground-glass opacities) [31].

Gefitinib, a TKI used for the first-line treatment of EGFR-mutated NSCLC for almost two decades, is known to aggravate pulmonary fibrosis inflicted by bleomycin [32–34]. Amphiregulin, a ligand of the EGFR, encoded by the AREG gene, is upregulated in many cancers, determining cell growth, proliferation and migration through major intracellular signaling pathways triggered by receptor binding. In murine models, silencing amphiregulin by siRNA or using EGFR-specific TKIs attenuated the fibrogenic effects of TGF-beta1, TGF-beta1 being known for its fibrosis-inducing characteristics [35]. In another study, it was shown that TGF-alpha-mediated fibrosis can be prevented by treating mice with gefitinib and erlotinib [36].

The available data are contradictory given that anti-EGFR TKIs can cause pulmonary fibrosis in humans while preventing pulmonary fibrosis in mice, so there can be several ways of explaining the difference. One explanation is that EGFR signaling can determine different downstream results, depending on the species it encounters. Another way of explaining this discrepancy is that the EGFR downstream signaling kinetics could be dysregulated and not necessarily dependent on the strength of the signal itself [37].

2.2. The Fibroblast Growth Factor Receptor

Fibroblast growth factor receptors (FGFRs) are TKRs that possess an important role in cell proliferation, migration and differentiation. The dysregulation of their expression can lead to the emergence of different tumors [38]. The FGFR family has four members: FGFR1-4, encoded by different genes but presenting high homology [39]. FGFR TKIs are becoming an important tool in the inhibition of cancer growth, with multiple clinical trials assessing the effectiveness of anti-FGFR TKIs [40].

FGFR can also be relevant in viral infections. FGF, bound to heparan sulfate molecules, interacts with FGFR, creating a trimolecular FGF–HS–FGFR complex, setting off subsequent FGFR activation [41]. FGFR1 was proven to be an important, indispensable cofactor in infection with adeno-associated virus 2. Viral invasion was thought to be regulated by heparan sulfate proteoglycans alone, but it was later understood that both HSPG and FGFR1 were implicated in the endocytosis of the virus [42]. FGFR was also relevant in influenza virus infection, being a cofactor necessary for the early stages of the infection [43].

In a study by Hardie et al., several human kinases were screened in order to identify those that could be linked to dengue fever replication. Of those explored, the study focused on the role of FGFR4, a member of the FGFR family. The study showed that dengue fever infection determines an impairment of FGFR phosphorylation. More interestingly, the inhibition of FGFRs via siRNA provided a decrease in the RNA replication of dengue virus, while simultaneously increasing its viral particle production, suggesting that the FGFR might play a regulatory role in the lifecycle of the virus, switching between the early and late stages [44]. In another study, FGF2 was blocked in a Zika-virus-infected human astrocyte cell culture to see how it affects viral replication. The study showed that treatment with the monoclonal antibody BGJ398 determined a decrease in viral replication and cell-to-cell transmission, mainly through the inhibition of the MAPK pathway, which is strongly linked to normal FGF/FGFR activity [45]. In another study, the association of Epstein–Barr virus (EBV) with nonkeratinizing nasopharyngeal cancer (NPC) was explained through the perspective of FGFR1 signaling in the LMP1 pathway. FGFR1 inhibition managed to suppress cell multiplication, migration and invasion in the NPC. Aerobic glycolysis and the epithelial cell transformation demonstrated the association between FGFR/FGF2 signaling present in the EBV activity and the NPC [46].

Another study that analyzed MERS-CoV-induced apoptosis in kidney and lung tissues discovered a correlation between FGFR2 inhibition and the degree of cell death induced by viral infection. By using a specific anti-FGFR TKI, tyrphostin AG1296, the authors observed a reduction in apoptosis by over 40%. However, an anti-EGFR tyrphostin, AG490, had no influence over MERS-CoV-induced apoptosis [47].

2.3. The Platelet-Derived Growth Factor Receptor

Platelet-derived growth factor receptors (PDGFRs) are TKRs with important functions in the development of connective tissue. The two types of receptors are PDGFRα and PDGFRβ. PDGF-PDGFR signaling is important in development, but in the adult age, its function remains relevant only in tissue repair and lesion healing [48]. The most mainstream PDGFR inhibitors are TKIs, with the vast majority of them being nonspecific, targeting additional structures involved in cancer development such as KIT and FLT3 [49].

It was shown that influenza virus entered the cell through the PDGFRβ/GM3 signaling pathway, and endocytosis was successfully inhibited with the TKIKi8751, which specifically targets PDGFRβ phosphorylation. [50]. Furthermore, it was also discovered that PDGFRα plays an important role in the entry of cytomegalovirus into fibroblasts. Through a genome-wide CRISPR screen, PDGFR was shown to have the most significant role in trimer-only human cytomegalovirus (HCMV) infection [51]. Moreover, in a similar study, the silencing of PDGRα reduced the spread of gH/gL/gO-positive HCMV, demonstrating PDGRα's essential role in cell endocytosis [52]. Contrastingly, in another study, it was shown that PDGFRα was not involved in the HMCV entry of the trimer, its silencing producing no effect on the virus endocytosis [53].

3. The Link between Antiviral and Anticancer Drugs

Anticancer drugs have consistently shown potential in the treatment of antiviral infections. During the SARS-CoV-2 pandemic, an important task for researchers has been to find a correlation between the antiviral and antineoplastic function of drugs in order to implement them most effectively in the treatment protocols of COVID-19 patients [54]. Even more so, oncological treatment during the SARS-CoV-2 pandemic is more difficult than ever, as cytotoxic therapies have side effects, such as leukopenia, which makes the organism highly susceptible to infections [55].

Ibrutinib, a powerful inhibitor of the Bruton tyrosine kinase (BTK), is a drug that has a possible anti-inflammatory effect best observed in the respiratory apparatus. Its ability to reduce lung damage, cytokine levels in the lung tissue and mortality have been documented in animal experiments using the H1N1 influenza virus strain. The animals that received ibrutinib survived and made a complete recovery [56].

The effect of ibrutinib was also tested in SARS-CoV-2 subjects. A total of 300 patients suffering from Waldenström's macroglobulinemia (WM) were included in a study in which they received BTK inhibitors. Six of these patients were diagnosed with a SARS-CoV-2 infection and received different doses of ibrutinib (five of the patients received 420 mg/day, while only one patient received 140 mg/day). Patients receiving the higher dose presented better evolution with easier symptoms and with hospitalization not being necessary. On the contrary, patients receiving the lower dose showed symptoms with increasing severity, which caused the necessity of hospitalization [57].

Acalabrutinib, another BTK inhibitor, was also successful in the treatment of several patients suffering from severe cases of COVID-19. The patients, 11 of whom received supplemental oxygen and 8 of whom were on mechanical ventilation, were administered acalabrutinib, with improved oxygenation being observed for the majority of them. This proved that BTK inhibitors are relevant for targeting excessive host inflammation in COVID-19 patients [58].

Selinexor, a selective inhibitor of nuclear export (SINE), is a drug approved for treating relapsed/refractory multiple myeloma [59]. SINEs are known to have the ability to reduce viral proliferation and thus were used in a clinical trial for patients suffering from COVID-19. The drug managed to inhibit important host–protein interactions for SARS-CoV-2 [60].

The Role of Tyrosine Kinase Inhibitors in the Treatment of Coronavirus Infections

TKIs are considered a potential treatment for COVID-19, as they are known to target specific host functions that are required by multiple viruses, including SARS-CoV-2 [61].

MAPK/ERK and PI3K/AKT/mTOR signaling responses have been shown to be relevant in MERS-CoV infection through bioinformatics analysis in vivo. Therefore, by suppressing these pathways, the replication was substantially inhibited in vitro [62].

For SARS-CoV, the potential for use of imatinib, an ABL 2 inhibitor approved for clinical practice 20 years ago, was attested due to the inhibition of the replication of SARS-CoV and MERS-CoV prior to RNA production. Thus, a correlation was found between Abl2 and the productive replication of SARS-CoV and MERS-CoV [63]. Imatinib can be useful for treating pneumonia associated with SARS-CoV-2 infection, as it has been proven to be efficient in treating pulmonary diseases [64]. It improved patients with pulmonary and systemic vascular leak [65], severe refractory asthma [66] and pulmonary artery hypertension [67]. On the contrary, it did not improve patients with idiopathic pulmonary fibrosis [68].

For the more recent SARS-CoV-2, imatinib's viral inhibiting properties have been tested in vitro and results showed potential for inhibition, acting especially on the spike protein and blocking the viral entry at the endosomal level [69]. Prostaglandin E2 stimulation and the deceleration of the increase in TNF-α, IL1-β and IL-6 were observed in the case of administering imatinib, thus reducing inflammation. Imatinib has been proven to interfere in the NF-κB signaling pathway, suppressing it [70]. This pathway is activated in SARS-CoV-2-infected patients and is believed to facilitate the activity of the virus [71]. There have been attempts for treatment with imatinib. In a study, imatinib was added to the treatment protocol at the same time with the interruption of ceftriaxone. Astonishingly, the fever disappeared, the supplementation with oxygen was ceased and pulmonary stability was radiologically confirmed [72].

The possible link between JAK inhibitors (JAKi) and SARS-CoV-2 has also been taken into account. JAKi are drugs that usually have a tendency to interfere with the immune system, increasing the infectious risk in patients. There have been three anecdotal cases of patients that tested positive for SARS-CoV-2 who are taking JAKi for alopecia areata. None of them had significant events but were nonetheless taken off JAKi. Thus, an important aspect during the pandemic is that doctors are careful what drugs they are prescribing to their patients, especially if those drugs have a potential influence on the evolution of COVID-19 [73].

4. SARS-CoV-2 and Viral Tumorigenicity: A Tale of a Two-Edged Sword

4.1. SARS-CoV-2-Induced Carcinogenesis via Tyrosine Kinase Receptors

The carcinogenic potential of viruses is a well-known and documented fact. Of the 219 viral species known to humans, almost 150 types of viruses have carcinogenic potential. Some, such as HPV, are exceptionally carcinogenic, being responsible for almost 95% of cervical cancer cases, while others, such as human herpesvirus 8, are linked to rarer types of cancer such as Kaposi's sarcoma [74]. While the number of coronaviruses is quite vast, very little information is available on their carcinogenic potential as of yet. A preclinical model analysis suggested that SARS-CoV-2 presents a very high affinity for EGFR, VEGFR and c-MET receptors present on glial cells, which are strongly related to gliomagenesis [75]. However, can SARS-CoV-2 penetrate the blood–brain barrier (BBB)? Given the large number of neurological symptoms described by COVID-19 patients, it was strongly suggested that the virus is capable of easily penetrating the BBB. In a study by Rhea et al., it was demonstrated that a radioiodinated S protein can freely traverse the BBB in murine models [76]. Another study has shown that SARS-CoV-2 is capable of infecting the choroid plexus, strongly disrupting the BBB [77].

Another relation between SARS-CoV-2 and TKR activity in cancer might be established between the large number of proinflammatory cytokines and chemokines, which are largely responsible for acute respiratory distress syndrome and the tumor microenvironment, which has a strong impact on carcinogenesis; more explicitly, a link between the IL-6/JAK/STAT3 pathway and the plethora of proinflammatory molecules found in patients suffering from COVID-19. For example, Zhang et al. observed that the mortality of bladder cancer patients suffering from COVID-19 was 10 times higher than that of other patients suffering from the virus [78]. This was theorized to be related to the activation of the IL-6/JAK/STAT3 pathway in the tumor microenvironment of bladder cancer patients, which further exacerbates the inflammation caused by COVID-19 [79]. Another example of the link between the tumor microenvironment and COVID-19 is found in ovarian cancer. Ovarian cancer is known to present increased levels of IL-2, IL-6, IL-12 and Il-13, while high levels of IL-6 are frequently encountered in COVID-19. The interaction between IL-6 and sIL-6R has been shown to promote ovarian cancer progression through the ERK, a TK that triggers increased cell survival, migration and invasion [80]. These mechanisms indicate that high levels of proinflammatory cytokines during the COVID-19 infection could act as a trigger for cancer development and progression, mediated by signals initiated through TKRs or downstream signaling pathways shared with TKRs.

4.2. Oncolytic Virotherapy Potential of the Coronavirus

Oncolytic virotherapy is becoming an attractive option for the treatment of patients with different forms of cancer. Several clinical studies have researched the use of viral therapy, providing promising results [81–83]. Coronaviruses have been researched in this particular subject as potentially capable of exerting an antitumoral effect. The virus has to be modified with an additional protein, or antibody, in order to direct them against the EGFR, thus creating a tumor-targeting virus. [84].

The infectious properties of mouse hepatitis coronavirus (MHCV) are obtained through the binding of its S protein to the murine carcinoembryonic antigen-related cell adhesion molecule 1 (CEACAM1a). Virus–cell membrane fusion is achieved through the induction of conformational changes after the binding of the N-terminal part of the cellular receptor (soR) and the S protein. In order to direct it against the EGFR, a single-chain monoclonal antibody 425 was bound to the soR, creating a bispecific adapter protein (soR-425). The soR-425 proved successful in targeting the EGFR in vitro, but the S protein fusion process was necessary for the virus entry. This research first demonstrated the potential of coronaviruses for tumor-targeting purposes [85].

A few years later, a similar experiment was performed and was again successful in vitro and subsequently continued with an in vivo study. A mouse previously exposed to a lethal intracranial tumor was treated with an MHCV soR-EGF (adaptor protein soR-EGF injected into the MHCV's genome) injection. This significantly prolonged its survival, stopping the recurrence of the tumor load [86].

5. Discussion

The EGFR and other TKRs seem to have a strong correlation with SARS-CoV-2, providing diverse insights into the treatment of COVID-19. The direct binding to the EGFR of TGEV shows that coronaviruses have an affinity for the EGFR, so a potential application for the future is blocking the endocytosis at this level by downregulating the signaling pathway that promotes it. TKIs, more specifically A9 (a tyrphostin-class TKI), produced a satisfactory response in vitro, partially inhibiting the endocytosis of the virus through the EGFR.

A common aspect between SARS-CoV-2 and the EGFR TKIs is that they promote interstitial lung disease, having a high similarity of symptoms and radiological showings. Furthermore, EGFR overexpression facilitates pulmonary fibrosis for a SARS-CoV-infected patient. Although many studies have offered the perspective that EGFR has antifibrotic properties, there has also been research demonstrating the opposite. The difference may come from the different species involved in the testing or from the fact that the signal's intensity/time is not relevant for the EGFR's activity in fibrosis.

Imatinib, a representative of the TKI class, has proven efficient in the inhibition of replication in SARS- and MERS-CoV, proving the implication of Abl2 in the replication. Imatinib was recently tested for SARS-CoV-2 and it successfully inhibited the endocytosis of the virus and also suppressed the NF-κB signaling pathway, which enhanced viral activity levels. The importance of JAKi for COVID-19 has been questioned on a theoretical level, JAKi being known to negatively affect the immune response. The purpose of this correlation is that during the pandemic, the prescription of drugs that may have an effect on the immune response of patients should be closely regulated in order to minimize the rampant advancement of the pandemic.

The EGFR and SARS-CoV-2 correlation is also relevant in the oncology field. The potential for oncolytic virotherapy is an important one. Coronaviruses can exert an antitumoral effect when attached to the cancer cell. The targeting of the cell is achieved through the modification of the virus. An additional protein or antibody is bound to the virus, making it prone to connect to the EGFR of the cancer cell.

Experiments have been conducted for coronaviruses (e.g., mouse hepatitis coronavirus) and proved successful in vitro and in vivo. The use of bispecific adapter proteins attached to the virus redirected its course towards the EGFR, and endocytosis occurred through S protein fusion.

SARS-CoV-2 can certainly be relevant in the oncolytic virotherapy approach, as per its similarity with the MHCV coronavirus, with tests and research required in order for SARS-CoV-2 to prove itself as an important candidate for effective tumor targeting and cancer treatment.

No treatment has proven successful in treating SARS-CoV-2 as of yet. With the race to implement an international immunization scheme through vaccination being strongly underway, it might prove wise to try to replicate different treatment strategies that proved effective for other types of viral agents. Additionally, with oncolytic viral therapy being a popular option in the last decade, SARS-CoV-2 might prove useful as a therapeutic agent for the treatment of different cancer forms.

Author Contributions: Conceptualization of the article was done by S.-A.A. and O.-S.P.; Research on the subject was done by E.B.; C.A.S.; the original draft was written by S.-A.A.; I.S.; S.D.; Tudorache, supervision and proofing of the manuscript was done by L.G.T. and A.D. All authors have read and agreed to the published version of the manuscript.

Funding: This research was funded by Grant PN-II-ID-PCE-3–1041 by the UEFISCDI Authority, Romania.

Conflicts of Interest: The authors declare no conflict of interest.

Abbreviations

TKR	Tyrosine kinase receptor
SARS-CoV	Severe acute respiratory syndrome coronavirus
MERS-CoV	Middle East respiratory syndrome coronavirus
ACE2	Angiotensin conversion enzyme 2
RBD	Receptor-binding domain
HR	Heptane repeats
PRRs	Pattern recognition receptors
PAMP	Pathogen-associated molecular patterns
TLR	Toll-like receptor
RLR-RIG	I-like receptor
NLR-NOD	Like receptor
CLmin-C	Type lectin-like receptors
MDA5	Cytosolic receptor melanoma differentiation-associated gene 5
cGAS	Nucleotidyltransferase cyclic GMP-AMP synthase
NF-κB	Transcription factor nuclear factor-κB
IRF3	Interferon regulatory factor 3
IFN	Interferon
GFR	Growth factor receptors
EGFR	Epithelial growth factor receptor
NSCLC	Non-small-cell lung cancer
HMCV	Human cytomegalovirus

TKI	Tyrosine kinase inhibitors
PI3K	Phosphoinositide 3 kinase
TGEV	Transmissible gastroenteritis virus
FGFR	Fibroblast growth factor receptor
EBV	Epstein–Barr virus
NPC	Non-keratinizing nasopharyngeal cancer
PDGFR	Platelet-derived growth factor receptors
BTK	Bruton tyrosine kinase
SINE	Selective inhibitor of nuclear export
MHCV	Mouse hepatitis coronavirus
CEACAM1a	Carcinoembryonic antigen-related cell adhesion molecule 1
JAKi	JAK inhibitors
BBB	Blood–brain barrier

References

1. Li, W.; Shi, Z.; Yu, M.; Ren, W.; Smith, C.; Epstein, J.H.; Wang, H.; Crameri, G.; Hu, Z.; Zhang, H.; et al. Bats Are Natural Reservoirs of SARS-Like Coronaviruses. *Science* **2005**, *310*, 676–679. [CrossRef]
2. Banerjee, A.; Kulcsar, K.; Misra, V.; Frieman, M.; Mossman, K. Bats and Coronaviruses. *Viruses* **2019**, *11*, 41. [CrossRef]
3. Zhou, P.; Yang, X.-L.; Wang, X.-G.; Hu, B.; Zhang, L.; Zhang, W.; Si, H.-R.; Zhu, Y.; Li, B.; Huang, C.-L.; et al. A pneumonia outbreak associated with a new coronavirus of probable bat origin. *Nature* **2020**, *579*, 270–273. [CrossRef]
4. Rothan, H.A.; Byrareddy, S.N. The epidemiology and pathogenesis of coronavirus disease (COVID-19) outbreak. *J. Autoimmun.* **2020**, *109*, 102433. [CrossRef] [PubMed]
5. Li, Q.; Guan, X.; Wu, P.; Wang, X.; Zhou, L.; Tong, Y.; Ren, R.; Leung, K.S.M.; Lau, E.H.Y.; Wong, J.Y.; et al. Early transmission dynamics in Wuhan, China, of Novel Coronavirus—Infected pneumonia. *N. Engl. J. Med.* **2020**, *382*, 1199–1207. [CrossRef]
6. Bai, Y.; Yao, L.; Wei, T.; Tian, F.; Jin, D.Y.; Chen, L.; Wang, M. Presumed Asymptomatic Carrier Transmission of COVID-19. *JAMA* **2020**, *323*, 1406–1407. [CrossRef] [PubMed]
7. Wang, D.; Hu, B.; Hu, C.; Zhu, F.; Liu, X.; Zhang, J.; Wang, B.; Xiang, H.; Cheng, Z.; Xiong, Y.; et al. Clinical Characteristics of 138 Hospitalized Patients With 2019 Novel Coronavirus–Infected Pneumonia in Wuhan, China. *JAMA* **2020**, *323*, 1061–1069. [CrossRef] [PubMed]
8. Guan, W.J.; Ni, Z.Y.; Hu, Y.; Liang, W.H.; Ou, C.Q.; He, J.X.; Liu, L.; Shan, H.; Lei, C.L.; Hui, D.S.C.; et al. Clinical Characteristics of Coronavirus Disease 2019 in China. *N. Engl. J. Med.* **2020**, *382*, 1708–1720. [CrossRef] [PubMed]
9. Tortorici, M.A.; Veesler, D. Structural Insights into Coronavirus Entry. In *Advances in Virus Research*; Academic Press: Cambridge, MA, USA, 2019; pp. 93–116.
10. Guo, Y.R.; Cao, Q.D.; Hong, Z.S.; Tan, Y.Y.; Chen, S.D.; Jin, H.J.; Tan, K.; Wang, D.; Yan, Y. The origin, transmission and clinical therapies on coronavirus disease 2019 (COVID-19) outbreak–an update on the status. *Mil. Med Res.* **2020**, *7*, 11. [CrossRef] [PubMed]
11. Vicenzi, E.; Canducci, F.; Pinna, D.; Mancini, N.; Carletti, S.; Lazzarin, A.; Bordignon, C.; Poli, G.; Clementi, M. Coronaviridae and SARS-associated coronavirus strain HSR1. *Emerg. Infect. Dis.* **2004**, *10*, 413. [CrossRef]
12. Zhao, X.; Chu, H.; Wong, B.H.-Y.; Chiu, M.C.; Wang, D.; Li, C.; Liu, X.; Yang, D.; Poon, V.K.-M.; Cai, J.; et al. Activation of C-Type Lectin Receptor and (RIG)-I-Like Receptors Contributes to Proinflammatory Response in Middle East Respiratory Syndrome Coronavirus-Infected Macrophages. *J. Infect. Dis.* **2019**, *221*, 647–659. [CrossRef] [PubMed]
13. Li, G.; Fan, Y.; Lai, Y.; Han, T.; Li, Z.; Zhou, P.; Pan, P.; Wang, W.; Hu, D.; Liu, X.; et al. Coronavirus infections and immune responses. *J. Med. Virol.* **2020**, *92*, 424–432. [CrossRef] [PubMed]
14. Wu, J.; Sun, L.; Chen, X.; Du, F.; Shi, H.; Chen, C.; Chen, Z.J. Cyclic GMP-AMP Is an Endogenous Second Messenger in Innate Immune Signaling by Cytosolic DNA. *Science* **2013**, *339*, 826–830. [CrossRef] [PubMed]
15. Takeuchi, O.; Akira, S. Innate immunity to virus infection. *Immunol. Rev.* **2008**, *227*, 75–86. [CrossRef]
16. Huang, C.; Wang, Y.; Li, X.; Ren, L.; Zhao, J.; Hu, Y.; Zhang, L.; Fan, G.; Xu, J.; Gu, X.; et al. Clinical features of patients infected with 2019 novel coronavirus in Wuhan, China. *Lancet* **2020**, *395*, 497–506. [CrossRef]
17. Thorne, S.H.; Hermiston, T.; Kirn, D. Oncolytic Virotherapy: Approaches to Tumor Targeting and Enhancing Antitumor Effects. In *Seminars in Oncology*; W.B. Saunders Ltd.: Philadelphia, PA, USA, 2005.
18. Liu, T.-C.; Galanis, E.; Kirn, D.H. Clinical trial results with oncolytic virotherapy: A century of promise, a decade of progress. *Nat. Clin. Pract. Oncol.* **2007**, *4*, 101–117. [CrossRef]
19. Carapancea, M.; Alexandru, O.; Fetea, A.S.; Dragutescu, L.; Castro, J.; Georgescu, A.; Popa-Wagner, A.; Bäcklund, M.L.; Lewensohn, R.; Dricu, A. Growth factor receptors signaling in glioblastoma cells: Therapeutic implications. *J. Neuro-Oncology* **2008**, *92*, 137–147. [CrossRef]
20. Popescu, A.M.; Alexandru, O.; Brindusa, C.; Purcaru, S.O.; Tache, D.E.; Tataranu, L.G.; Taisescu, C.; Dricu, A. Targeting the VEGF and PDGF signaling pathway in glioblastoma treatment. *Int. J. Clin. Exp. Pathol.* **2015**, *8*, 7825–7837.

21. Hondermarck, H.; Bartlett, N.W.; Nurcombe, V. The role of growth factor receptors in viral infections: An opportunity for drug repurposing against emerging viral diseases such as COVID-19? *FASEB BioAdvances* **2020**, *2*, 296–303. [CrossRef]
22. Carapancea, M.; Cosaceanu, D.; Budiu, R.; Kwiecinska, A.; Tataranu, L.G.; Ciubotaru, V.; Alexandru, O.; Banita, M.; Pisoschi, C.; Bäcklund, M.L.; et al. Dual targeting of IGF-1R and PDGFR inhibits proliferation in high-grade gliomas cells and induces radiosensitivity in JNK-1 expressing cells. *J. Neuro-Oncology* **2007**, *85*, 245–254. [CrossRef]
23. Schlessinger, J. Receptor Tyrosine Kinases: Legacy of the First Two Decades. *Cold Spring Harb. Perspect. Biol.* **2014**, *6*, a008912. [CrossRef] [PubMed]
24. Yarden, Y.; Pines, G. The ERBB network: At last, cancer therapy meets systems biology. *Nat. Rev. Cancer* **2012**, *12*, 553–563. [CrossRef] [PubMed]
25. Liu, X.; Wang, P.; Zhang, C.; Ma, Z. Epidermal growth factor receptor (EGFR): A rising star in the era of precision medicine of lung cancer. *Oncotarget* **2017**, *8*, 50209–50220. [CrossRef] [PubMed]
26. Hu, W.; Zhang, S.; Shen, Y.; Yang, Q. Epidermal growth factor receptor is a co-factor for transmissible gastroenteritis virus entry. *Virology* **2018**, *521*, 33–43. [CrossRef] [PubMed]
27. Hu, W.; Zhu, L.; Yang, X.; Lin, J.; Yang, Q. The epidermal growth factor receptor regulates cofilin activity and promotes transmissible gastroenteritis virus entry into intestinal epithelial cells. *Oncotarget* **2016**, *7*, 12206–12221. [CrossRef] [PubMed]
28. Dong, W.; Xie, W.; Liu, Y.; Sui, B.; Zhang, H.; Liu, L.; Tan, Y.; Tong, X.; Fu, Z.F.; Yin, P.; et al. Receptor tyrosine kinase inhibitors block proliferation of TGEV mainly through p38 mitogen-activated protein kinase pathways. *Antivir. Res.* **2020**, *173*, 104651. [CrossRef] [PubMed]
29. Venkataraman, T.; Coleman, C.M.; Frieman, M.B. Overactive Epidermal Growth Factor Receptor Signaling Leads to Increased Fibrosis after Severe Acute Respiratory Syndrome Coronavirus Infection. *J. Virol.* **2017**, *91*. [CrossRef]
30. Shah, R.R. Tyrosine Kinase Inhibitor-Induced Interstitial Lung Disease: Clinical Features, Diagnostic Challenges, and Therapeutic Dilemmas. *Drug Saf.* **2016**, *39*, 1073–1091. [CrossRef]
31. Chang, H.-L.; Chen, Y.-H.; Taiwan, H.-C.; Yang, C.-J. EGFR Tyrosine Kinase Inhibitor–Associated Interstitial Lung Disease during the Coronavirus Disease 2019 Pandemic. *J. Thorac. Oncol.* **2020**, *15*, e129–e131. [CrossRef] [PubMed]
32. Kato, T.; Nishio, K. Clinical aspects of epidermal growth factor receptor inhibitors: Benefit and risk. *Respirology* **2006**, *11*, 693–698. [CrossRef]
33. Totura, A.L.; Whitmore, A.C.; Agnihothram, S.; Schäfer, A.; Katze, M.G.; Heise, M.T.; Baric, R.S. Toll-Like Receptor 3 Signaling via TRIF Contributes to a Protective Innate Immune Response to Severe Acute Respiratory Syndrome Coronavirus Infection. *MBio* **2015**, *6*. [CrossRef] [PubMed]
34. Suzuki, H.; Aoshiba, K.; Yokohori, N.; Nagai, A. Epidermal growth factor receptor tyrosine kinase inhibition augments a murine model of pulmonary fibrosis. *Cancer Res.* **2003**, *63*, 5054–5059. [PubMed]
35. Zhou, Y.; Lee, J.-Y.; Lee, C.-M.; Cho, W.-K.; Kang, M.-J.; Koff, J.L.; Yoon, P.-O.; Chae, J.; Park, H.-O.; Elias, J.A.; et al. Amphiregulin, an Epidermal Growth Factor Receptor Ligand, Plays an Essential Role in the Pathogenesis of Transforming Growth Factor-β-induced Pulmonary Fibrosis. *J. Biol. Chem.* **2012**, *287*, 41991–42000. [CrossRef] [PubMed]
36. Hardie, W.D.; Davidson, C.; Ikegami, M.; Leikauf, G.D.; Le Cras, T.D.; Prestridge, A.; Whitsett, J.A.; Korfhagen, T.R. EGF receptor tyrosine kinase inhibitors diminish transforming growth factor-α-induced pulmonary fibrosis. *Am. J. Physiol.-Lung Cell. Mol. Physiol.* **2008**, *294*, L1217–L1225. [CrossRef] [PubMed]
37. Venkataraman, T.; Frieman, M.B. The role of epidermal growth factor receptor (EGFR) signaling in SARS coronavirus-induced pulmonary fibrosis. *Antivir. Res.* **2017**, *143*, 142–150. [CrossRef]
38. Ornitz, D.M.; Itoh, N. Fibroblast growth factors. *Genome Biol.* **2001**, *2*, 1–12. [CrossRef]
39. Itoh, N.; Ornitz, D.M. Evolution of the Fgf and Fgfr gene families. *Trends Genet.* **2004**, *20*, 563–569. [CrossRef]
40. Porta, R.; Borea, R.; Coelho, A.; Khan, S.; Araújo, A.; Reclusa, P.; Franchina, T.; Van Der Steen, N.; Van Dam, P.; Ferri, J.; et al. FGFR a promising druggable target in cancer: Molecular biology and new drugs. *Crit. Rev. Oncol. Hematol.* **2017**, *113*, 256–267. [CrossRef]
41. Ornitz, D.M. FGFs, heparan sulfate and FGFRs: Complex interactions essential for development. *BioEssays* **2000**, *22*, 108–112. [CrossRef]
42. Qing, K.; Mah, C.; Hansen, J.; Zhou, S.; Dwarki, V.; Srivastava, A. Human fibroblast growth factor receptor 1 is a co-receptor for infection by adeno-associated virus 2. *Nat. Med.* **1999**, *5*, 71–77. [CrossRef]
43. König, R.; Stertz, S.; Zhou, Y.; Inoue, A.; Hoffmann, H.-H.; Bhattacharyya, S.; Alamares, J.G.; Tscherne, D.M.; Ortigoza, M.B.; Liang, Y.; et al. Human host factors required for influenza virus replication. *Nature* **2010**, *463*, 813–817. [CrossRef] [PubMed]
44. Cortese, M.; Kumar, A.; Matula, P.; Kaderali, L.; Scaturro, P.; Erfle, H.; Acosta, E.G.; Buehler, S.; Ruggieri, A.; Chatel-Chaix, L.; et al. Reciprocal Effects of Fibroblast Growth Factor Receptor Signaling on Dengue Virus Replication and Virion Production. *Cell Rep.* **2019**, *27*, 2579–2592.e6. [CrossRef] [PubMed]
45. Limonta, D.; Jovel, J.; Kumar, A.; Lu, J.; Hou, S.; Airo, A.M.; Lopez-Orozco, J.; Wong, C.P.; Saito, L.; Branton, W.; et al. Fibroblast Growth Factor 2 Enhances Zika Virus Infection in Human Fetal Brain. *J. Infect. Dis.* **2019**, *220*, 1377–1387. [CrossRef] [PubMed]
46. Lo, A.K.-F.; Dawson, C.W.; Young, L.S.; Ko, C.-W.; Hau, P.-M.; Lo, K.-W. Activation of the FGFR1 signalling pathway by the Epstein-Barr virus-encoded LMP1 promotes aerobic glycolysis and transformation of human nasopharyngeal epithelial cells. *J. Pathol.* **2015**, *237*, 238–248. [CrossRef]

47. Yeung, M.-L.; Yao, Y.; Jia, L.; Chan, J.F.W.; Chan, K.-H.; Cheung, K.-F.; Chen, H.; Poon, V.K.M.; Tsang, A.K.L.; To, K.K.; et al. MERS coronavirus induces apoptosis in kidney and lung by upregulating Smad7 and FGF2. *Nat. Microbiol.* **2016**, *1*, 16004. [CrossRef] [PubMed]
48. Chen, P.-H.; Chen, X.; He, X. Platelet-derived growth factors and their receptors: Structural and functional perspectives. *Biochim. Biophys. Acta—Proteins Proteom.* **2013**, *1834*, 2176–2186. [CrossRef] [PubMed]
49. Levitzki, A. PDGF receptor kinase inhibitors for the treatment of PDGF driven diseases. *Cytokine Growth Factor Rev.* **2004**, *15*, 229–235. [CrossRef] [PubMed]
50. Vrijens, P.; Noppen, S.; Boogaerts, T.; Vanstreels, E.; Ronca, R.; Chiodelli, P.; Laporte, M.; Vanderlinden, E.; Liekens, S.; Stevaert, A.; et al. Influenza virus entry via the GM3 ganglioside-mediated platelet-derived growth factor receptor β signalling pathway. *J. Gen. Virol.* **2019**, *100*, 583–601. [CrossRef]
51. Wu, K.; Oberstein, A.; Wang, W.; Shenk, T. Role of PDGF receptor-α during human cytomegalovirus entry into fibroblasts. *Proc. Natl. Acad. Sci. USA* **2018**, *115*, E9889–E9898. [CrossRef]
52. Wu, Y.; Prager, A.; Boos, S.; Resch, M.; Brizic, I.; Mach, M.; Wildner, S.; Scrivano, L.; Adler, B. Human cytomegalovirus glycoprotein complex gH/gL/gO uses PDGFR-α as a key for entry. *PLoS Pathog.* **2017**, *13*, e1006281. [CrossRef]
53. Liu, J.; Jardetzky, T.S.; Chin, A.L.; Johnson, D.C.; Vanarsdall, A.L. The Human Cytomegalovirus Trimer and Pentamer Promote Sequential Steps in Entry into Epithelial and Endothelial Cells at Cell Surfaces and Endosomes. *J. Virol.* **2018**, *92*. [CrossRef]
54. Allegra, A.; Pioggia, G.; Tonacci, A.; Musolino, C.; Gangemi, S. Cancer and SARS-CoV-2 Infection: Diagnostic and Therapeutic Challenges. *Cancers* **2020**, *12*, 1581. [CrossRef] [PubMed]
55. Anil, I.; Arnold, R.; Benkwitz-Beford, S.; Branford, S.; Campton, N.; Cazier, J.-B.; Cheng, V.; Curley, H.; D'Costa, J.; Edmondson, A.; et al. The UK Coronavirus Cancer Monitoring Project: Protecting patients with cancer in the era of COVID-19. *Lancet Oncol.* **2020**, *21*, 622–624. [CrossRef]
56. Florence, J.M.; Krupa, A.; Booshehri, L.M.; Davis, S.A.; Matthay, M.A.; Kurdowska, A.K. Inhibiting Bruton's tyrosine kinase rescues mice from lethal influenza-induced acute lung injury. *Am. J. Physiol.-Lung Cell. Mol. Physiol.* **2018**, *315*, L52–L58. [CrossRef] [PubMed]
57. Treon, S.P.; Castillo, J.J.; Skarbnik, A.P.; Soumerai, J.D.; Ghobrial, I.M.; Guerrera, M.L.; Meid, K.E.; Yang, G. The BTK inhibitor ibrutinib may protect against pulmonary injury in COVID-19–infected patients. *Blood* **2020**, *135*, 1912–1915. [CrossRef] [PubMed]
58. Roschewski, M.; Lionakis, M.S.; Sharman, J.P.; Roswarski, J.; Goy, A.; Monticelli, M.A.; Roshon, M.; Wrzesinski, S.H.; Desai, J.V.; Zarakas, M.A.; et al. Inhibition of Bruton tyrosine kinase in patients with severe COVID-19. *Sci. Immunol.* **2020**, *5*, eabd0110. [CrossRef] [PubMed]
59. Allegra, A.; Innao, V.; Allegra, A.G.; Leanza, R.; Musolino, C. Selective inhibitors of nuclear export in the treatment of hematologic malignancies. *Clin. Lymphoma Myeloma Leuk.* **2019**, *19*, 689–698. [CrossRef] [PubMed]
60. Agree, I. Karyopharm to Evaluate Low Dose Selinexor as a Potential Treatment for Hospitalized Patients with COVID-19. Available online: https://www.globenewswire.com/news-release/2020/04/07/2012711/0/en/Karyopharm-to-Evaluate-Low-Dose-Selinexor-as-a-Potential-Treatment-for-Hospitalized-Patients-with-COVID-19.html (accessed on 1 May 2021).
61. Schor, S.; Einav, S. Repurposing of Kinase Inhibitors as Broad-Spectrum Antiviral Drugs. *DNA Cell Biol.* **2018**, *37*, 63–69. [CrossRef] [PubMed]
62. Kindrachuk, J.; Ork, B.; Hart, B.J.; Mazur, S.; Holbrook, M.R.; Frieman, M.B.; Traynor, D.; Johnson, R.F.; Dyall, J.; Kuhn, J.H.; et al. Antiviral Potential of ERK/MAPK and PI3K/AKT/mTOR Signaling Modulation for Middle East Respiratory Syndrome Coronavirus Infection as Identified by Temporal Kinome Analysis. *Antimicrob. Agents Chemother.* **2015**, *59*, 1088–1099. [CrossRef]
63. Coleman, C.M.; Sisk, J.M.; Mingo, R.M.; Nelson, E.A.; White, J.M.; Frieman, M.B. Abelson Kinase Inhibitors Are Potent Inhibitors of Severe Acute Respiratory Syndrome Coronavirus and Middle East Respiratory Syndrome Coronavirus Fusion. *J. Virol.* **2016**, *90*, 8924–8933. [CrossRef] [PubMed]
64. Assaad, H.S.; Assaad-Khalil, S. Imatinib a Tyrosine Kinase Inhibitor: A potential treatment for SARS- COV-2 induced pneumonia. *Alex. J. Med.* **2020**, *56*, 68–72. [CrossRef]
65. Overbeek, M.J.; Amerongen, G.P.V.N.; Boonstra, A.; Smit, E.F.; Vonk-Noordegraaf, A. Possible role of imatinib in clinical pulmonary veno-occlusive disease. *Eur. Respir. J.* **2008**, *32*, 232–235. [CrossRef]
66. Cahill, K.N.; Katz, H.; Cui, J.; Lai, J.; Kazani, S.; Crosby-Thompson, A.; Garofalo, D.; Castro, M.; Jarjour, N.N.; DiMango, E.; et al. Effect of KIT Inhibition by Imatinib on Airway Mast Cells in Patients with Severe Refractory Asthma (KIA). *J. Allergy Clin. Immunol.* **2017**, *139*, AB169. [CrossRef]
67. Hoeper, M.M.; Barst, R.J.; Bourge, R.C.; Feldman, J.; Frost, A.E.; Galié, N.; Gómez-Sánchez, M.A.; Grimminger, F.; Grünig, E.; Hassoun, P.M.; et al. Imatinib mesylate as add-on therapy for pulmonary arterial hypertension: Results of the randomized IMPRES study. *Circulation* **2013**, *127*, 1128–1138. [CrossRef] [PubMed]
68. Daniels, C.E.; Lasky, J.A.; Limper, A.H.; Mieras, K.; Gabor, E.; Schroeder, D.R. Imatinib treatment for idiopathic pulmonary fibrosis: Randomized placebo-controlled trial results. *Am. J. Respir. Crit. Care Med.* **2010**, *181*, 604–610. [CrossRef] [PubMed]
69. Mulgaonkar, N.S.; Wang, H.; Mallawarachchi, S.; Ruzek, D.; Martina, B.; Fernando, S. Bcr-Abl tyrosine kinase inhibitor imatinib as a potential drug for COVID-19. *bioRxiv* **2020**. [CrossRef]
70. Rizzo, A.N.; Sammani, S.; Esquinca, A.E.; Jacobson, J.R.; Garcia, J.G.N.; Letsiou, E.; Dudek, S.M. Imatinib attenuates inflammation and vascular leak in a clinically relevant two-hit model of acute lung injury. *Am. J. Physiol.-Lung Cell. Mol. Physiol.* **2015**, *309*, L1294–L1304. [CrossRef]

71. Dosch, S.F.; Mahajan, S.D.; Collins, A.R. Collins, SARS coronavirus spike protein-induced innate immune response occurs via activation of the NF-κB pathway in human monocyte macrophages in vitro. *Virus Res.* **2009**, *142*, 19–27. [CrossRef]
72. Morales-Ortega, A.; Bernal-Bello, D.; Llarena-Barroso, C.; Frutos-Pérez, B.; Duarte-Millán, M.Á.; de Viedma-García, V.G.; Farfán-Sedano, A.I.; Canalejo-Castrillero, E.; Ruiz-Giardín, J.M.; Ruiz-Ruiz, J.; et al. Imatinib for COVID-19: A case report. *Clin. Immunol.* **2020**, *218*, 108518. [CrossRef]
73. Peterson, D.; Damsky, W.; King, B. The use of Janus kinase inhibitors in the time of severe acute respiratory syndrome coronavirus 2 (SARS-CoV-2). *J. Am. Acad. Dermatol.* **2020**, *82*, e223–e226. [CrossRef] [PubMed]
74. McLaughlin-Drubin, M.E.; Munger, K. Viruses associated with human cancer. *Biochim. Biophys. Acta—Mol. Basis Dis.* **2008**, *1782*, 127–150. [CrossRef] [PubMed]
75. Khan, I.; Hatiboglu, M.A. Can COVID-19 induce glioma tumorogenesis through binding cell receptors? *Med. Hypotheses* **2020**, *144*, 110009. [CrossRef] [PubMed]
76. Rhea, E.M.; Logsdon, A.F.; Hansen, K.M.; Williams, L.M.; Reed, M.J.; Baumann, K.K.; Holden, S.J.; Raber, J.; Banks, W.A.; Erickson, M.A.; et al. The S1 protein of SARS-CoV-2 crosses the blood-brain barrier in mice. *Nat. Neurosci.* **2021**, *24*, 368–378. [CrossRef]
77. Pellegrini, L.; Albecka, A.; Mallery, D.L.; Kellner, M.J.; Paul, D.; Carter, A.P.; James, L.C.; Lancaster, M.A. SARS-CoV-2 Infects the Brain Choroid Plexus and Disrupts the Blood-CSF Barrier in Human Brain Organoids. *Cell Stem Cell* **2020**, *27*, 951–961. [CrossRef] [PubMed]
78. Zhang, L.; Zhu, F.; Xie, L.; Wang, C.; Wang, J.; Chen, R.; Jia, P.; Guan, H.Q.; Peng, L.; Chen, Y.; et al. Clinical characteristics of COVID-19-infected cancer patients: A retrospective case study in three hospitals within Wuhan, China. *Ann. Oncol.* **2020**, *31*, 894–901. [CrossRef] [PubMed]
79. Goulet, C.R.; Champagne, A.; Bernard, G.; Vandal, D.; Chabaud, S.; Pouliot, F.; Bolduc, S. Cancer-associated fibroblasts induce epithelial–mesenchymal transition of bladder cancer cells through paracrine IL-6 signalling. *BMC Cancer* **2019**, *19*, 1–13. [CrossRef] [PubMed]
80. Lo, C.-W.; Chen, M.-W.; Hsiao, M.; Wang, S.; Chen, C.-A.; Hsiao, S.-M.; Chang, J.-S.; Lai, T.-C.; Rose-John, S.; Kuo, M.-L.; et al. IL-6 Trans-Signaling in Formation and Progression of Malignant Ascites in Ovarian Cancer. *Cancer Res.* **2010**, *71*, 424–434. [CrossRef]
81. Ring, C.J.A. Cytolytic viruses as potential anti-cancer agents. *J. Gen. Virol.* **2002**, *83*, 491–502. [CrossRef]
82. Artene, S.-A.; Turcu-Stiolica, A.; Ciurea, M.E.; Folcuti, C.; Tataranu, L.G.; Alexandru, O.; Purcaru, O.S.; Tache, D.E.; Boldeanu, M.V.; Silosi, C.; et al. Comparative effect of immunotherapy and standard therapy in patients with high grade glioma: A meta-analysis of published clinical trials. *Sci. Rep.* **2018**, *8*, 11800. [CrossRef]
83. Artene, S.-A.; Turcu-Stiolica, A.; Hartley, R.; Ciurea, M.E.; Daianu, O.; Brindusa, C.; Alexandru, O.; Tataranu, L.G.; Purcaru, Ş.O.; Dricu, A. Dendritic cell immunotherapy versus bevacizumab plus irinotecan in recurrent malignant glioma patients: A survival gain analysis. *OncoTargets Ther.* **2016**, *9*, 6669–6677. [CrossRef] [PubMed]
84. Verheije, M.H.; Rottier, P.J.M. Retargeting of Viruses to Generate Oncolytic Agents. *Adv. Virol.* **2011**, *2012*, 1–15. [CrossRef] [PubMed]
85. Würdinger, T.; Verheije, M.H.; Broen, K.; Bosch, B.J.; Haijema, B.J.; De Haan, C.A.M.; Van Beusechem, V.W.; Gerritsen, W.R.; Rottier, P.J.M. Soluble Receptor-Mediated Targeting of Mouse Hepatitis Coronavirus to the Human Epidermal Growth Factor Receptor. *J. Virol.* **2005**, *79*, 15314–15322. [CrossRef] [PubMed]
86. Verheije, M.H.; Lamfers, M.L.M.; Würdinger, T.; Grinwis, G.C.M.; Gerritsen, W.R.; Van Beusechem, V.W.; Rottier, P.J.M. Coronavirus Genetically Redirected to the Epidermal Growth Factor Receptor Exhibits Effective Antitumor Activity against a Malignant Glioblastoma. *J. Virol.* **2009**, *83*, 7507–7516. [CrossRef] [PubMed]

International Journal of
Molecular Sciences

MDPI

Article

Assessment of Epinephrine and Norepinephrine in Gastric Carcinoma

Alina Maria Mehedințeanu [1,†], Veronica Sfredel [2,†], Puiu Olivian Stovicek [3], Michael Schenker [1], Georgică Costinel Târtea [2,*], Octavian Istrătoaie [4,*], Ana-Maria Ciurea [1] and Cristin Constantin Vere [5]

1 Department of Oncology, University of Medicine and Pharmacy of Craiova, 200349 Craiova, Romania; alina.maria591@gmail.com (A.M.M.); michael.schenker@umfcv.ro (M.S.); amciurea14@gmail.com (A.-M.C.)
2 Department of Physiology, University of Medicine and Pharmacy of Craiova, 200349 Craiova, Romania; veronica.sfredel@umfcv.ro
3 Department of Pharmacology, Faculty of Nursing, Târgu Jiu Subsidiary, Titu Maiorescu University, 04317 Bucharest, Romania; olivian_sfn@yahoo.com
4 Department of Cardiology, University of Medicine and Pharmacy of Craiova, 200349 Craiova, Romania
5 Department of Gastroenterology, University of Medicine and Pharmacy of Craiova, 200349 Craiova, Romania; cristin.vere@umfcv.ro
* Correspondence: georgetartea@gmail.com (G.C.T.); octavian.istratoaie@umfcv.ro (O.I.)
† These authors contributed equally to this work.

Abstract: The aim of our study was to assess the sympathetic nervous system's involvement in the evolution of gastric carcinoma in patients by analyzing the mediators of this system (epinephrine and norepinephrine), as well as by analyzing the histological expression of the norepinephrine transporter (NET). We conducted an observational study including 91 patients diagnosed with gastric carcinoma and an additional 200 patients without cancer between November 2017 and October 2018. We set the primary endpoint as mortality from any cause in the first two years after enrolment in the study. The patients were monitored by a 24-h Holter electrocardiogram (ECG) to assess sympathetic or parasympathetic predominance. Blood was also collected from the patients to measure plasma free metanephrine (Meta) and normetanephrine (N-Meta), and tumor histological samples were collected for the analysis of NET expression. All of this was performed prior to the application of any antineoplastic therapy. Each patient was monitored for two years. We found higher heart rates in patients with gastric carcinoma than those without cancer. Regarding Meta and N-Meta, elevated levels were recorded in the patients with gastric carcinoma, correlating with the degree of tumor differentiation and other negative prognostic factors such as tumor invasion, lymph node metastasis, and distant metastases. Elevated Meta and N-Meta was also associated with a poor survival rate. All these data suggest that the predominance of the sympathetic nervous system's activity predicts increased gastric carcinoma severity.

Keywords: gastric carcinoma; norepinephrine transporter; plasma free metanephrines and normetanephrines

Citation: Mehedințeanu, A.M.; Sfredel, V.; Stovicek, P.O.; Schenker, M.; Târtea, G.C.; Istrătoaie, O.; Ciurea, A.-M.; Vere, C.C. Assessment of Epinephrine and Norepinephrine in Gastric Carcinoma. *Int. J. Mol. Sci.* **2021**, *22*, 2042. https://doi.org/10.3390/ijms22042042

Academic Editor: Hidekazu Suzuki

Received: 1 February 2021
Accepted: 14 February 2021
Published: 18 February 2021

1. Introduction

Gastric cancer is a malignant disease with a high degree of lethality; according to Globocan 2018, it ranks fifth in terms of the incidence of malignancies, with 1,033,701 cases annually (5.7%). It is also the third leading cause of cancer mortality, with an annual death toll of approximately 782,685 (8.2%) [1]. This condition is much more common among men, occupying third place for the total number of neoplasms, whereas this condition is ranked in fifth place regarding incidence for women [1]. The factors leading to this disease have not been identified exactly, but strong correlations have been found between its occurrence and diet (i.e., a diet rich in salty and smoked foods), *Helicobacter pylori* infection, vitamin deficiency, a low consumption of fruits and vegetables, smoking, a family history of gastric cancer, stress, and long-term stomach inflammation [2].

The evolution of gastric cancer is unpredictable, and because of the extensive nature of the diagnostic methods, as well as diagnosis often occurring in advanced stages of the disease, the disease only becomes symptomatic in the advanced stages for the vast majority of patients [3]. The involvement of the autonomic nervous system in the development and evolution of gastric cancer has not been fully elucidated, but numerous studies have shown that there is a close relationship. Not only does the autonomic nervous system innervate the digestive tract but the tumor cells secrete growth factors and exhibit elevated levels of catecholamines and various receptors [4–6]. The involvement of the vegetative nervous system was first demonstrated by Batsakis approximately 30 years ago when he described the presence of nerves located in the vicinity of human epithelial carcinomas, such as gastric, head and neck, or prostate cancers [7,8]. These nerves have been described as directly involved in metastatic dissemination through a process called perineural invasion (PNI) in which neoplastic cancer cells are able to invade and migrate into, around, and through the nerves, with PNI frequently being associated with poor clinical results [9].

In this study, we wanted to evaluate the involvement of the sympathetic nervous system in the evolution of patients with gastric carcinoma by analyzing the mediators of this system (epinephrine and norepinephrine) as well as the histological expression of the norepinephrine transporter (NET). The norepinephrine transporter is a monoamine transporter responsible for capturing extracellular norepinephrine (N-Meta), also known as noradrenaline. The latter has an inhibitory role in the gastrointestinal tract. This is also true for epinephrine, which mainly enters the gastrointestinal tract through the bloodstream after being secreted by the adrenal medulla directly into the circulation. It is known that this transporter is also involved in the uptake of extracellular dopamine; the reuptake of the two neurotransmitters plays an important role in regulating their concentrations in the synaptic terminals [10–13].

In order to obtain homogeneous data, we chose to analyze the influence of epinephrine and norepinephrine only on gastric tumors because of the physiological particularities of the stomach's innervation. It is necessary to mention that the stomach is much more dependent on extrinsic neural inputs, represented by nuclei located in the caudal brainstem, from which sympathetic and parasympathetic pathways start or are controlled. In contrast to the stomach, the small and large intestines have a high degree of independent neuronal control and can function even if they lack extrinsic neural inputs [14].

2. Results

2.1. Assessment of the Heart Rates of the Patients Included in the Study

To assess the predominance of the sympathetic or parasympathetic autonomic nervous system, we evaluated 91 patients suffering from gastric cancer using a Holter electrocardiogram (ECG) for approximately 24 h, both at the time of diagnosis and before starting any antineoplastic therapy, by calculating the average heart rate (HR) during the day, during the night, and for 24 h. For the controls, 200 patients without gastric carcinoma who belonged to the same age group as the gastric carcinoma patients were evaluated using a Holter ECG. We observed (Figure 1A,B) that during the day, higher heart rates predominated in the group of patients with gastric carcinoma than in the control group (HR during the day = 90.76 $\pm$ 13.64 beats per minute (bpm) in the group of patients with gastric carcinoma versus 82.29 $\pm$ 7.86 bpm in the control group; $p = 0.0012$). These differences were maintained overnight (HR at night = 65.74 $\pm$ 16.44 bpm in the gastric carcinoma patients versus 58.21 $\pm$ 5.11 bpm in the control group; $p = 0.0015$) and for the entire monitoring period of approximately 24 h (HR for the 24 h period = 78.25 $\pm$ 14.01 bpm in the gastric carcinoma patients versus 68.96 $\pm$ 6.80 bpm in the control group; $p < 0.0000$). Another observation was that the patients with gastric carcinoma had no significant difference in heart rate during the day compared to that during the night, as shown in patients without cancer (Figure 1C,D). All these data suggest an increased predominance of sympathetic nervous system influences in the patients with gastric carcinoma versus patients without cancer.

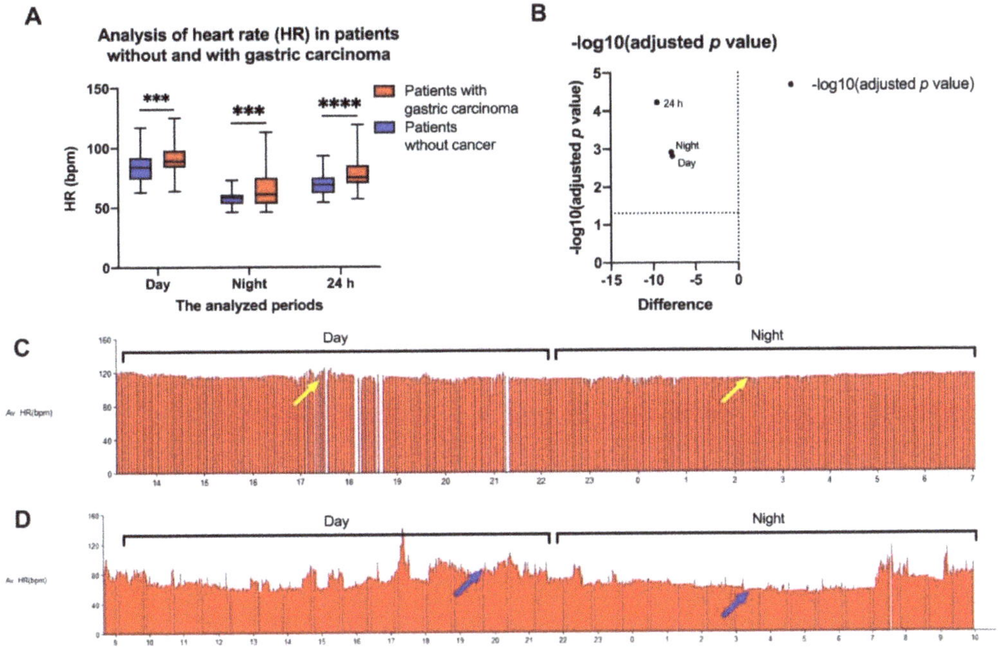

Figure 1. Evaluation of heart rates (HRs) by Holter electrocardiogram (ECG) of the patients included in the study. (**A**) Minimum, maximum, and average HRs in patients with gastric carcinoma and patients without cancer, with (**B**) a volcano association plot. (**C**) Representative histogram of HR for a 24 h period of a patient with gastric carcinoma; approximately the same HRs can be observed during the day and night (yellow arrows). (**D**) Representative histogram of HR for a 24 h period of a cancer-free patient; it was higher during the day (as normal) and lower at night (blue arrows). Av HR, atrioventricular heart rate; bpm, beats per minute. Student's *t*-test, *** $p < 0.001$ and **** $p < 0.0001$.

2.2. Relationship between Norepinephrine Transporter Expression and Clinicopathological Features

The expression of the norepinephrine transporter was analyzed in samples from patients with gastric carcinoma (N = 91) as well as from the 200 patients (controls) who required gastric resection for benign reasons (Figure 2A–D). Using multispectral microscopy (Figure 3A–D), we analyzed the expression of the norepinephrine transporter, calculating the integrated optical density (IOD) only for the target color signal. Depending on the tumor grading, we observed an increase in IOD from well-differentiated (G1) to moderately differentiated (G2) and poorly differentiated (G3) tumors (Figure 4A and Supplementary Table S1). The norepinephrine transporter expression was higher in the patients with gastric carcinoma in those aged < 60 years ($p = 0.0115$) and in those with localization of the tumor in the gastric body or pyloric area ($p = 0.0033$), with tumor invasion T_{3-4} ($p = 0.0093$), with lymph node metastasis $N_{\geq 2}$ ($p = 0.0371$), and with TNM classification of malignant tumors (TNM) stages T_{III-IV} ($p = 0.003$) (Figure 5a and Supplementary Table S2).

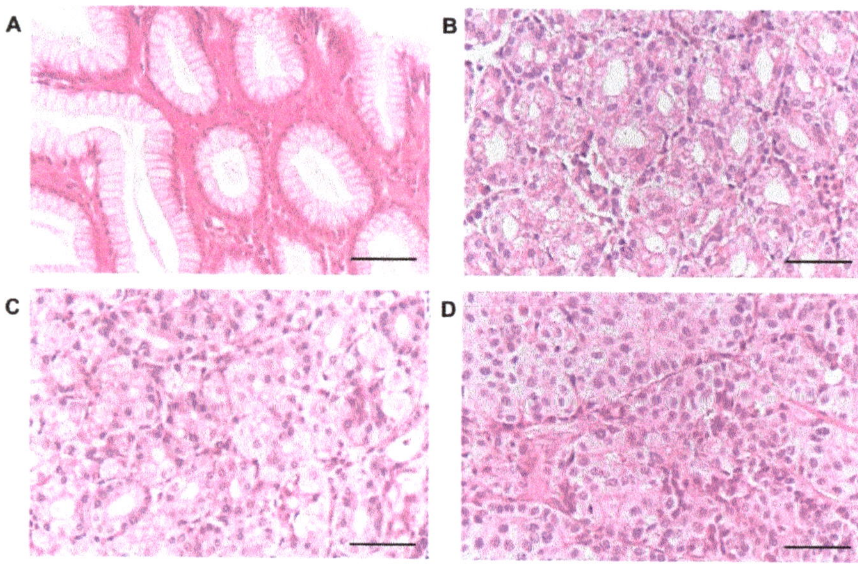

Figure 2. Examples of histological samples collected from the patients included in the study. (**A**) Normal gastric mucosa. (**B**) Gastric carcinoma with a high degree of cell differentiation—G1. (**C**) Gastric carcinoma with a moderate degree of cell differentiation—G2. (**D**) Gastric carcinoma with a poor degree of cell differentiation—G3. Hematoxylin and eosin staining. Magnification, 400×. Scale bars represent 20 µm.

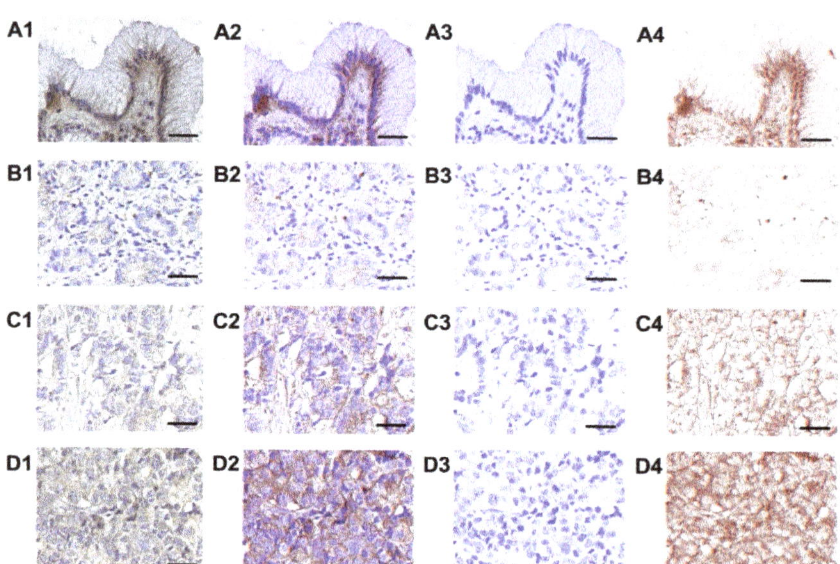

Figure 3. Assessment of the expression of norepinephrine transporters in normal (**A**) and tumor gastric tissue (**B–D**) by spectral unmixing microscopy: (**B**) well-differentiated (G1), (**C**) moderately differentiated (G2), and (**D**) poorly differentiated (G3). (1) Images from the optical microscopy; (2) slides immuno-stained for norepinephrine transporters with 3,3′-diaminobenzidines (DAB) and counterstained with hematoxylin (signals are shown overlapping); (3) images with pure hematoxylin only (nuclei and cell membranes); (4) images with pure DAB only (only the signal for norepinephrine transporters). Magnification, 400×. Scale bars represent 20 µm.

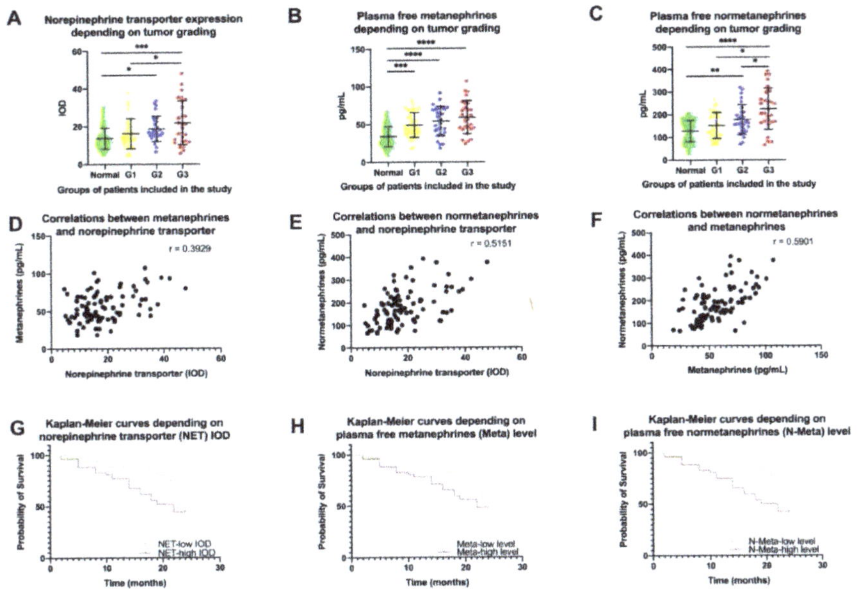

Figure 4. (**A**) Norepinephrine transporter expression quantified according to the integrated optical density (IOD) in cancer-free patients and in different tumor differentiation gradings in patients with gastric carcinoma. Plasma levels of free metanephrine (**B**) and normetanephrine (**C**) in cancer-free patients and in different stages of tumor differentiation in patients with gastric carcinoma. Correlations between metanephrine and norepinephrine transporter (**D**), normetanephrine and normetanephrine transporter (**E**), and normetanephrine and metanephrine (**F**). Kaplan–Meier curves depending on the norepinephrine transporter IOD (**G**), on plasma free metanephrine (**H**), and on plasma free normetanephrine (**I**). One-way ANOVA; * $p < 0.05$, ** $p < 0.01$, *** $p < 0.001$, and **** $p < 0.0001$.

2.3. Relationship between Plasma Free Metanephrine (Meta) and Normetanephrine and Clinicopathological Features

Another analysis performed in our study was measuring the free metanephrine and normetanephrine in the plasma for all the patients included before starting any therapy. Both the plasma free metanephrine and plasma free normetanephrine were higher in the patients with gastric carcinoma than those without. The cancer-free patients had plasma free metanephrine (Meta) values of 34.05 ± 13.23 pg/mL and plasma free normetanephrine (N-Meta) values of 128.3 ± 46.70 pg/mL. In the patients with gastric carcinoma, increased plasmatic levels of metanephrine and normetanephrine were correlated with tumor grading, increasing from well-differentiated (Meta = 48.09 ± 16.45 pg/mL and N-Meta = 152.1 ± 57.05 pg/mL) to moderately differentiated (Meta = 54.14 ± 19.59 pg/mL and N-Meta = 178.1 ± 65.53 pg/mL) and poorly differentiated (Meta = 59.13 ± 21.88 pg/mL and N-Meta = 225.4 ± 91.22 pg/mL) tumors (Figure 4B,C). In terms of the clinicopathological features concerned, we observed that higher free metanephrine could be found in patients with gastric carcinoma according to the histological type, i.e., adenocarcinoma (Ad.c.) as opposed to mixed carcinoma/signet-ring-cell carcinoma (M.c./S.r.c.c.) ($p = 0.0004$); the location of the tumor in the gastric body or pyloric area ($p = 0.0047$); tumor invasion $T_{3–4}$ ($p = 0.0165$); lymph node metastasis $N_{\geq 2}$ ($p = 0.0473$); and TNM stage $T_{III–IV}$ ($p = 0.0148$) (Figure 5B and Supplementary Table S3). We also observed higher plasma free normetanephrine in patients with gastric carcinoma with a tumor size ≥5 cm ($p = 0.0217$), histological type Ad.c as opposed to M.c./S.r.c.c. ($p = 0.0253$), a location of the tumor in the gastric body or pyloric area ($p = 0.0132$), a tumor invasion $T_{3–4}$ ($p = 0.0177$), a lymph node metastasis $N_{\geq 2}$ ($p = 0.0127$), and a TNM stage $T_{III–IV}$ ($p = 0.0275$) (Figure 5C and Supplementary Table S4). It should be noted that patients who had the signet-ring-cell gastric cancer type, despite having a low degree of differentiation (G3), had low plasma levels of free metanephrine and normetanephrine.

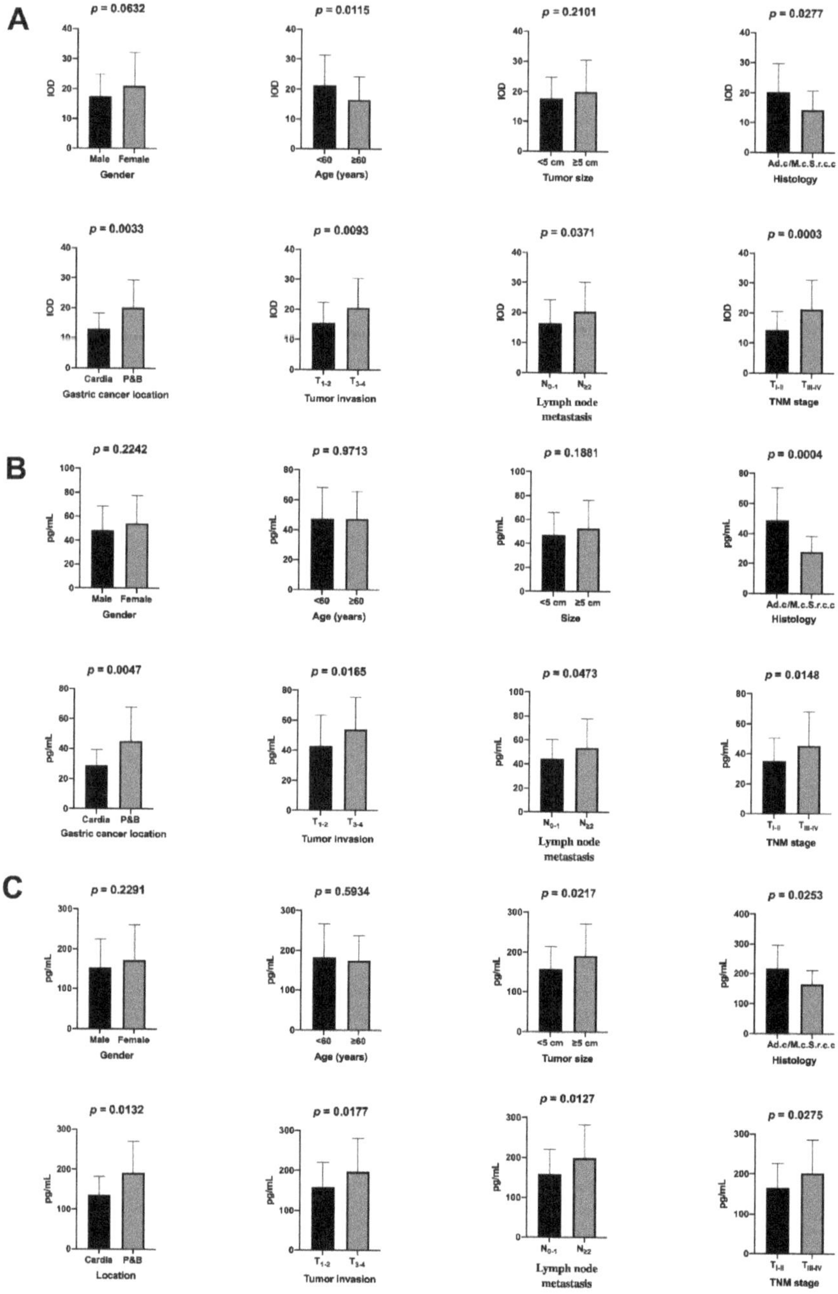

Figure 5. (**A**) Relationship between norepinephrine transporter expression and clinicopathological features. Relationship between plasma free metanephrine (**B**) and normetanephrine (**C**) and clinicopathological features. Ad.c., adenocarcinoma; M.c./S.r.c.c., mixed carcinoma/signet-ring-cell carcinoma; P&B, gastric body or pyloric area.

2.4. Correlation between Norepinephrine Transporter Expression and Plasma Free Metanephrine and Normetanephrine

We observed a moderate positive correlation between the IOD for norepinephrine transporter expression and plasma free metanephrine (r = 0.3929; 95% confidence interval = 0.2034–0.5540; R-squared = 0.1544), and a strong positive correlation between the IOD for norepinephrine transporter expression and plasma free normetanephrine (r = 0.5151; 95% confidence interval = 0.3459–0.6519; R-squared = 0.2654). There was also a strong positive correlation between the plasma free metanephrine and plasma free normetanephrine (r = 0.5901; 95% confidence interval = 0.4373–0.7098; R-squared = 0.3842). These data are summarized in Figure 4D–F.

2.5. Univariate Analysis of Prognostic Factors

The 91 patients with gastric carcinoma were divided into a low-NET group (N = 36/91) and a high-NET group (N = 55/91) based on the median IOD for the norepinephrine transporter. Depending on cut-off levels for plasma free metanephrine (65 pg/mL) and normetanephrine (196 pg/mL), the patients with gastric carcinoma were divided into low-(N = 39/91) and high-Meta groups (N = 52/91) and into low-(N = 33/91) and high-N-Meta groups (N = 58/91). Univariate analysis with a log-rank test indicated that the high-NET patients had a significantly poorer survival rate at two years after inclusion in the study than the low-NET patients (44.23% vs. 63.88%; $p = 0.0358$; hazard ratio and its reciprocal = 1.956 and 0.5140, respectively; Figure 4G). Lower survival rates were also observed in the patients with higher metanephrine but without statistical significance (48.07% vs. 58.33%; $p = 0.1487$; hazard ratio and its reciprocal = 1.462 and 0.6838, respectively; Figure 4H), as well as in those with high free normetanephrine with statistical significance (42.30% vs. 66.67%; $p = 0.0104$; hazard ratio and its reciprocal = 2.289 and 0.4369, respectively; Figure 4I).

3. Discussion

In this study, we analyzed the correlations between the clinicopathological aspects of patients diagnosed with gastric cancer (gastric adenocarcinoma) and the involvement of the autonomic nervous system in the carcinogenesis process by identifying certain features of the sympathetic nervous system and the norepinephrine transporter, identified in neoplastic cells. We attempted to support the claim that the vegetative nervous system can influence the development and evolution of gastric cancer.

The sympathetic nervous system is part of the vegetative nervous system and responsible for the fight reaction, also known as the sympathetic–adrenal response. It secretes adrenaline (epinephrine) and noradrenaline (norepinephrine), catecholamines that are subsequently released into the blood [15]. The action exerted by the sympathetic nervous system causes a series of reactions in various organs of the body, such as an increased heart rate, decreased motility in the large intestine, reduced secretions by salivary glands, and vasoconstriction [16]. Free plasma metanephrine and normetanephrine are metabolites of catecholamines, the latter being considered a hormone that is released into the blood, especially during periods of physical or emotional stress, depression, or anxiety, causing both psychological and endocrine changes [17]. These catecholamines produced from the precursor tyrosine can, on the one hand, alter the immune response and, on the other hand, promote several biological signaling pathways involved in tumor initiation, growth, and metastasis [16–19].

Regarding the involvement of the sympathetic nervous system in the development and evolution of gastric cancer, the primary pathway is mediated by the action of neurotransmitters on β2-adrenergic receptors, activating an intracellular signaling cascade via adenylyl cyclase [18,19]. Studies have highlighted the impact of the main neurotransmitter of the sympathetic nervous system, norepinephrine, on vascular endothelial growth factor (VEGF) and matrix metalloproteinase 2/9 (MMP-2 and MMP-9) [20]. Other studies have also shown an important role of norepinephrine in epithelial–mesenchymal transition

(EMT). For example, Shan et al. demonstrated that norepinephrine causes, in gastric carcinoma, a decrease in E-cadherin expression and an increase in vimentin expression; both changes increase cell motility and confer the ability of tumor invasion [21]. This mechanism can occur through the β2-adrenergic receptor (AR)–hypoxia-inducible factor-1-alpha axis, which is also involved in the promotion of tumor progression by chronic stress in animal cancer models [22]. EMT can also be initiated in gastric cancer by the β2-AR–metalloproteinase (MMP)-7 pathway through the activation of AP-1 and signal transducer and transcriptional activator 3 (STAT3) [23,24].

We found that the highest plasma values of serum metanephrine and normetanephrine were increased in patients with poorly differentiated gastric adenocarcinoma. They varied according to the degree of differentiation, and increased values were found among those with localization in the gastric body or pyloric area and with histopathological aspects of adenocarcinoma, as well as among patients who had metastases in regional lymph nodes or distant metastasis. In this regard, a recent study that evaluated the activity of periostin, which mediates the critical steps in gastric carcinoma, showed that it is expressed in the stroma of gastric carcinoma but not in normal gastric tissue, and this is strongly correlated with the expression of alpha-smooth muscle actin (SMA) [25]. Isoprenaline causes an increase in periostin expression in gastric cancer, with the activation of the previously mentioned axis, but it can also promote angiogenesis by stimulating VEGF secretion and the upregulation of VEGFR2 and plexin-A1 [26,27].

Another aspect highlighted by our study is the implications of psychological stress for the initiation and progression of gastric carcinoma. Psychological stress initiates a response of the hypothalamic–pituitary–adrenal axis, which raises catecholamine levels; catecholamines interact with certain biological components of tumor cells through certain signaling pathways. This can lead to the progression of certain cancers, such as those of the ovaries, nasopharynx, or pancreas [27,28].

Regarding the influence of catecholamines on therapy for severe gastric cancer, it has been observed that the stimulation of gastric cancer cells with catecholamines in vitro increases trastuzumab resistance by not only activating STAT3 and extracellular signal-regulated kinases (ERKs) but also by upregulating mucin 4 (MUC4) expression [29]. However, these cellular signaling mechanisms induced by catecholamines may become possible therapeutic targets. For example, propranolol, a non-selective adrenergic blocker, can cause cell cycle arrest and induce apoptosis in gastric carcinoma cells by blocking nuclear factor-kB (NF-kB), MMP2/9, VEGF, and cyclooxygenase-2 (COX-2) [30,31].

It is well known that the sympathetic nervous system influences cardiac activity, causing an increase in heart rate, as was shown in our patients. Most of the patients with gastric cancer in our study had increased heart rates directly proportional to the plasma levels of free metanephrine and normetanephrine. Shi et al. recently reported that the severity of gastric cancer in diagnosed patients can also be predicted by perturbations in the nonlinear dynamic patterns of heart rate variability (HRV) [32].

In other primary tumors, the activity of the sympathetic nervous system has been evaluated in both preclinical and clinical studies. The plasma norepinephrine and epinephrine concentrations are significantly higher in patients with oral and oropharyngeal squamous cell carcinoma (SCC) than in non-cancer patients [33]. In epithelial ovarian cancer, norepinephrine reduces cisplatin's efficacy and can affect DNA integrity [34]. Epinephrine increases the phosphorylation of p38 MAPK in breast cancer cells and, thereby, enhances the malignancy of this type of cancer [35]. A chemical sympathectomy markedly reduces the incidence of fibrosarcoma and significantly prolongs survival in rats [36]. Regarding other tumors of the gastrointestinal tract, it has been found that norepinephrine facilitates tumor growth in pancreatic cancer [37], induces hepatocellular carcinoma invasion and anoikis resistance through β2-AR-mediated epidermal growth factor receptor transactivation [38], and facilitates cell proliferation in esophageal squamous cell carcinoma [39]. Catecholamines have also been shown to promote metastasis and tumor progression in prostate and lung cancers and melanoma [40].

4. Materials and Methods

4.1. Patients

This was an observational study, in which 91 patients diagnosed with gastric carcinoma with different degrees of tumor differentiation were consecutively included, following surgery or upper digestive endoscopy, at the Emergency County Hospital of Craiova, Romania, between November 2017 and October 2018. For the controls, we delimited a group of 200 patients without cancer of the same age group and gender as the cancer patients. We set the primary endpoint as mortality from any cause in the first two years after enrolling in the study. The design of the study is shown in Figure 6.

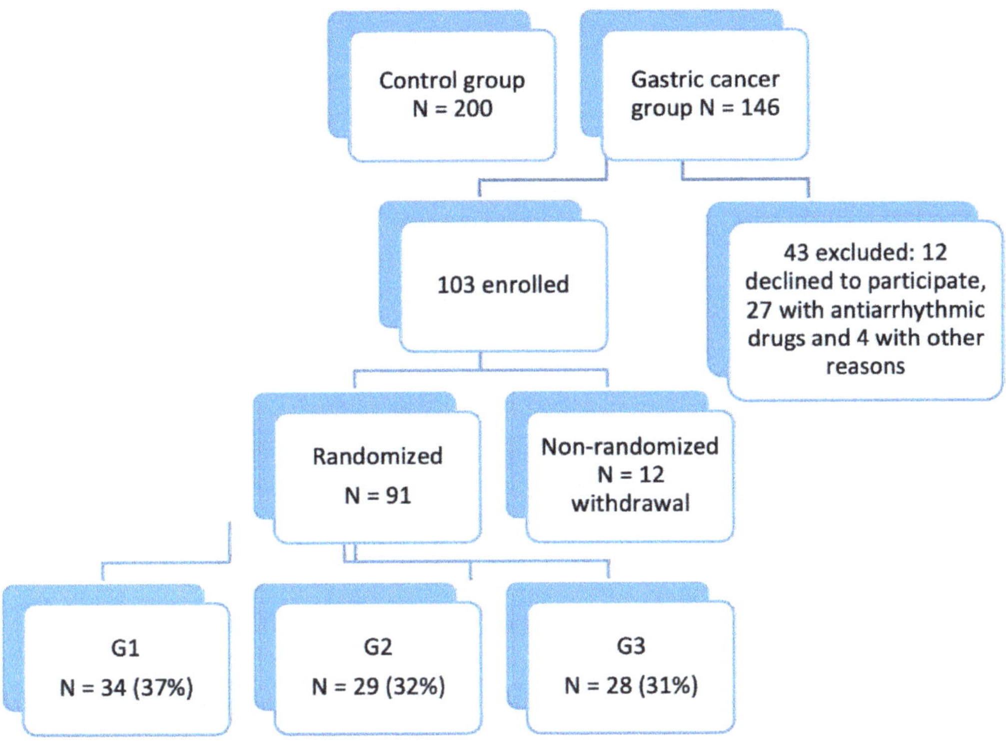

Figure 6. Design of the study.

All of the stages of the study were explained to the patients before commencement, and participation was possible only after providing written consent, with the patients being informed about the confidentiality of personal data and the procedures being performed in accordance with current regulations, without negatively influencing normal diagnostic or treatment procedures.

This study was approved by the Ethics Committee of the University of Medicine and Pharmacy of Craiova (No. 71/02.04.2017), respecting the ethical principles underlying the Declaration of Helsinki and the University Code of Ethics on Good Conduct of Research and the codes of practice established by the Code of Medical Ethics.

Patients diagnosed with gastric cancer, confirmed by postoperative histopathological examination or biopsy, had blood samples taken for the measurement of serum metanephrine and normetanephrine, after which they underwent Holter ECGs, before starting chemotherapeutic treatment, to avoid possible post-drug interactions.

4.2. Assessment of Heart Rate

To assess the predominance of the sympathetic or parasympathetic nervous system, the patients were also monitored using a Holter ECG TLC5000 (Contec Medical Systems, Qinhuangdao, Hebei Province, China), together with a former analysis of heart rate variability in terms of both frequency and the field of time. Ten patch electrodes were applied to the patients' chests, through which 12 ECG leads were connected. The important parameters in our study were the minimum, average, and maximum heart rate, not only over 24 h but also during the night and day. The ECG Holter parameters were interpreted according to the recommendations of the American College of Cardiology/American Heart Association (ACC/AHA) ambulatory electrocardiography guide [41]. It should be noted that the patients were examined by this method after the diagnosis of gastric adenocarcinoma was confirmed. Another important criterion was a lack of previous medication that could have influenced the heart rhythm (especially betablockers, calcium channel blockers, current blockers, or other antiarrhythmics). All the patients in our study performed at least 150 min of moderate-intensity aerobic physical activity per week. There were no differences between the groups included in the study regarding the degree of physical activity. Since all the patients underwent gastrectomy and since this intervention involved damage to local nerve plexuses, we unfortunately did not consider it appropriate to analyze the heart rate variability postoperatively. Moreover, postoperative stress is another factor that inevitably changes heart rate variability.

4.3. Histopathological Examination

A histopathological examination was performed following surgery or tumor biopsy via upper digestive endoscopy. The biological samples were introduced into 10% formic aldehyde solution, in which a neutral pH was created by adding calcium bicarbonate to neutralize the formic acid for fixation. After fixation, the biological samples were washed with water and then paraffin embedded as follows. The samples were completely dehydrated by passing them through ethyl alcohol of different concentrations. They were then clarified by removing the alcohol from the tissue, and the samples were passed through successive paraffin baths, incorporated into paraffin blocks, and solidified. Finally, the paraffin blocks were sectioned, then the sectioned tissues were glued to slides and stained with hematoxylin–eosin, after which histopathological diagnoses were established with certainty.

4.4. Immunohistochemistry

The paraffin-embedded tissues were sectioned into 3 μm serial sections using the HM350 rotary microtome (Thermo Fisher Scientific, Waltham, MA, USA) equipped with a section transfer system in a cold-water bath with a Peltier cooling mode. These slides were then transferred to a water bath heated to 40 °C to be stretched and evened out. The slides were recovered from the bath on blades covered with poly-lysine with a positively charged amino acid residue to increase the adhesion of the sections on the blades. The poly-lysine-coated slides were dried in a thermostat at 37 °C for 24 h.

The next day, the slides were first deparaffined by passing them through three successive xylene baths, for 15 min each, and then rehydrated in alcohol solutions with decreasing concentrations; any final traces of alcohol were removed by washing with distilled water. For antigenic recovery, the slides were boiled for 21 min in successive cycles of 3 min each, in a microwave oven, at a power of 600 W, in a solution of sodium citrate at a pH of 6. This was followed by cooling the slides for 30 min and then washing them with tap water and distilled water for 15 min. To block endogenous peroxidase activity, the slides were incubated in 1% hydrogen peroxide and distilled water for 30 min at room temperature, stored for another 30 min in 3% skimmed-milk powder in phosphate-buffered saline (PBS), and then incubated with primary antibodies at 4 °C for 18 h.

The primary antibodies used in this work were norepinephrine transporter monoclonal antibody (CL3063)/NBP2-62704 (dilution 1:20; Novus Biological, Abingdon, UK).

Finally, the signal was identified via 3,3′-diaminobenzidines (DAB) (Dako, Glostrup, Denmark). Subsequently, the slides were cover-slipped in DPX (Sigma-Aldrich, St. Louis, MO, USA) after hematoxylin and eosin staining. All the slides stained for each primary antibody were processed at the same time to observe the protocol correlation, along with the control sections, which were stained with either DAB or hematoxylin–eosin in order to obtain the pure spectrum for those colors. Negative controls were obtained by omitting primary antibodies.

4.5. Acquisition and Image Processing

For the quantification of the target immunohistochemical signal, and taking into account the histopathological aspect, light microscopy images were obtained using a Nikon Eclipse 90i motorized microscope (Apidrag, Bucharest, Romania). This microscope was equipped with a Nuance FX multispectral camera as well as the Nuance imaging analysis software (Perkin Elmer, Hopkinton, MA, USA). An optical microscopy image was initially obtained, followed by a mixed image (the color spectra for hematoxylin and DAB were separately superimposed on this image). In another step, separate images were obtained for hematoxylin and DAB (Supplementary Figure S1). The unmixed DAB signal was quantified by randomly obtaining 10 images captured with a 20× lens. The color signal was quantitatively analyzed based on the integrated optical density, using the Image-Pro Plus AMS 7 image analysis software (Media Cybernetics, Bethesda, MD, USA). With the help of this software, regions of interest were defined where we evaluated the color signals and calculated the IOD, while the stroma was manually excluded from the obtained images.

4.6. Dosage of Plasma Free Metanephrine and Normetanephrine

Blood samples were taken from fasted patients who had avoided alcohol and caffeine for 24 h before sampling. They were also informed to avoid certain drugs that may influence serum metanephrine and normetanephrine, such as acetaminophen, tricyclic antidepressants, phenoxybenzamine, alpha-agonists, or monoaminoxidase inhibitors. For patients undergoing treatment with these drugs, their medication was discontinued for at least five days before sampling.

Venous blood was collected in a pre-cooled ethylenediaminetetraacetic acid (EDTA) K3 vacutainer, which was then gently shaken, overturned, and placed on ice; it was transported to the laboratory within 2 h. The samples were processed no later than 2 h after sampling.

The serum metanephrine and normetanephrine were quantified via the competitive enzyme-linked immunosorbent assay (ELISA), and the values were determined after a precipitation stage. Their reference values were the following: metanephrine, <65 pg/mL, and normetanephrine, <196 pg/mL, with detection limits of 5 and 10 pg/mL, respectively [42].

4.7. Statistical Analysis

The data obtained with the Image-Pro Plus AMS 7 image analysis software were exported to Microsoft Excel 2010 (Microsoft Corporation, Redmond, WA, USA) and analyzed using GraphPad Prism 8 (San Diego, CA, USA). All the results are reported as the means and standard deviations. To compare the means of two groups, we used Student's *t*-test. To compare the means of more than two groups, we used an ANOVA. To examine the correlations between the different categories of data, we used Pearson's correlation test. To analyze whether there was a link between a variable and survival time, we used the log-rank test. $p < 0.05$ was considered to indicate a statistically significant difference between the compared means from the various groups.

5. Conclusions

The predominance of the sympathetic nervous system's activity in patients with gastric cancer, through increased heart rates, elevated plasma free metanephrine and normetanephrine, and increased expression of the norepinephrine transporter in tumor cells, is a negative prognostic factor for these patients. These observations may highlight future therapeutic or prognostic targets.

Supplementary Materials: The supplementary materials can be found at https://www.mdpi.com/1422-0067/22/4/2042/s1: Figure S1, example of spectral unmixing for the series of slides; Table S1, norepinephrine transporter expression quantified according to the integrated optical density in cancer-free patients and in different tumor differentiation gradings in patients with gastric carcinoma; Table S2, norepinephrine transporter expression quantified depending on clinicopathological features; Table S3, plasma free metanephrine level (pg/mL) depending on the clinicopathological features; Table S4, plasma free normetanephrine level (pg/mL) depending on the clinicopathological features.

Author Contributions: Conceptualization, A.M.M., V.S., G.C.T., and C.C.V.; methodology, V.S. and O.I.; software, G.C.T.; validation, V.S., O.I., M.S., and A.-M.C.; formal analysis, A.-M.C. and P.O.S.; investigation, A.M.M., P.O.S., M.S., and A.-M.C.; writing—original draft preparation, G.C.T. and O.I.; writing—review and editing, A.M.M. and V.S.; visualization, P.O.S., M.S., and A.-M.C.; supervision, C.C.V. All authors have read and agreed to the published version of the manuscript.

Funding: This research received no external funding.

Institutional Review Board Statement: The study was conducted according to the guidelines of the Declaration of Helsinki, and approved by the Ethics Committee of UNIVERSITY OF MEDICINE AND PHARMACY OF CRAIOVA, ROMANIA (No. 71/02.04.2017).

Informed Consent Statement: Informed consent was obtained from all subjects involved in the study.

Acknowledgments: The authors would like to especially thank Pirici Daniel from the Department of Histology, University of Medicine and Pharmacy of Craiova, Romania, for the processing of the histological samples and for the help given in the multispectral evaluation of the microscopic images.

Conflicts of Interest: The authors declare no conflict of interest.

Abbreviations

bpm	Beats per minute
DAB	3,3′-diaminobenzidine
EDTA	Ethylenediaminetetraacetic acid
ELISA	Enzyme-linked immunosorbent assay
HRV	Heart rate variability
IOD	Integrated optical density
Meta	Metanephrine
N-meta	Normetanephrine
NET	Norepinephrine transporter
STAT3	Signal transducer and activator of transcription 3
VEGF	Vascular endothelial growth factor
VEGFR	Vascular endothelial growth factor receptor

References

1. Global Cancer Observatory. Available online: https://gco.iarc.fr/today/online-analysis-pie?v=2018&mode=cancer&mode_population=continents&population=900&populations=900&key=total&sex=0&cancer=39&type=0&statistic=5&prevalence=0&population_group=0&ages_group%5B%5D=0&ages_group%5B%5D=17&nb_items=7&group_cancer=1&include_nmsc=1&include_nmsc_other=1&half_pie=0&donut=0&population_group_globocan_id= (accessed on 15 September 2020).
2. Kim, J.; Cho, Y.A.; Choi, W.J.; Jeong, S.H. Gene-diet interactions in gastric cancer risk: A systematic review. *World J. Gastroenterol.* **2014**, *20*, 9600–9610. [CrossRef] [PubMed]
3. Entschladen, F.; Palm, D.; Lang, K.; Drell IV, T.L.; Zaenker, K.S. Neoneurogenesis: Tumors may initiate their own innervation by the release of neurotrophic factors in analogy to lymphangiogenesis and neoangiogenesis. *Med. Hypotheses* **2006**, *67*, 33–35. [CrossRef] [PubMed]
4. Gao, J.P.; Xu, W.; Liu, W.T.; Yan, M.; Zhu, Z.G. Tumor heterogeneity of gastric cancer: From the perspective of tumor-initiating cell. *World J. Gastroenterol.* **2018**, *24*, 2567–2581. [CrossRef]
5. Entschladen, F.; Palm, D.; Niggemann, B.; Zaenker, K.S. The cancer's nervous tooth: Considering the neruronal crosstalk within tumors. *Semin. Cancer Biol.* **2008**, *18*, 171–175. [CrossRef]

6. Schuller, H.M.; Al-Wadei, H.A.; Majidi, M. GABA B receptor is a novel drug target for pancreatic cancer. *Cancer* **2008**, *112*, 767–778. [CrossRef]
7. Batsakis, J.G. Nerves and neurotropic carcinomas. *Ann. Otol. Rhinol. Laryngol.* **1985**, *94*, 426–427.
8. Rodin, A.E.; Larson, D.L.; Roberts, D.K. Nature of the perineural space invaded by prostatic carcinoma. *Cancer* **1967**, *20*, 1772–1779. [CrossRef]
9. Anderson, P.R.; Hanlon, A.L.; Patchefsky, A.; Al-Saleem, T.; Hanks, G.E. Perineural invasion and Gleason 7–10 tumors predict increased failure in prostate cancer patients with pretreatment PSA. *Int. J. Radiat. Oncol. Biol. Phys.* **1998**, *41*, 1087–1092. [CrossRef]
10. Schroeter, S.; Apparsundaram, S.; Wiley, R.G.; Miner, L.H.; Sesack, S.R.; Blakely, R.D. Immunolocalization of the cocaine- and antidepressant-sensitive l-norepinephrine transporter. *J. Comp. Neurol.* **2000**, *420*, 211–232. [CrossRef]
11. Tellioglu, T.; Robertson, D. Genetic or acquired deficits in the norepinephrine transporter: Current understanding of clinical implications. *Expert Rev. Mol. Med.* **2001**, *3*, 1–10. [CrossRef] [PubMed]
12. Kristensen, A.S.; Andersen, J.; Jørgensen, T.N.; Sørensen, L.; Eriksen, J.; Loland, C.J.; Strømgaard, K.; Gether, U. SLC6 neurotransmitter transporters. Structure, function and regulation. *Pharmacol. Rev.* **2011**, *63*, 585–640. [CrossRef]
13. Torres, G.E.; Gainetdinov, R.R.; Caron, M.G. Plasma membrane monoamine transporters: Structure, regulation and function. *Nat. Rev. Neurosci.* **2003**, *4*, 13–25. [CrossRef]
14. Browning, K.N.; Travagli, R.A. Central nervous system control of gastrointestinal motility and secretion and modulation of gastrointestinal functions. *Compr. Physiol.* **2014**, *4*, 1339–1368. [CrossRef]
15. Wang, K.; Zhao, X.H.; Liu, J.; Zhang, R.; Li, J.P. Nervous system and gastric cancer. *Biochim. Biophys. Acta (BBA) Rev. Cancer* **2020**, *1873*, 188313. [CrossRef]
16. Repasky, E.A.; Eng, J.; Hylander, B.L. Stress, metabolism and cancer: Integrated pathways contributing to immune suppression. *Cancer J. (Sudbury Mass.)* **2015**, *21*, 97–103. [CrossRef]
17. Peters, L.J.; Kelly, H. The influence of stress and stress hormones on the transplantability of a non-immunogenic syngeneic murine tumor. *Cancer* **1977**, *39*, 1482–1488. [CrossRef] [PubMed]
18. Lin, X.; Luo, K.; Lv, Z.; Huang, J. Beta-adrenoceptor action on pancreatic cancer cell proliferation and tumor growth in mice. *Hepato Gastroenterol.* **2012**, *59*, 584–588. [CrossRef]
19. Ciurea, R.N.; Rogoveanu, I.; Pirici, D.; Târtea, G.C.; Streba, C.T.; Florescu, C.; Cătălin, B.; Puiu, I.; Târtea, E.A.; Vere, C.C. B2 adrenergic receptors and morphological changes of the enteric nervous system in colorectal adenocarcinoma. *World J. Gastroenterol.* **2017**, *23*, 1250–1261. [CrossRef]
20. Yang, E.V.; Sood, A.K.; Chen, M.; Li, Y.; Eubank, T.D.; Marsh, C.B.; Jewell, S.; Flavahan, N.A.; Morrison, C.; Yeh, P.E.; et al. Norepinephrine up-regulates the expression of vascular endothelial growth factor, matrix metalloproteinase (MMP)-2, and MMP-9 in nasopharyngeal carcinoma tumor cells. *Cancer Res.* **2006**, *66*, 10357–10364. [CrossRef]
21. Shan, T.; Cui, X.; Li, W.; Lin, W.; Li, Y.; Chen, X.; Wu, T. Novel regulatory program for norepinephrine-induced epithelial-mesenchymal transition in gastric adenocarcinoma cell lines. *Cancer Sci.* **2014**, *105*, 847–856. [CrossRef]
22. Hu, H.T.; Ma, Q.Y.; Zhang, D.; Shen, S.G.; Han, L.; Ma, Y.D.; Li, R.F.; Xie, K.P. HIF-1alpha links beta-adrenoceptor agonists and pancreatic cancer cells under normoxic condition. *Acta Pharmacol. Sin.* **2010**, *31*, 102–110. [CrossRef]
23. Shi, M.; Liu, D.; Duan, H.; Han, C.; Wei, B.; Qian, L.; Chen, C.; Guo, L.; Hu, M.; Yu, M.; et al. Catecholamine up-regulates MMP-7 expression by activating AP-1 and STAT3 in gastric cancer. *Mol. Cancer* **2010**, *9*, 269. [CrossRef] [PubMed]
24. Lu, Y.J.; Geng, Z.J.; Sun, X.Y.; Li, Y.H.; Fu, X.B.; Zhao, X.Y.; Wei, B. Isoprenaline induces epithelial-mesenchymal transition in gastric cancer cells. *Mol. Cell. Biochem.* **2015**, *408*, 1–13. [CrossRef]
25. Liu, G.X.; Xi, H.Q.; Sun, X.Y.; Geng, Z.J.; Yang, S.W.; Lu, Y.J.; Wei, B.; Chen, L. Isoprenaline Induces Periostin Expression in Gastric Cancer. *Yonsei Med. J.* **2016**, *57*, 557–564. [CrossRef]
26. Lu, Y.; Xu, Q.; Zuo, Y.; Liu, L.; Liu, S.; Chen, L.; Wang, K.; Lei, Y.; Zhao, X.; Li, Y. Isoprenaline/β2-AR activates Plexin-A1/VEGFR2 signals via VEGF secretion in gastric cancer cells to promote tumor angiogenesis. *BMC Cancer* **2017**, *17*, 875. [CrossRef]
27. Lutgendorf, S.K.; Lamkin, D.M.; Jennings, N.B.; Arevalo, J.M.; Penedo, F.; DeGeest, K.; Langley, R.R.; Lucci, J.A.; Cole, S.W.; Lubaroff, D.M.; et al. Biobehavioral influences on matrix metalloproteinase expression in ovarian carcinoma. *Clin. Cancer Res.* **2008**, *14*, 6839–6846. [CrossRef] [PubMed]
28. Sood, A.K.; Armaiz-Pena, G.N.; Halder, J.; Nick, A.M.; Stone, R.L.; Hu, W.; Carroll, A.R.; Spannuth, W.A.; Deavers, M.T.; Allen, J.K.; et al. Adrenergic modulation of focal adhesion kinase protects human ovarian cancer cells from anoikis. *J. Clin. Investig.* **2010**, *120*, 1515–1523. [CrossRef]
29. Shi, M.; Yang, Z.; Hu, M.; Liu, D.; Hu, Y.; Qian, L.; Zhang, W.; Chen, H.; Guo, L.; Yu, M.; et al. Catecholamine-Induced β2-adrenergic receptor activation mediates desensitization of gastric cancer cells to trastuzumab by upregulating MUC4 expression. *J. Immunol.* **2013**, *190*, 5600–5608. [CrossRef] [PubMed]
30. Liao, X.; Che, X.; Zhao, W.; Zhang, D.; Bi, T.; Wang, G. The β-adrenoceptor antagonist, propranolol, induces human gastric cancer cell apoptosis and cell cycle arrest via inhibiting nuclear factor κB signaling. *Oncol. rep.* **2010**, *24*, 1669–1676. [CrossRef]
31. Takahashi, K.; Kaira, K.; Shimizu, A.; Sato, T.; Takahashi, N.; Ogawa, H.; Yoshinari, D.; Yokobori, T.; Asao, T.; Takeyoshi, I.; et al. Clinical significance of β2-adrenergic receptor expression in patients with surgically resected gastric adenocarcinoma. *Tumour Biol. J. Int. Soc. Oncodev. Biol. Med.* **2016**, *37*, 13885–13892. [CrossRef]
32. Shi, B.; Wang, L.; Yan, C.; Chen, D.; Liu, M.; Li, P. Nonlinear heart rate variability biomarkers for gastric cancer severity: A pilot study. *Sci. Rep.* **2019**, *9*, 13833. [CrossRef]

33. Bastos, D.B.; Sarafim-Silva, B.; Sundefeld, M.; Ribeiro, A.A.; Brandão, J.; Biasoli, É.R.; Miyahara, G.I.; Casarini, D.E.; Bernabé, D.G. Circulating catecholamines are associated with biobehavioral factors and anxiety symptoms in head and neck cancer patients. *PLoS ONE* **2018**, *13*, e0202515. [CrossRef]
34. Lamboy-Caraballo, R.; Ortiz-Sanchez, C.; Acevedo-Santiago, A.; Matta, J.; NA Monteiro, A.; N Armaiz-Pena, G. Norepinephrine-Induced DNA Damage in Ovarian Cancer Cells. *Int. J. Mol. Sci.* **2020**, *21*, 2250. [CrossRef] [PubMed]
35. Ouyang, X.; Zhu, Z.; Yang, C.; Wang, L.; Ding, G.; Jiang, F. Epinephrine increases malignancy of breast cancer through p38 MAPK signaling pathway in depressive disorders. *Int. J. Clin. Exp. Pathol.* **2019**, *12*, 1932–1946.
36. Lackovicova, L.; Banovska, L.; Bundzikova, J.; Janega, P.; Bizik, J.; Kiss, A.; Mravec, B. Chemical sympathectomy suppresses fibrosarcoma development and improves survival of tumor-bearing rats. *Neoplasma* **2011**, *58*, 424–429. [CrossRef] [PubMed]
37. Renz, B.W.; Takahashi, R.; Tanaka, T.; Macchini, M.; Hayakawa, Y.; Dantes, Z.; Maurer, H.C.; Chen, X.; Jiang, Z.; Westphalen, C.B.; et al. β2 Adrenergic-Neurotrophin Feedforward Loop Promotes Pancreatic Cancer. *Cancer Cell* **2018**, *33*, 75–90.e7. [CrossRef]
38. Li, J.; Yang, X.M.; Wang, Y.H.; Feng, M.X.; Liu, X.J.; Zhang, Y.L.; Huang, S.; Wu, Z.; Xue, F.; Qin, W.X.; et al. Monoamine oxidase A suppresses hepatocellular carcinoma metastasis by inhibiting the adrenergic system and its transactivation of EGFR signaling. *J. Hepatol.* **2014**, *60*, 1225–1234. [CrossRef] [PubMed]
39. Liu, X.; Wu, W.K.; Yu, L.; Sung, J.J.; Srivastava, G.; Zhang, S.T.; Cho, C.H. Epinephrine stimulates esophageal squamous-cell carcinoma cell proliferation via beta-adrenoceptor-dependent transactivation of extracellular signal-regulated kinase/cyclooxygenase-2 pathway. *J. Cell. Biochem.* **2008**, *105*, 53–60. [CrossRef]
40. Jiang, S.H.; Hu, L.P.; Wang, X.; Li, J.; Zhang, Z.G. Neurotransmitters: Emerging targets in cancer. *Oncogene* **2020**, *39*, 503–515. [CrossRef]
41. Crawford, M.H.; Bernstein, S.J.; Deedwania, P.C.; DiMarco, J.P.; Ferrick, K.J.; Garson, A., Jr.; Green., L.A.; Greene, H.L.; Silka, M.J.; Stone, P.H.; et al. ACC/AHA guidelines for ambulatory electrocardiography: Executive summary and recommendations. A report of the American College of Cardiology/American Heart Association task force on practice guidelines (committee to revise the guidelines for ambulatory electrocardiography). *Circulation* **1999**, *100*, 886–893. [CrossRef]
42. Eisenhofer, G.; Peitzsch, M. Laboratory evaluation of pheochromocytoma and paraganglioma. *Clin. Chem.* **2014**, *60*, 1486–1499. [CrossRef] [PubMed]

International Journal of *Molecular Sciences*

MDPI

Review

Updated Insights on EGFR Signaling Pathways in Glioma

Alexandru Oprita [1,†], Stefania-Carina Baloi [1,†], Georgiana-Adeline Staicu [1], Oana Alexandru [2], Daniela Elise Tache [1], Suzana Danoiu [3], Elena Simona Micu [4,*] and Ani-Simona Sevastre [5,*]

[1] Department of Biochemistry, Faculty of Medicine, University of Medicine and Pharmacy, 200349 Craiova, Romania; opr.alexandru@gmail.com (A.O.); carina_baloi@yahoo.com (S.-C.B.); adstaicu@gmail.com (G.-A.S.); dodolita@yahoo.it (D.E.T.)

[2] Department of Neurology, University of Medicine and Pharmacy of Craiova and Clinical, Hospital of Neuropsychiatry, 200349 Craiova, Romania; oanale@hotmail.com

[3] Department of Pathophysiology, Faculty of Medicine, University of Medicine and Pharmacy, 200349 Craiova, Romania; suzdanoiu@gmail.com

[4] Department 4—Medical Specialities, Faculty of Medicine, University of Medicine and Pharmacy, 200349 Craiova, Romania

[5] Department of Pharmaceutical Technology, Faculty of Pharmacy, University of Medicine and Pharmacy, 200349 Craiova, Romania

* Correspondence: gvs.2005@yahoo.com (E.S.M.); anifetea_umf@yahoo.com (A.-S.S.); Tel.: +40-351-443-500 (A.-S.S.)

† These authors contributed equally to this work.

Abstract: Nowadays, due to recent advances in molecular biology, the pathogenesis of glioblastoma is better understood. For the newly diagnosed, the current standard of care is represented by resection followed by radiotherapy and temozolomide administration, but because median overall survival remains poor, new diagnosis and treatment strategies are needed. Due to the quick progression, even with aggressive multimodal treatment, glioblastoma remains almost incurable. It is known that epidermal growth factor receptor (EGFR) amplification is a characteristic of the classical subtype of glioma. However, targeted therapies against this type of receptor have not yet shown a clear clinical benefit. Many factors contribute to resistance, such as ineffective blood–brain barrier penetration, heterogeneity, mutations, as well as compensatory signaling pathways. A better understanding of the EGFR signaling network, and its interrelations with other pathways, are essential to clarify the mechanisms of resistance and create better therapeutic agents.

Keywords: glioma; pathways; EGFR; clinical trials

Citation: Oprita, A.; Baloi, S.-C.; Staicu, G.-A.; Alexandru, O.; Tache, D.E.; Danoiu, S.; Micu, E.S.; Sevastre, A.-S. Updated Insights on EGFR Signaling Pathways in Glioma. *Int. J. Mol. Sci.* **2021**, *22*, 587. https://doi.org/10.3390/ijms22020587

Received: 15 November 2020
Accepted: 5 January 2021
Published: 8 January 2021

1. Introduction

With an overall survival of less than 35% in five years [1], malignant primary brain tumors are the most difficult to treat cancers. Of those, the most common type is represented by gliomas. Based on the expression patterns' differences, glioblastomas are divided into three subtypes as follows: classical, proneural, and mesenchymal [2]. Because glioblastoma multiforme (GBM), a grade IV glioma [3], is one of the most aggressive primary brain tumors, recent studies and reviews have focused on deepening our understanding of the disease [4–9].

At present, GBM's pathogenesis is better understood due to recent advances in molecular biology. For newly diagnosed glioblastoma, the current standard of care is represented by resection, followed by radiotherapy and temozolomide (TMZ) administration [10], but the median overall survival (OS) is not fully improved; therefore, new diagnosis and treatment strategies are needed [11,12].

Glioblastoma is the most common and the most deleterious glioma [13]. The 2011–2015 Statistical Report of the Central Brain Tumor Registry of the United States (CBTRUS) showed that glioblastoma represents 48% of the malignant brain and central nervous system tumors, with an incidence rate in the United States 1.58 times higher in males compared to females [14]. Due to the quick progression, even with aggressive multimodal treatment, glioblastoma remains almost incurable [15,16].

Nowadays, chemotherapy has a significant role in glioblastoma's treatment strategies, with numerous research studies aimed to develop more efficient chemotherapeutic drugs [17]. Understanding the disease's pathogenesis has a key role in identifying disease biomarkers and developing new potential chemotherapeutic drugs. We present some of the most promising signaling pathways involved in pathogenesis, with their specific targeting components.

GBM is characterized by nuclear atypia, cellular pleomorphism, mitotic activity, anaplasia, and rapid proliferation alternated with an aggressive invasion of the surrounding brain tissue. In its microenvironment, glioma cells are faced with many challenges such as acidity, hypoxia, and low nutrient availability. To maintain rapid growth, they need to modulate metabolic activity [18,19].

In multicellular organisms, tyrosine phosphorylation is involved in signal transduction, leading to differentiation, proliferation, migration, and survival [20,21].

Receptor tyrosine kinases (RTKs) are activated by binding their extracellular domain to corresponding ligands determining their oligomerization. This process activates the intracellular domain, facilitating the recruitment of proteins that start a signaling cascade, integrating numerous signaling pathways that lead to specific cellular responses [22]. Among all RTKs, epidermal growth factor receptor (EGFR) is the most amplified in GBM [23]. EGFR amplification is observed in the classical subtype of glioma [2]. EGFR gene amplification is detected in 57.4% of primary GBM patients, leading to high levels of EGFR protein, contributing to tumorigenesis and progression [24].

However, targeted therapies against this type of receptor have not yet shown a clear clinical benefit. Many factors contribute to resistance, such as ineffective blood–brain barrier penetration, heterogeneity, mutations, and compensatory signaling pathways. A better understanding of the EGFR signaling network and its interrelations with other pathways are essential to improve drug activity, clarify the mechanisms of resistance, and develop better therapeutic agents.

2. Understanding EGFR Features

The transmembrane receptor of tyrosine kinase epidermal growth factor (EGFR), also known as HER (human EGFR related) 1 or ErbB1, along with HER2/neu (ErbB-2), HER3 (ErbB-3), and HER4 (ErbB-4), is a member of the ErbB family and it is located on chromosome band 7p12 [25].

Like all RTKs, EGFR has an extracellular region, a single transmembrane domain, an intracellular juxtamembrane domain, a tyrosine kinase, and a C-terminal region. The ligands of ErbB receptors are divided into two main groups: EGFR activators called EGF agonists, and neuregulins that bind to ErbB3 and ErbB4 [26,27].

The extracellular region of EGFR has two homologous domains (I and III) that bind ligands and two cysteine rich domains (II and IV) [28].

The juxtamembrane region tethers inactive EGFRs to the plasma membrane cytosolic surface, which contributes to EGFR activation [29]. Structural studies highlight the functional importance for certain regions, such as the structure of the first 30 amino acids from the intracellular juxtamembrane region of EGFR and the C-terminal 190 amino acids [27].

There are more than 40 EGFR ligands that control its signaling. They can be divided into high-affinity ligands, such as epithelial growth factor (EGF), heparin-binding EGF-like growth factor (HB-EGF), Transforming growth factor alpha (TGF-α), betacellulin (BTC), and low-affinity ligands, such as epiregulin (EREG), amphiregulin (AREG), and epigen (EPGN).

Expression of EGF-family proteins and activation of EGFR are features of cardiac disease [30,31]. Moreover, molecular alterations of EGFR include overexpression, deletion, or amplification, in different types of cancer. In GBM, EGFR amplification promotes invasion, proliferation, and drug resistance to radio- and chemotherapy [32].

Several trials on EGFR targeted therapy have failed to produce conclusive evidence, maybe because of the EGFR molecular heterogeneity in GBM, of the low specificity of the designed drugs, as well as because of low brain penetration [33]. Despite all this, the detection of EGFR alterations is still used as a prognostic marker for GBM because 24–67% of GBMs are characterized by a mutated gene, 40% by amplification, and 60% by EGFR overexpression [34].

In recent years, studies have proved that EGFR has pro-survival kinase-independent functions in malignant cells. This fact has offered a different perspective of understanding EGFR implications in cancer, with new ideas of EGFR targeted cancer therapy [35–37].

3. Mechanisms of EGFR Pathway Activation

There are several different mechanisms of EGFR pathway activation, such as increased ligand production or overexpression/defective inactivation/mutation of the receptor. Many studies focused on the EGFR signaling mechanism in recent years, trying to conclude how the extracellular EGFR-ligand binding propagates through the single transmembrane helix (TM) to trigger intracellular kinase activation [38–40].

3.1. EGFR Activation Mechanisms in Normal Physiologic Status

The expression of EGFR in normal cells is about 4×10^4–10×10^4 receptors/cell [41], whereas, in cancer cells, more than 10^6 receptors/cell are observed [42]

The EGFR RNA expression is increased by stimulating the EGFR-specific transcription factor (ETF). The receptor expression is regulated by epidermal growth factor (EGF) itself and other proteins such as E1A, Sp1, and AP2 [36].

Like all RTKs, EGFR is activated by ligands featuring receptor-specificity. Briefly, ligand binding leads to a dimeric active conformation of EGFR by homodimerization (complexed with another EGFR) or heterodimerization (complex with another ErbB member). The tyrosine residues from other RTKs are autophosphorylated after ligand stimulation, and phenylalanine substitutions significantly impair the kinase signaling and the downstream signaling. Differently, EGFR Tyr-845 phosphorylation is not a required mechanism for ligand-induced EGFR activation, but it may represent the main mechanism for EGFR transactivation [43,44].

Proteins that express a proto-oncogene tyrosine-protein kinase (Src) homology domain 2 (SH2) region bind to the activated receptor, areactivated, and forward the signal to the downstream effectors, propagating critical cellular signaling pathways [45]. EGFR can simultaneously activate several signal transduction pathways such as phosphatidylinositide 3 kinase (PI3K) and serine–threonine kinase (AKT) and RAS/MAPK pathways [46].

3.1.1. Extracellular Domain Activation

For EGFR, the dimerization is completely receptor-mediated, with no physical interaction between two activating ligands. In normal physiologic status, the receptors are in a dynamic monomer–dimer equilibrium. In the absence of ligands, the extracellular domain presents a tethered configuration (intra-molecular links entirely block the dimerization arm), and the intracellular tyrosine kinase domain (TKD) is inactive. Ligand binding leads to a conformational change that exposes the buried dimerization arm, and the extracellular domain dimerizes, inducing conformational changes of the intracellular domain and enabling kinase activation [45].

A recent study by Chung et al. described physiological EGFR activation as being due to a ligand-mediated extracellular domain dimerization that stabilizes the *N*-terminal transmembrane dimer and disrupts autoinhibition, allowing the C-terminal juxtamembrane (JM-B) segment to stabilize the asymmetric kinase domain (KD) dimer, resulting in activation of EGFR signaling. They also concluded that the stimulus stabilizes the active KD conformation in pathological states and further the asymmetric KD dimerization. The inside–out coupling is weaker than the physiological outside–in coupling, suggesting that the extracellular (EC) dimer is linked through the *N*-terminal TM dimer with the asymmetric oncogenic KD dimer [38].

3.1.2. Intracellular Domains Activation

By ligand-induced dimerization, the cis-autoinhibition is released, and through a unique allosteric mechanism, the kinase activity of EGFR isactivated. It is well known that this mechanism consists of physical interaction between the C-terminal tail of the activator kinase and the other kinase *N*-terminal tail (receiver kinase) of the dimer pair, inducing conformational changes of the *N*-lobe of receiver kinase and trans-phosphorylation C-terminal tail of the activator [47].

3.1.3. Downstream Signaling of EGFR

EGFR activation and autophosphorylation result in the recruitment of downstream signaling proteins. Almost all autophosphorylation sites are binding sites for Src Homology 2-(SH2) or Phosphotyrosine binding-(PTB) signaling proteins. The SH2- proteins may be bound directly to the receptor, or indirectly through docking proteins using PTB domains [48]. EGFR can recruit and regulate many signaling pathways such as PI-3 K/AKT, RAS/MAPK, and JAK2/STAT. Therefore, EGFR functions as a hub involved in regulating various cellular processes [21,23], as shown in Figure 1.

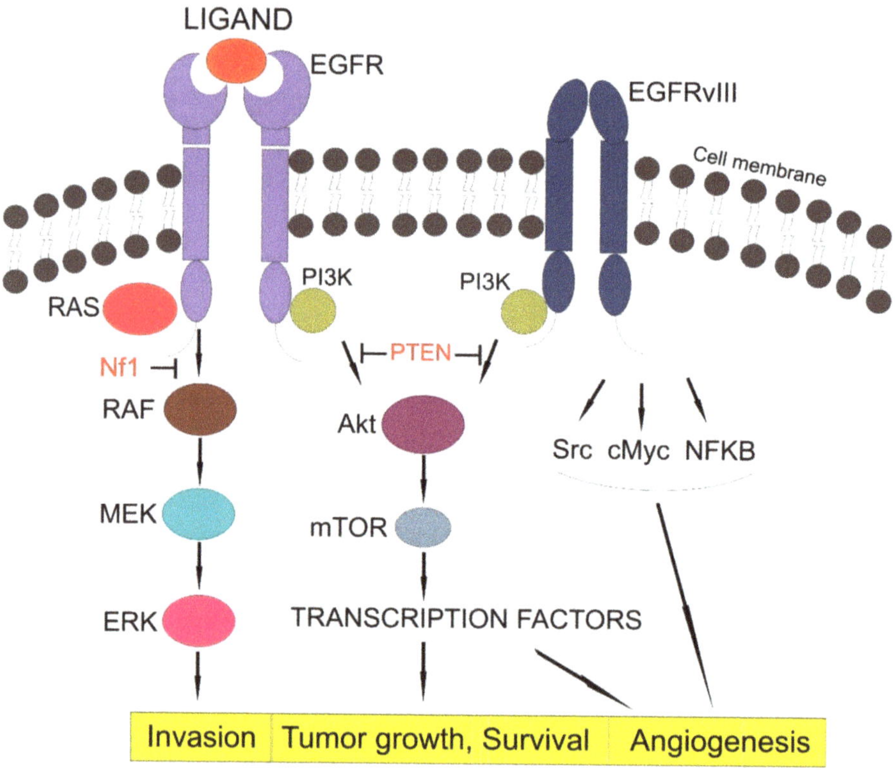

Figure 1. EGFR signaling pathway (EGFR—epithelial growth factor receptor, EGFRvIII—Epidermal growth factor receptor variant III, Pi3K—Phosphoinositide 3-kinase, RAS—family of genes involving cellular signal transduction, PTEN—Phosphatase and tensin homolog, NF1—Neurofibromatosis type 1, RAF—serine/threonine-specific protein kinases, MEK—Mitogen-activated protein kinase, ERK—extracellular signal-regulated kinase, AkT—Protein kinase B, mTOR—mammalian target of rapamycin, Src—Proto-oncogene tyrosine-protein kinase, cMyc—c proto-oncogene, NFKB—nuclear factor kappa-light-chain-enhancer of activated B cells, Block arrow—inhibition activity, Point arrow—pathway flow).

The PI-3K/AKT signaling pathway involves PI3K, an enzyme with SH2-signal transducer and its downstream effector AKT, regulating apoptosis and cell survival. Once EGFR is activated and phosphorylated, PI3K is brought to the cell membrane, and it phosphorylates phosphatidylinositol 4,5-bisphosphate (PIP2), forming phosphatidylinositol (3,4,5)-trisphosphate (PIP3). AKT reacts with PIP3, and it is phosphorylated at Threonin308 by phosphoinosite-dependent protein kinase-1 (PDK1) and at Serine 473 by the mammalian target of rapamycin complex 2 (mTORC2), reaching full activity. The phosphatase and tensin homolog (PTEN) negatively regulate the PI3K/AKT pathway by dephosphorylating and delocalizing PIP3 from the cellular membrane, resulting in the relocalization of AKT in the cytoplasm, where it is unable to be reactivated [49,50].

Class IA is one of the three different classes of PI3Ks featuring subunits with regulatory activity such as p85. Active EGFR achieves association with regulatory p85 through dimerization with human HER3, or via the docking protein GRB2-associated binder 1 (GAB1), relieving the inhibitory effect of p85 [51]. GAB1 is a scaffolding protein involved in recruiting additional signaling proteins such as PI3K, SHP2, and p120RasGap. It is involved in many EGFR signaling outputs, and is the predominant mechanism linking EGFR to PI3K/Akt signaling [52,53].

Due to its increasing importance in different human cancers, GAB1 may represent an emerging potential therapeutic target.

The RAS/MAPK signaling pathway involves the growth-factor-receptor bound-2 (GRB2), which forms a complex with Son of Sevenless (SOS), a guanine-nucleotide exchange factor (GEF) and activates the RAS G-protein by exchanging guanosine diphosphate (GDP) with guanosine triphosphate (GTP) [54]. Consequently, RAS and mitogen-activated protein kinases (MAPKs) initiate a downstream signaling cascade to phosphorylate the nuclear protein Jun. Jun creates complexes with different nuclear proteins leading to the key transcription factor activator protein 1 (AP-1), responsible for translation and transcription of proteins involved in the growth and division of cells. Activated RAS is negatively regulated by GTPase activating proteins (GAPs), such as the tumor suppressor neurofibromin 1 (NF1) [55].

Signal transduction and activator of transcription 3 (STAT3) is tyrosine-phosphorylated or activated as pSTAT3 due to EGFR-regulation of interleukin-6 (IL-6) expression. This mechanism leads to a feed-forward in the IL-6/Janus kinase (JAK)/STAT3 loop [21,56–58].

3.2. *Oncogenic Status and EGFR Activation*

The EGFR is one of the most frequently altered oncogenes in brain cancers. Except for hematopoietic cells, the majority of cell types express ErbB family members [35].

In glioblastoma cells, the EGFR tyrosine kinase activity may be dysregulated by multiple oncogenic mechanisms, such as gene mutation, overexpression of EGFR protein, increased gene copy number, rearrangements of chromosomes, and activation by autocrine function [59].

3.2.1. Mutations of Cell Signaling Regulators

The EGFR gene is located on chromosome 7p11.2 and consists of 28 exons encoding a transmembrane protein receptor composed of 464 amino acids. Exons 5–7 and 13–16 encode the ligand binding domain, and exons 18–24 encode the tyrosine kinase domain. The region encoded by exons 25–28 is the site of autophosphorylation.

Although EGFR is one of the most important drug targets in cancer therapies, its mutations present an organ–site asymmetry, depending on the cancer's organ of origin [60]. Although mutations occur in the kinase domain (KD) in other tumors, in gliomas, heterogeneous mutations and deletions are focused on the ligand-binding ectodomain (ECD). This tissue-specific feature leads to type-II tyrosine kinase inhibitors (TKIs) with high sensitivity for the inactive symmetric KD dimer (sKD), when administered in GBM mutations [61]. However, both intra- and extracellular GBM mutations result in ligand-independent oncogenic activation.

Almost 50% of the tumors characterized by EGFR amplification are positive for the mutant EGFRvIII and EGFR single nucleotide variants (SNVs). Due to this tumor-specific feature, novel therapeutic agents are currently under development to target the overexpressed EGFR or EGFRvIII proteins. An in-frame deletion of exons 2–7 characterizes the EGFRvIII, which results in overexpression of a truncated receptor that lacks some significant parts of the ECD. This prototypic oncoprotein is unable to bind ligands, and it is constitutively active. Several studies examined the effect of the EGFRvIII constitutive activity on the wtEGFR and ErbB2 protein levels. For example, one study evaluated the effect of Tyrphostin AG1478 on the protein levels and demonstrated that its administration increased protein levels of wtEGFR and erbB2 in vIIIA1 cells, due to the catalytic activity of EGFRvIII, while in its absence, the levels were reduced [62]. Furthermore, the unique peptide sequence of EGFRvIII generated by the fusion of exons 1 and 8 may serve as a tumor-specific target in immunotherapy [63], although subsequent phase III trial results are not as promising as initially anticipated [64].

A meta-analysis performed in 2017 by Felsberg et al. proved that EGFRvIII and EGFR SNVs do not represent prognostic keys in EGFR-amplified glioma patients. However, the amplification of EGFR is retained in recurrent glioma [63], although improved long-term survival by EGFRvIII therapy has been reported in glioblastoma patients [65].

Nevertheless, the research on EGFRvIII continues, producing inconclusive results. For example, Struve et al. just published in early 2020 the results of a study focused on the effect of EGFRvIII in regulating DNA mismatch repair. They tested if EGFRvIII influences temozolomide's sensitivity and demonstrated that, under standard treatment with temozolomide, EGFRvIII expression leads to prolonged survival only in patients with tumors with O6-methylguanine-DNA methyltransferase (MGMT) methylated promoter. Their results showed that EGFRvIII sensitizes a type of GBM to the current standard of care treatment with temozolomide through the upregulation of DNA mismatch repair (MMR) [65]. However, patients with tumors that have both EGFRvIII and MGMT methylation are very uncommon, and the conclusion that EGFRvIII status was associated with increased survival had a $p = 0.06$. This level would not normally be considered significant, especially not in this sort of multivariate analysis [66].

3.2.2. Overexpression and Gene Amplification

The EGFR gene is amplified in approximately 40% of glioblastomas. The primary and secondary GBM differ in genetic profiles and primary GBMs have a higher prevalence of EGFR gene amplification and overexpression than secondary GBMs [67]. In a study performed by Watanabe et al., EGFR gene amplification was associated with protein overexpression in most tumor cells, but 10% of GBM with overexpression of EGFR protein lacked EGFR gene amplification [68]. However, previous studies have stated that EGFR overexpression or activation does not necessarily cause a simple amplification of its downstream signals, but dose-dependent changes in oncogene-induced downstream signaling and biological responses have been reported [69].

3.2.3. Rearrangements of Chromosomes

Breakpoint sequence analyses proved different types of chromosomal rearrangements and mechanisms of DNA repair. Analyses of single nucleotide polymorphisms suggested that different deletions may appear from amplified non-vIII EGFR precursor [70].

In a study performed in 2018 on glioma tumor samples by Tomoyuki et al., complex chromosomal rearrangements involving chromosome 7 were observed [70].

A study performed by Lopez-Gines et al. showed that trisomy/polysomy 7 and monosomy 10 were frequently associated with glioma. The combination of these anomalies is important in glioblastoma's tumorigenesis. Moreover, the association seems to be independent of EGFR gene amplification [71].

3.2.4. Activation by Autocrine Function

It is well known that wild-type EGFR ligands such as transforming growth factor-alpha (TGF-alpha) and heparin-binding EGF (HB-EGF) are often increased in glioblastoma leading to an autocrine loop resulting in the autonomy growth of glioma cells [72]. GBM expresses an EGFR mutant (EGFRvIII) that signals constitutively, does not bind ligand, and is considered to have more tumorigenicity than wild-type EGFR. In a U251-MG glioma cell line, the expression of EGFRvIII may result in specific up-regulation of some genes (TGF-α, EPHA2, HB-EGF, IL8, FOSL1, MAP4K4, DUSP6, and EMP1) influencing signaling pathways involved in oncogenesis. TGF-α and HB-EGF (EGFR ligands) induce the expression of EGFRvIII, suggesting that EGFRvIII has a role in creating an autocrine loop with wild-type EGFR. By inhibiting HB-EGF activity with neutralizing antibodies, EGFRvIII-induced cell proliferation may be reduced, suggesting that EGFRvIII-HB-EGF-wild-type EGFR autocrine loop has a major role in signal transduction in glioblastoma cells [73]. Furthermore, studies have demonstrated that the expression of the EGFR alone has a poor transformation effect on cells. Though, coexpression of TGF-α ligand leads to a significant increase in transformation and therapies based on neutralizing the ligands have demonstrated the decreased growth of cells that harbor such loops [74,75].

4. Applied Theory—Therapies Targeting EGFR

The distribution of EGFR in cancer cells is the basic pillar of many targeted strategies pursued to inhibit its signaling pathway [76,77].

EGFR activity may be controlled by binding to the tyrosine kinase domain or binding to the extracellular component. There are three generations of tyrosine kinase inhibitors approved for clinical use. The first mechanism targets signal transduction and is characteristic of the tyrosine kinase inhibitors (RTKIs, TKIs), which bind to the tyrosine kinase domain of EGFR and inhibit its activity. First-generation TKIs, inhibit the receptor by competitive binding with ATP. Subsequent generations of TKIs were created to overcome drug resistance. Second-generation TKIs irreversibly inhibit all four ERBB (originally named because of the homology with the erythroblastoma viral gene product, v-erbB) receptors, whereas the third-generation TKI are specifically designed to target the T790M resistance mutation [78]. As first-generation inhibitors, active drugs include: erlotinib, gefitinib, lapatinib and vandetanib. Afatinib, dacomitinib, and tesevatinib are examples of second-generation small molecule EGFR inhibitors. Osimertinib is the first third generation RTKI. The monoclonal antibodies act as receptor blockers by binding to the extracellular component of the EGFR and block it from binding to its ligands. Cetuximab, necitumumab, and panitumumab are examples of biological therapy targeting EGFR [21]. An overview of clinical trials focused on anti-EGFR strategies used in GBM is provided in Table 1.

Table 1. Epidermal growth factor receptor (EGFR)-targeted therapies for adult high-grade gliomas currently in investigational and/or clinical use.

Therapeutic Agent	Mechanism	Results	Reference
Afatinib (Tovok, BIBW2992)	Second-generation EGFR inhibitor	As a single agent, Afatinib proved good safety, but limited activity on GBM patients. It was promising in combination with TMZ in a case report.6 months progression-free survival (PFS) worse than TMZ: Afatinib alone 3% vs. Afatinib + TMZ 10% vs. TMZ alone 23% Ongoing clinical trials: NCT02423525	[79]
Cetuximab (Erbitux, DTXSID70142901)	Antibody targeting the L2 domain of EGFR	Cetuximab was not very effective in GBM clinical trials. 6-month PFS was 33%, and median PFS was 16 weeks Ongoing clinical trials: NCT02800486 NCT02861898	[80]
Dacomitinib (Vizimpro, PF299804)	Second-generation EGFR inhibitor	Dacomitinib proved to be promising in pre-clinical models. 6-months PFS was 10.6% with a median PFS of 2.7 months Ongoing clinical trials: NCT01112527 NCT01520870	[81]

Table 1. *Cont.*

Therapeutic Agent	Mechanism	Results	Reference
Erlotinib (Tarceva, OSI-774)	First-generation EGFR inhibitor	Erlotinib showed poor results in GBM clinical trials. The median PFS: 1.8 months Erlotinib vs. 2.4 months TMZ/BCNU (bis-chloroethylnitrosourea) Ongoing clinical trials: NCT01257594 NCT02239952	[82]
Gefitinib (Iressa, ZD1839)	First-generation EGFR inhibitor	Gefitinib showed poor results in GBM clinical trials. The median overall survival time from treatment initiation was 39.4 weeks	[83]
Lapatinib (Tykerb, GSK 572016)	First-generation EGFR inhibitor	Lapatinib demonstrated poor results in GBM clinical trials. The studies lacked objective responses, with early progression rate of 76%. Ongoing clinical trials: NCT01591577 NCT02101905	[84]
Nimotuzumab (OSAG101)	Antibody targeting the L2 domain of EGFR	Nimotuzumab in addition to standard treatment is well tolerated and has increased survival rates in EGFR positive expression newly diagnosed GBM patients. The PFS and OS rates were 49.3% and 83.3% for 1-year and 29.0% and 51.1% for 2-year. Ongoing clinical trials: NCT03620032	[85]
Osimertinib (AZD9291)	Third-generation EGFR inhibitor	Osimertinib is in phase I/II clinical trial. Compared to other EGFR-TKIs, AZD9291 demonstrated improved ability to inhibit GBM cells proliferation.Complete response of left frontal lobe tumor after 4 weeks of osimertinib.	[86–88]
Panitumumab (Vectibix, ABX-EGF)	Antibody targeting the L2 domain of EGFR	Panitumumab was not very effective in GBM clinical trials.Panitumumab-IRDye800CW specificities for tumor core and margin were slightly higher than those of 5-ALA. Ongoing clinical trials: NCT03510208	[89]
Rindopepimut (CDX110)	Vaccine	When co-administrated with Bevacizumab, Rindopepimut significantly prolonged patient survival. 6 months PFS was 28% (rindopepimut), compared with 16% (control)Phase II trial (NCT00458601) was completed in 2018.	[90]
Vandetanib (Caprelsa, ZD6474)	Second-generation EGFR inhibitor	Vandetanib was a moderately tolerated drug, with no significant activity as a single agent in patients with recurrent malignant glioma. Median overall survival was 6.3 months. Ongoing clinical trials: NCT02239952	[91]
Tesevatinib (KD019)	Second-generation EGFR inhibitor	Tesevatinib is in Phase II study in patients with recurrent glioblastoma, with no results posted. Ongoing clinical trials: NCT02844439	[92]
bscEGFRvIIIxCD3	Antibody BisAbs	Fully human bispecific single chain antibody fragments bi-scFv (EGFRvIII:CD3 bi-scFv) was recently developed with the aim to redirect CD3-expressing T cells to target malignant EGFRvIII-expressing glioma.	[93]
mAB806	Antibody targeting the EGFRvIII-specific sequence	Structural extracellular mutations lead to a similar intermediate conformation, that can be synergistically targeted intra- and extracellularly by mAb806 antibody. Lapatinib co-treatment sensitized unresponsive wild type (WT)-EGFR to mAb806.	[94]
^{125}I mAB425	Antibody toxin or radioactive isotope conjugated	Single or in combination with TMZ, ^{125}I mAB425 prolonged patient survival (median survival of 20.4 months, compared to 14.5 months for ^{125}I mAB425 alone), with minimal toxicity in normal tissue.	[95]

Table 1. *Cont.*

Therapeutic Agent	Mechanism	Results	Reference
Chimeric anti-gen receptor T cell therapy (CAR-T cells)	Chimeric antigen receptor therapy (CARs) targeting EGFRvIII	Chimeric antigen receptor (CAR) T cells are in phase I clinical trials in high-grade glioma (HGG) patients. Pre-clinical models proved to be promising. Ongoing clinical trials: NCT02331693 NCT02844062 NCT02209376 NCT01454596 NCT02664363	[96,97]
Antisense oligonucleotides, siRNA, ribozymes, and miRNA-based therapy	RNA-based therapies	Feasibility of RNA-based therapies must be further evaluated using pre-clinical models.	[98,99]

Other strategies consist of radio-immunotherapy, docking molecule conjugate toxins, chimeric antigen receptor T cells (CAR-T cells), RNA-based therapies, oncolytic viruses, exosomes, and nanoparticles [100]. EGFR-targeted nanoparticles may be combined with focused ultrasound to achieve local drug delivery [101]. Studies have shown that magnetic nanoparticles' superparamagnetic properties allow them to be guided by an external magnet. However, their therapeutic use is limited in treating in vivo brain pathologies due to insufficient local ability to cross the blood–brain barrier. So, focused ultrasound combined with magnetic targeting synergistically delivers drug-loaded magnetic nanoparticles at the target tissue [102]. Boronated EGFR binding compounds are under investigation in so-called boron neutron capture therapy (BNCT). To improve the unsatisfactory bioavailability of large molecules or viruses due to low blood–brain barrier permeability, the convection technique (CED) is also being investigated [103]. Studies have shown that magnetic nanoparticles' superparamagnetic properties allow them to be guided by an external magnet. However, their therapeutic use is limited in treating in vivo brain pathologies due to insufficient local ability to cross the blood–brain barrier. So, focused ultrasound combined with magnetic targeting synergistically delivers drug-loaded magnetic nanoparticles at the target tissue [101,102].

For example, a study using Cetuximab conjugated magnetic iron oxide nanoparticles showed a significantly enhanced anti-tumor activity compared tocetuximab alone. This was due to improved cellular targeting and uptake, EGFR internalization, EGFR signaling alterations, and apoptosis induction in glioma stem-like cells and tumor non-stem cells that expressed EGFR [104]. In Figure 2, the EGFR-based therapies used in glioblastoma are mentioned.

4.1. *Small Molecule Receptor Tyrosine Kinase Inhibitors*

Although several compounds are approved for various diseases, none are approved for glioblastoma due to numerous negative clinical trials. Trials have not shown efficacy either alone or in combination for the oldest small molecule kinase inhibitors: gefitinib, erlotinib, lapatinib, and afatinib [105].

Gefitinib (Iressa, ZD1839) is the first approved EGFR-targeted small-molecule. Initial results from the clinical studies proved that gefitinib was safe when administered for lung carcinoma. However, responses were observed only in a subset of patients featuring chemotherapy–refractory advanced NSCLC (Nonsmall-cell lung carcinoma). The specific mutations of the EGFR gene explained this. It was suggested that these mutations stabilize the interaction of ATP and gefitinib with EGFR. Nevertheless, in the Phase II trial for recurrent GBM, gefitinib did not show improved overall survival [106], neither in the Phase I/II trial when combined with radiation in newly diagnosed GBM [107].

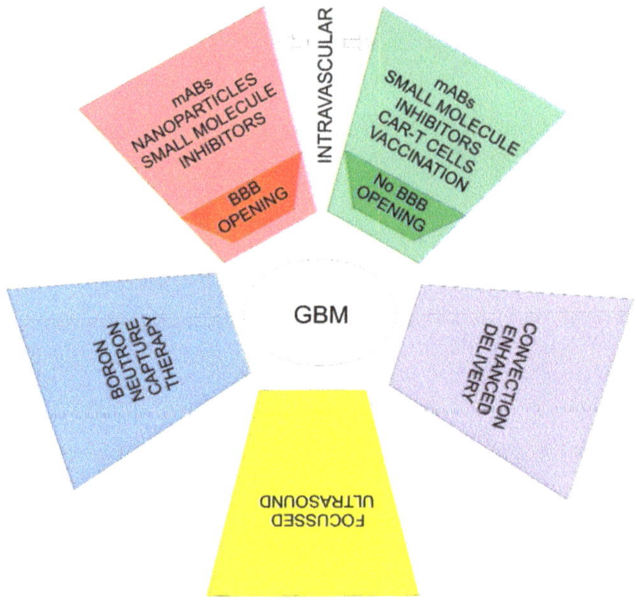

Figure 2. EGFR-based therapies in glioblastoma. (BBB—blood brain barrier, mABs—monoclonal antibodies, CAR-T—Chimeric anti-gen receptor T cell therapy)

Erlotinib (Tarceva, OSI-774) proved to prolong NSCLC patients' survival rate upon chemotherapy [108]. These results cannot be achieved in GBM, because EGFR mutations occur in the extracellular domain in GBM, whereas in lung cancers, they are typically observed in the kinase domain. Therefore, unlike NSCLC, GBMs are not sensitive to first-generation EGFR inhibitors [103]. As a single agent, it showed no efficacy in newly diagnosed GBM [109] and later studies that co-administered temsirolimus or bevacizumab were also unsuccessful [110,111].

Lapatinib (Tykerb, GSK 572016) had minimal efficacy alone or in combination with pazopanib in recurrent glioblastoma [112].

Afatinib (Tovok, BIBW2992) had limited efficacy as a single agent in one clinical trial in recurrent glioblastoma [79].

One of the drawbacks of the small molecule inhibitors is their brain penetrance. A study performed by Liu et al. showed that erlotinib could be distributed inside an intracranial U87 xenograft [113]. In another clinical trial, Gefitinib tissue concentration was two- to three-fold plasma concentrations, which was not the cause of insufficient efficacy [114].

Tesevatinib is another second-generation RTKI that is currently under evaluation in patients with recurrent GBM [92]. The first results should be published this year. This trial investigates the drug activity in EGFRvIII positive and negative GBM, with or without EGFR amplification.

Dacomitinib (Vizimpro, PF299804) is a second-generation EGFR inhibitor. Despite its poor global results in a phase II trial in recurrent GBM, Dacomitinib significantly benefited some patients [115].

A recent study investigating Osimertinib (AZD9291), a third-generation EGFR inhibitor, showed that it inhibits with high potency (<100 nM) the constitutive activity of EGFRvIII tyrosine kinase while also inhibiting its downstream signaling. Furthermore, Chagoya et al. proved that osimertinib inhibited the in vitro growth of the D317 cell line and heterotopic and orthotopic xenograft models [89].

To date, there have been eight completed clinical studies involving glioma and Vandetanib (Caprelsa, ZD6474), a second-generation EGFR inhibitor. They all investigated vandetanib's effect together with other therapies (radiotherapy or therapeutic agents), but the results were not satisfactory [116].

4.2. Monoclonal Antibodies

Cetuximab (Erbitux, DTXSID70142901) is a monoclonal antibody (mAb) targeting the L2 domain of EGFR, preventing dimerization and subsequent cross-activation, thus interrupting downstream signal transduction. It has been approved for the treatment of colorectal, head, and neck cancers [117]. In the progressive high-grade glioma (HGG) patient population, the drug was well tolerated but had limited activity and failed to demonstrate benefit [118]. In new research, the photo-immunoconjugate nanoparticle (PIC-NP) significantly enhanced the photosensitizers in cancer cells and increased the light-activated cytotoxicity in U87 cells overexpressing EGFR [119,120].

Nimotuzumab (OSAG101) is another antibody targeting the L2 domain of EGFR. It was tested in clinical trials for its efficiency in adults with glioblastoma, but the results were not satisfactory. Currently, there is an ongoing clinical trial investigating the effect of nimotuzumab co-administered with temozolomide and radiotherapy. Preliminary results show that nimotuzumab was well tolerated, with an increased survival rate in newly diagnosed GBM patients [86].

Another antibody targeting the L2 domain of EGFR is Panitumumab (Vectibix, ABX-EGF). In combination with irinotecan, it was not very effective in solid tumors [121]. Panitumumab -IRDye800 is currently under investigation in Phase II trials as a GBM diagnostic agent [122].

bscEGFRvIIIxCD3 is a bispecific T-cell engager antibody (BiTEs) that binds to the CD3 T-cell coreceptor and recruits cytotoxic T cells. It was designed to redirect the T-cells towards tumors expressing EGFRvIII. Used in vitro and in vivo on mice, bscEGFRvIIIxCD3 showed the potent killing of GBM expressing EGFRvIII [123].

mAB806 is an antibody targeting the EGFRvIII-specific sequence. The antibody mAb806 is under investigation for glioblastoma treatment, although its mechanism of action remains unknown [124]. It was shown to potentiate the sensitivity of glioma xenotransplants to radiotherapy [125].

4.3. Targeted Isotopes

The isotope 125I mAB425 is a radioactive isotope conjugated with a specific antibody. Emrich et al. demonstrated that the administration of 125IMAb425 and intensive medical management led to a significant increase in median survival in patients with high-grade gliomas [126]. However, the results of subsequent studies failed expectations.

4.4. Immunotherapy

4.4.1. CAR-T Cells Targeting EGFRvIII

A new technology developed in cancer therapy is the engineering of T cells to recognize their target by expressing a chimeric antigen receptor (CAR). Glioblastomas express the EGFRvIII, with its unique site of antigenicity. Therefore, these chimeric antigen receptor (CAR)-T cells were engineered to recognize the vIII-receptor mutation through a humanized single-chain antibody fragment (scFv) fused with some key constituents of T-cell receptor intra-cytoplasmic signaling domains [127].

This strategy is currently in early clinical trials. Sahin et al. developed a third-generation chimeric antigen receptor (CAR), specific for EGFRvIII (G3-EGFRvIII), that expresses CD28 (Cluster of Differentiation 28) and CD134 (Cluster of Differentiation 134). Their findings suggest that G3-EGFRvIII CAR represents a potential anti-glioblastoma strategy [128]. A Phase 1 pilot study that investigated the safety and feasibility of CAR-T-EGFRvIII in treating patients with EGFRvIII+ glioblastoma just terminated in 2019, and results are expected to be published [129]—currently, 15 trials are still recruiting.

4.4.2. EGFR as an Immunologic Target—Vaccination

EGFRvIII represents the most common mutation of EGFR. It creates a tumor-specific antigen detectable in almost 30% of human GBM. Deleting the EGFR exons 2–7 results in EGFRvIII with a truncated extracellular domain, resulting in a unique, GBM cell-specific, antibody-reactive antigen. An EGFRvIII-specific peptide conjugated to a keyhole limpet hemocyanin represents the structure of a vaccine called Rindopepimut (CDX110). The latest Phase III clinical trial showed that rindopepimut did not increase the survival rate in newly diagnosed glioblastoma patients [130–132].

4.5. *Targeting the Regulation of EGFR Gene Expression*

This strategy consists of using antisense oligonucleotides, siRNA, ribozymes, and miRNA. In glioblastoma, the microRNAs control the post-transcriptional gene expression of receptor tyrosine kinase (RTK) signaling pathways by blocking or accelerating the mRNA. Recent work demonstrates that extracellular vesicles (EVs) can carry and transfer EGFR [133] and that cell communication through EVs enhances glioblastoma's intratumoral heterogeneity [134]. Bronisz et al. showed that miR-1 could interact with a major EV protein Annexin A2 (ANAXA2) to reduce glioblastoma tumorigenicity [135]. Furthermore, one recent study of Liao et al. showed that the extracellular EC domain methylation using protein arginine methyltransferase 1 (PRMT1) increased EGF binding and dimerization, with enhanced receptor activation counteracting the effect of cetuximab in a mouse model of colon cancer [136]. The expression of the EGFRvIII is also under investigation. Unfortunately, none of the strategies targeting EGFR gene expression regulation have yet had any preclinical development.

4.6. *Nanoparticles*

In order to efficiently deliver the therapeutic agent, novel pharmaceutical formulations are currently used. It is well known that the bioavailability of drugs may be low because of the blood-brain barrier (BBB) permeability. Nanoparticles are vesicular carriers able to increase the bioavailability due to targeted drug release while protecting their content. EGFR is an ideal molecule for glioblastoma tumor targeting and numerous agents have been entrapped in a variety of nanoparticles [137].

There is only limited experience in early clinical trials for cetuximab conjugated liposomes [138]. A Phase II study just terminated in March 2020 investigated the effect of combining Temozolomide and a nanocomplex called SGT-53 (systemic gene therapy—53) to treat recurrent glioblastoma [139]. Previous results showed prolonged survival rates in glioblastoma mouse models [140].

Regardless of strategy, all therapeutic agents face the main problem of delivery across the blood–brain barrier, often cited as the explanation for EGFR targeting failure in glioblastoma [103].

5. Facing a Real Challenge—Drug Resistance

There is evidence that targeted therapy towards mutations responsible for cancer growth and progression is effective in different tumor types. For GBM, the responses to EGFR-pathway inhibitors were not as expected, and they are mainly explained by drug resistance [141]. Two major mechanisms could explain the EGFR therapy resistance.

The first resistance mechanism involves target independence. In this situation, glioma cells without EGFR protein expression experience no negative impact from EGFR inhibitors. For this type of resistance, the loss of extra-chromosomally encoded EGFR is a frequent mechanism. Target independence may occur after small molecule therapy. The dynamic EGFRvIII expression regulation by small circular fragments of extra-chromosomal DNA is involved in the resistance to EGFR inhibition. Some studies demonstrated that GBM cells treated with erlotinib reversibly suppressed mutant EGFR by producing extra-chromosomal DNA, making the GBM cells resistant to EGFR inhibition. After withdrawing erlotinib, the mutations re-emerged on extra-chromosomal DNA, leading to the upregulation of EGFRvIII with consequent re-sensitization of GBM cells [142].

The second mechanism regards target compensation. In this situation, glioma cells fight back against EGFR inhibition by activating compensatory pathways independent of EGFR signaling. Insulin-like growth factor 1 (IGFR1), platelet-derived growth factor β (PDGFβ), mesenchymal-epidermal transition (cMET), and their downstream targets are involved in these compensatory pathways [77].

Given these two mechanisms, rational strategies should include multi-target therapies targeting truncation mutations for the first mechanism and multi-target therapies targeting compensatory proteins for the second mechanism. Resistance may be overcome by dosing/epigenetic therapy, targeting truncation mutations, or through multitarget therapy.

For the first strategy, glioma cells' re-sensitization may be achieved by using drug scheduling [143]. Pulsatile intermittent drug therapy with EGFR-inhibitors in high doses can lead to better inhibition of the target, delay of therapy resistance, and toxicity reduction compared to continuous dosing [144].

Targeting truncated mutations is a suitable strategy for glioma treatment due to their high frequency in this pathology. As candidates for co-targeting, the following deserve discussion: PTEN with PI3K as a molecular target and cyclin-dependent kinase inhibitor 2 A (CDKN2A) with cyclin-dependent kinase (CDK)4/6 as a molecular target. Some PI3K inhibitors are currently undergoing clinical trials, such as GDC-0084, PX-866, pilaralisib, buparlisib, and XL765 [145]. Abemaciclib, palbociclib, and ribocyclib are examples of FDA (Food and Drug Administration)-approved oral drugs with good BBB permeability [146,147] that may be investigated for targeting CDK4/6.

In many types of cancer, multitarget therapy is a preferred option. Therapies may become more efficient by combining EGFR inhibitors with other downstream blocking agents. Several glioma specific epitopes such as IL13RA2 and EphA2 are under investigation for poly-target therapy with antibody drug conjugates (ADCs) and CAR-T cells [148–151].

In a recent 2020 study, Meng et al. proved that the cross-activation of EGFR and MET signaling pathways contributes to temozolomide resistance in glioblastoma patients. To simultaneously diminish both EGFR and MET activation, they developed a nanoinhibitor with double functionalized brain-target (BIP-MPC-NP) by conjugating cMBP and Inherbin3 modified poli-2-methacryloyloxyethyl phosphorylcholine (MPC)-nanoparticles. The study shows that DNA damage repair is reduced, and sensitivity is augmented by downregulating the E2F1 transcription factor in temozolomide resistant glioma in mice. These results demonstrate that the abovementioned nanoinhibitor is a suitable candidate for overcoming drug resistance in glioma [152].

Furthermore, patient mutations affecting the trafficking of therapeutic antibodies is another potential mechanism contributing to therapeutic resistance [153,154].

6. Conclusions

The global understanding of EGFR signaling has dramatically advanced in the last ten years. However, extensive work is still required in order to understand all signaling pathways and their implications fully. The application of EGFR-targeted therapy for glioma treatment has been less successful than expected. Gliomas require a complex signaling network that dictates the tumor sensitivity of EGFR-targeted therapies. Low BBB penetration, as well as tumor heterogeneity, secondary mutations, and compensatory signaling pathways, contribute to resistance. The development of new combinatorial therapies may improve patient quality of life through personalized, tailored choices of appropriate therapeutic strategies. An integrated approach is required to offer a complete view of this critical receptor by combining cellular, biochemical, structural, and genetic modeling techniques.

Author Contributions: Conceptualization, A.-S.S.; methodology, A.O., S.-C.B., D.E.T., S.D.; investigation, A.O., S.-C.B.; resources, G.-A.S.; writing—original draft preparation, O.A.; writing—review and editing, A.O., E.S.M., A.-S.S.; visualization, A.-S.S.; supervision, A.-S.S. All authors have read and agreed to the published version of the manuscript.

Funding: This research received no external funding.

Institutional Review Board Statement: Not applicable.

Informed Consent Statement: Not applicable.

Data Availability Statement: The data presented in this study are openly available.

Conflicts of Interest: The authors declare no conflict of interest.

References

1. Zhang, A.S.; Ostrom, Q.T.; Kruchko, C.; Rogers, L.; Peereboom, D.M.; Barnholtz-Sloan, J.S. Complete prevalence of malignant primary brain tumors registry data in the United States compared with other common cancers 2010. *Neuro Oncol.* **2017**, *19*, 726–735. [CrossRef] [PubMed]
2. Verhaak, R.G.; Hoadley, K.A.; Purdom, E.; Wang, V.; Qi, Y.; Wilkerson, M.D.; Miller, C.R.; Ding, L.; Golub, T.; Mesirov, J.P.; et al. Integrated genomic analysis identifies clinically relevant subtypes of glioblastoma characterized by abnormalities in PDGFRA, IDH1, EGFR, and NF1. *Cancer Cell.* **2010**, *17*, 98–110. [CrossRef] [PubMed]
3. Louis, D.N.; Perry, A.; Reifenberger, G.; von Deimling, A.; Figarella-Branger, D.; Cavenee, W.K.; Ohgaki, H.; Wiestler, O.D.; Kleihues, P.; Ellison, D.W. The 2016 World Health Organization Classification of Tumors of the Central Nervous System: A summary. *Acta Neuropathol.* **2016**, *131*, 803–820. [CrossRef] [PubMed]
4. Armstrong, T.S.; Dirven, L.; Arons, D.; Bates, A.; Chang, S.M.; Coens, C.; Espinasse, C.; Gilbert, M.R.; Jenkinson, D.; Kluetz, P.; et al. Glioma patient-reported outcome assessment in clinical care and research: A Response Assessment in Neuro-Oncology collaborative report. *Lancet Oncol.* **2020**, *21*, e97–e103. [CrossRef]
5. Alexandru, O.; Sevastre, A.S.; Castro, J.; Artene, S.A.; Tache, D.E.; Purcaru, O.S.; Sfredel, V.; Tataranu, L.G.; Dricu, A. Platelet-Derived Growth Factor Receptor and Ionizing Radiation in High Grade Glioma Cell Lines. *Int. J. Mol. Sci.* **2019**, *20*, 4663. [CrossRef]
6. Alexandru, O.; Horescu, C.; Sevastre, A.S.; Cioc, C.E.; Baloi, C.; Oprita, A.; Dricu, A. Receptor tyrosine kinase targeting in glioblastoma: Performance, limitations and future approaches. *Współczesna Onkol.* **2020**, *24*, 55–66. [CrossRef]
7. Sevastre, A.-S.; Horescu, C.; Carina Baloi, S.; Cioc, C.E.; Vatu, B.I.; Tuta, C.; Artene, S.A.; Danciulescu, M.M.; Tudorache, S.; Dricu, A. Benefits of Nanomedicine for Therapeutic Intervention in Malignant Diseases. *Coatings* **2019**, *9*, 628. [CrossRef]
8. Novak, M.; KoprivnikarKrajnc, M.; Hrastar, B.; Breznik, B.; Majc, B.; Mlinar, M.; Rotter, A.; Porčnik, A.; Mlakar, J.; Stare, K.; et al. CCR5-Mediated Signaling Is Involved in Invasion of Glioblastoma Cells in Its Microenvironment. *Int. J. Mol. Sci.* **2020**, *21*, 4199. [CrossRef]
9. Alexandru, O.; Dragutescu, L.; Tătăranu, L.; Ciubotaru, V.; Sevastre, A.; Georgescu, A.M.; Purcaru, O.; Dănoiu, S.; Bäcklund, L.M.; Dricu, A. Helianthin induces antiproliferative effect on human glioblastoma cells in vitro. *J. Neuro Oncol.* **2011**, *102*, 9–18. [CrossRef]
10. Fernandes, C.; Costa, A.; Osório, L.; Lago, R.C.; Linhares, P.; Carvalho, B.; Caeiro, C. *Glioblastoma [Internet]. Current Standards of Care in Glioblastoma Therapy*; De Vleeschouwer, S., Ed.; Codon Publications: Brisbane, Australia, 2017; Chapter 11. Available online: https://www.ncbi.nlm.nih.gov/books/NBK469987/ (accessed on 22 June 2020). [CrossRef]
11. Rajaratnam, V.; Islam, M.M.; Yang, M.; Slaby, R.; Ramirez, H.M.; Mirza, S.P. Glioblastoma: Pathogenesis and Current Status of Chemotherapy and Other Novel Treatments. *Cancers* **2020**, *12*, 937. [CrossRef]
12. Horescu, C.; Cioc, C.E.; Tuta, C.; Sevastre, A.S.; Tache, D.E.; Alexandru, O.; Artene, S.A.; Danoiu, S.; Dricu, A.; Purcaru, S.O. The effect of temozolomide in combination with doxorubicin in glioblastoma cells in vitro. *J. Immunoass. Immunochem.* **2020**, *6*, 1–11. [CrossRef] [PubMed]
13. Stöppler, M.C.; Shiel, W.C.; Credo Reference (Firm); WebMD (Firm). *Webster's New World Medical Dictionary*, 3rd ed.; Credo Reference: Boston, MA, USA; Wiley: Hoboken, NJ, USA, 2014; p. 1.
14. Ostrom, Q.T.; Gittleman, H.; Truitt, G.; Boscia, A.; Kruchko, C.; Barnholtz-Sloan, J.S. CBTRUS Statistical Report: Primary Brain and Other Central Nervous System Tumors Diagnosed in the United States in 2011–2015. *Neuro Oncol.* **2018**, *20*, iv1–iv86. [CrossRef] [PubMed]
15. Tirosh, I.; Suvà, M.L. Tackling the Many Facets of Glioblastoma Heterogeneity. *Cell Stem Cell.* **2020**, *26*, 303–304. [CrossRef] [PubMed]
16. Carapancea, M.; Alexandru, O.; Fetea, A.S.; Dragutescu, L.; Castro, J.; Georgescu, A.; Popa-Wagner, A.; Bäcklund, M.L.; Lewensohn, R.; Dricu, A. Growth factor receptors signaling in glioblastoma cells: Therapeutic implications. *J. Neurooncol.* **2009**, *92*, 137–147. [CrossRef] [PubMed]
17. Oprita, A.; Sevastre, A.S. New pharmaceutical dosage forms used in the treatment of breast cancer. Polymeric micelles. *Med. Oncol.* **2020**, *1*, 38–52.

18. Mao, H.; Lebrun, D.G.; Yang, J.; Zhu, V.F.; Li, M. Deregulated signaling pathways in glioblastoma multiforme: Molecular mechanisms and therapeutic targets. *Cancer Investig.* **2012**, *30*, 48–56. [CrossRef]
19. Alexandru, O.; Ciubotaru, V.; Tataranu, L.; Fetea, S.; Badea, P.; Dricu, A. The relationship between cognitive function, tumour histology and surgical treatment in patients with primary brain tumours. Communications of the European Neurological Society. *J. Neurol.* **2008**, *255*, 151–156. [CrossRef]
20. Schlessinger, J. Cell signaling by receptor tyrosine kinases. *Cell* **2000**, *103*, 211–225. [CrossRef]
21. Eskilsson, E.; Røsland, G.V.; Solecki, G.; Wang, Q.; Harter, P.N.; Graziani, G.; Verhaak, R.G.W.; Winkler, F.; Bjerkvig, R.; Miletic, H. EGFR heterogeneity and implications for therapeutic intervention in glioblastoma. *Neuro Oncol.* **2018**, *20*, 743–752. [CrossRef]
22. Hunter, T. Tyrosine phosphorylation: Thirty years and counting. *Curr. Opin. Cell Biol.* **2009**, *21*, 140–146. [CrossRef]
23. An, Z.; Aksoy, O.; Zheng, T.; Fan, Q.W.; Weiss, W.A. Epidermal growth factor receptor and EGFRvIII in glioblastoma: Signaling pathways and targeted therapies. *Oncogene* **2018**, *37*, 1561–1575. [CrossRef] [PubMed]
24. Brennan, C.W.; Verhaak, R.G.; McKenna, A.; Campos, B.; Noushmehr, H.; Salama, S.R.; Zheng, S.; Chakravarty, D.; Sanborn, J.Z.; Berman, S.H.; et al. The somatic genomic landscape of glioblastoma. *Cell* **2013**, *155*, 462–477. [CrossRef] [PubMed]
25. Zhang, H.; Berezov, A.; Wang, Q.; Zhang, G.; Drebin, J.; Murali, R.; Greene, M.I. ErbB receptors: From oncogenes to targeted cancer therapies. *J. Clin. Investig.* **2007**, *117*, 2051–2058. [CrossRef] [PubMed]
26. Wood, E.R.; Truesdale, A.T.; McDonald, O.B.; Yuan, D.; Hassell, A.; Dickerson, S.H.; Ellis, B.; Pennisi, C.; Horne, E.; Lackey, K.; et al. A unique structure for epidermal growth factor receptor bound to GW572016 (Lapatinib): Relationships among protein conformation, inhibitor off-rate, and receptor activity in tumor cells. *Cancer Res.* **2004**, *64*, 6652–6659. [CrossRef]
27. Ferguson, K.M. Structure-based view of epidermal growth factor receptor regulation. *Annu. Rev. Biophys.* **2008**, *37*, 353–373. [CrossRef]
28. Hubbard, S.R.; Till, J.H. Protein tyrosine kinase structure and function. *Annu. Rev. Biochem.* **2000**, *69*, 373–398. [CrossRef]
29. Arkhipov, A.; Shan, Y.; Das, R.; Endres, N.; Eastwood, M.; Wemmer, D.; Kuriyan, J.; Shaw, D. Architecture and Membrane Interactions of the EGF Receptor. *Cell* **2013**, *152*, 557–569. [CrossRef]
30. Forrester, S.J.; Kawai, T.; O'Brien, S.; Thomas, W.; Harris, R.C.; Eguchi, S. Epidermal Growth Factor Receptor Transactivation: Mechanisms, Pathophysiology, and Potential Therapies in the Cardiovascular System. *Annu. Rev. Pharmacol. Toxicol.* **2016**, *56*, 627–653. [CrossRef]
31. Makki, N.; Thiel, K.W.; Miller, F.J., Jr. The epidermal growth factor receptor and its ligands in cardiovascular disease. *Int. J. Mol. Sci.* **2013**, *14*, 20597–20613. [CrossRef]
32. Alexandru, O.; Purcaru, S.O.; Tataranu, L.G.; Lucan, L.; Castro, J.; Folcuţi, C.; Artene, S.A.; Tuţă, C.; Dricu, A. The Influence of EGFR Inactivation on the Radiation Response in High Grade Glioma. *Int. J. Mol. Sci.* **2018**, *19*, 229. [CrossRef]
33. Kwatra, M.M. A Rational Approach to Target the Epidermal Growth Factor Receptor in Glioblastoma. *Curr. Cancer Drug Targets* **2017**, *17*, 290–296. [CrossRef] [PubMed]
34. Muñoz-Hidalgo, L.; San-Miguel, T.; Megías, J.; Monleón, D.; Navarro, L.; Roldán, P.; Cerdá-Nicolás, M.; López-Ginés, C. Somatic copy number alterations are associated with EGFR amplification and shortened survival in patients with primary glioblastoma. *Neoplasia* **2020**, *22*, 10–21. [CrossRef] [PubMed]
35. Thomas, R.; Weihua, Z. Rethink of EGFR in Cancer with Its Kinase Independent Function on Board. *Front. Oncol.* **2019**, *9*, 800. [CrossRef] [PubMed]
36. Sigismund, S.; Avanzato, D.; Lanzetti, L. Emerging functions of the EGFR in cancer. *Mol. Oncol.* **2018**, *12*, 3–20. [CrossRef] [PubMed]
37. Jureczek, J.; Feldmann, A.; Bergmann, R.; Arndt, C.; Berndt, N.; Koristka, S.; Loureiro, L.R.; Mitwasi, N.; Hoffmann, A.; Kegler, A.; et al. Highly Efficient Targeting of EGFR-Expressing Tumor Cells with UniCAR T Cells via Target Modules Based on Cetuximab®. *Onco Targets Ther.* **2020**, *13*, 5515–5527. [CrossRef]
38. Tsai, C.-J.; Nussinov, R. Emerging Allosteric Mechanism of EGFR Activation in Physiological and Pathological Contexts. *Biophys. J.* **2019**, *117*, 5–13. [CrossRef]
39. Hsu, P.C.; Jablons, D.M.; Yang, C.T.; You, L. Epidermal Growth Factor Receptor (EGFR) Pathway, Yes-Associated Protein (YAP) and the Regulation of Programmed Death-Ligand 1 (PD-L1) in Non-Small Cell Lung Cancer (NSCLC). *Int. J. Mol. Sci.* **2019**, *20*, 3821. [CrossRef]
40. Vitiello, P.P.; Cardone, C.; Martini, G.; Ciardiello, D.; Belli, V.; Matrone, N.; Barra, G.; Napolitano, S.; Della Corte, C.; Turano, M.; et al. Receptor tyrosine kinase-dependent PI3K activation is an escape mechanism to vertical suppression of the EGFR/RAS/MAPK pathway in KRAS-mutated human colorectal cancer cell lines. *J. Exp. Clin. Cancer Res.* **2019**, *38*, 41. [CrossRef]
41. Carpenter, G.; Cohen, S. Epidermal growth factor. *Annu. Rev. Biochem.* **1979**, *48*, 193–216. [CrossRef]
42. Gullick, W.J.; Marsden, J.J.; Whittle, N.; Ward, B.; Bobrow, L.; Waterfield, M.D. Expression of Epidermal Growth Factor Receptors on Human Cervical, Ovarian, and Vulval Carcinomas. *Cancer Res.* **1986**, *46*, 285–292.
43. Parsons, J.T.; Parsons, S.J. Src family protein tyrosine kinases: Cooperating with growth factor and adhesion signaling pathways. *Curr. Opin. Cell Biol.* **1997**, *9*, 187–192. [CrossRef]
44. Biscardi, J.S.; Maa, M.C.; Tice, D.A.; Cox, M.E.; Leu, T.H.; Parsons, S.J. c-Src-mediated phosphorylation of the epidermal growth factor receptor on Tyr845 and Tyr1101 is associated with modulation of receptor function. *J. Biol. Chem.* **1999**, *274*, 8335–8343. [CrossRef] [PubMed]

45. Dawson, J.P.; Berger, M.B.; Lin, C.C.; Schlessinger, J.; Lemmon, M.A.; Ferguson, K.M. Epidermal growth factor receptor dimerization and activation require ligand-induced conformational changes in the dimer interface. *Mol. Cell. Biol.* **2005**, *25*, 7734–7742. [CrossRef] [PubMed]
46. Hervieu, A.; Kermorgant, S. The Role of PI3K in Met Driven Cancer: A Recap. *Front. Mol. Biosci.* **2018**, *5*, 86. [CrossRef]
47. Lemmon, M.A.; Schlessinger, J. Cell signaling by receptor tyrosine kinases. *Cell* **2010**, *141*, 1117–1134. [CrossRef]
48. Brummer, T.; Schmitz-Peiffer, C.; Daly, R.J. Docking proteins. *FEBS J.* **2010**, *277*, 4356–4369. [CrossRef]
49. Kharbanda, A.; Walter, D.; Gudiel, A.; Schek, N.; Feldser, D.; Witze, E. Blocking EGFR palmitoylation suppresses PI3K signaling and mutant KRAS lung tumorigenesis. *Sci. Signal.* **2020**, *13*, eaax2364. [CrossRef]
50. Mayer, I.A.; Arteaga, C.L. The PI3K/AKT Pathway as a Target for Cancer Treatment. *Annu. Rev. Med.* **2016**, *67*, 11–28. [CrossRef]
51. Mattoon, D.R.; Lamothe, B.; Lax, I.; Schlessinger, J. The docking protein Gab1 is the primary mediator of EGF-stimulated activation of the PI-3K/Akt cell survival pathway. *BMC Biol.* **2004**, *2*, 24. [CrossRef]
52. Mulcahy, E.Q.X.; Colón, R.R.; Abounader, R. HGF/MET Signaling in Malignant Brain Tumors. *Int. J. Mol. Sci.* **2020**, *21*, 7546. [CrossRef]
53. Kiyatkin, A.; Aksamitiene, E.; Markevich, N.I.; Borisov, N.M.; Hoek, J.B.; Kholodenko, B.N. Scaffolding protein Grb2-associated binder 1 sustains epidermal growth factor-induced mitogenic and survival signaling by multiple positive feedback loops. *J. Biol. Chem.* **2006**, *281*, 19925–19938. [CrossRef] [PubMed]
54. Pawson, T. Specificity in signal transduction: From phosphotyrosine SH2 domain interactions to complex cellular systems. *Cell* **2004**, *116*, 191–203. [CrossRef]
55. Ward, A.F.; Braun, B.S.; Shannon, K.M. Targeting oncogenic Ras signaling in hematologic malignancies. *Blood* **2012**, *120*, 3397–3406. [CrossRef] [PubMed]
56. Gao, S.P.; Mark, K.G.; Leslie, K.; Pao, W.; Motoi, N.; Gerald, W.L.; Travis, W.D.; Bornmann, W.; Veach, D.; Clarkson, B.; et al. Mutations in the EGFR kinase domain mediate STAT3 activation via IL-6 production in human lung adenocarcinomas. *J. Clin. Investig.* **2007**, *117*, 3846–3856. [CrossRef] [PubMed]
57. Gao, S.P.; Chang, Q.; Mao, N.; Daly, L.A.; Vogel, R.; Chan, T.; Liu, S.H.; Bournazou, E.; Schori, E.; Zhang, H.; et al. JAK2 inhibition sensitizes resistant EGFR-mutant lung adenocarcinoma to tyrosine kinase inhibitors. *Sci. Signal.* **2016**, *9*, ra33. [CrossRef]
58. Padfield, E.; Ellis, H.P.; Kurian, K.M. Current Therapeutic Advances Targeting EGFR and EGFRvIII in Glioblastoma. *Front. Oncol.* **2015**, *5*, 5. [CrossRef]
59. Jorissen, R.N.; Walker, F.; Pouliot, N.; Garrett, T.P.; Ward, C.W.; Burgess, A.W. Epidermal growth factor receptor: Mechanisms of activation and signalling. *Exp. Cell Res.* **2003**, *284*, 31–53. [CrossRef]
60. Huang, P.H.; Xu, A.M.; White, F.M. Oncogenic EGFR signaling networks in Glioma. *Sci. Signal.* **2009**, *2*, 1–13. [CrossRef]
61. Furnari, F.B.; Cloughesy, T.F.; Cavenee, W.K.; Mischel, P.S. Heterogeneity of epidermal growth factor receptor signalling networks in glioblastoma. *Nat. Rev. Cancer* **2015**, *15*, 302–310. [CrossRef]
62. Zeineldin, R.; Ning, Y.; Hudson, L.G. The constitutive activity of epidermal growth factor receptor vIII leads to activation and differential trafficking of wild-type epidermal growth factor receptor and erbB2. *J. Histochem. Cytochem.* **2010**, *58*, 529–541. [CrossRef]
63. Felsberg, J.; Hentschel, B.; Kaulich, K.; Gramatzki, D.; Zacher, A.; Malzkorn, B.; Kamp, M.; Sabel, M.; Simon, M.; Westphal, M.; et al. Epidermal Growth Factor Receptor Variant III (EGFRvIII) Positivity in EGFR-Amplified Glioblastomas: Prognostic Role and Comparison between Primary and Recurrent Tumors. *Clin. Cancer Res.* **2017**, *23*, 6846–6855. [CrossRef] [PubMed]
64. Platten, M. EGFRvIII vaccine in glioblastoma-InACT-IVe or not ReACTive enough? *Neuro Oncol.* **2017**, *19*, 1425–1426. [CrossRef] [PubMed]
65. Struve, N.; Binder, Z.A.; Stead, L.F.; Brend, T.; Bagley, S.J.; Faulkner, C.; Ott, L.; Müller-Goebel, J.; Weik, A.-S.; Hoffer, K.; et al. EGFRvIII upregulates DNA mismatch repair resulting in increased temozolomide sensitivity of MGMT promoter methylated glioblastoma. *Oncogene* **2020**, *39*, 3041–3055. [CrossRef] [PubMed]
66. Brito, C.; Azevedo, A.; Esteves, S.; Marques, A.R.; Martins, C.; Costa, I.; Mafra, M.; Bravo Marques, J.M.; Roque, L.; Pojo, M. Clinical insights gained by refining the 2016 WHO classification of diffuse gliomas with: EGFR amplification, TERT mutations, PTEN deletion and MGMT methylation. *BMC Cancer* **2019**, *19*, 968. [CrossRef] [PubMed]
67. Ohgaki, H.; Kleihues, P. Genetic pathways to primary and secondary glioblastoma. *Am. J. Pathol.* **2007**, *170*, 1445–1453. [CrossRef] [PubMed]
68. Watanabe, K.; Tachibana, O.; Sata, K.; Yonekawa, Y.; Kleihues, P.; Ohgaki, H. Overexpression of the EGF receptor and p53 mutations are mutually exclusive in the evolution of primary and secondary glioblastomas. *Brain Pathol.* **1996**, *6*, 217–223. [CrossRef]
69. Sarkisian, C.J.; Keister, B.A.; Stairs, D.B.; Boxer, R.B.; Moody, S.E.; Chodosh, L.A. Dose-dependent oncogene-induced senescence in vivo and its evasion during mammary tumorigenesis. *Nat. Cell Biol.* **2007**, *9*, 493–505. [CrossRef]
70. Koga, T.; Li, B.; Figueroa, J.M.; Ren, B.; Chen, C.C.; Carter, B.S.; Furnari, F.B. Mapping of genomic EGFRvIII deletions in glioblastoma: Insight into rearrangement mechanisms and biomarker development. *Neuro Oncol.* **2018**, *20*, 1310–1320. [CrossRef]
71. Lopez-Gines, C.; Cerda-Nicolas, M.; Gil-Benso, R.; Pellin, A.; Lopez-Guerrero, J.A.; Callaghan, R.; Benito, R.; Roldan, P.; Piquer, J.; Llacer, J.; et al. Association of chromosome 7, chromosome 10 and EGFR gene amplification in glioblastoma multiforme. *Clin. Neuropathol.* **2005**, *24*, 209–218.
72. Singh, B.; Carpenter, G.; Coffey, R.J. EGF receptor ligands: Recent advances. *F1000Research* **2016**, *5*, 2270. [CrossRef]

73. Ramnarain, D.B.; Park, S.; Lee, D.Y.; Hatanpaa, K.J.; Scoggin, S.O.; Out, H.; Libermann, T.A.; Raisanen, J.M.; Ashfaq, R.; Wong, E.T.; et al. Differential gene expression analysis reveals generation of an autocrine loop by a mutant epidermal growth factor receptor in glioma cells. *Cancer Res.* **2006**, *66*, 867–874. [CrossRef] [PubMed]
74. Tang, P.; Steck, P.A.; Yung, W.K. The autocrine loop of TGF-α/EGFR and brain tumors. *J. Neurooncol.* **1997**, *35*, 303–314. [CrossRef] [PubMed]
75. Filmus, J.; Shi, W.; Spencer, T. Role of transforming growth factor α (TGF-α) in the transformation of ras-transfected rat intestinal epithelial cells. *Oncogene* **1993**, *8*, 1017–1022.
76. Arteaga, C.L.; Engelman, J.A. ErbB receptors: From oncogene discovery to basic science to mechanism-based cancer therapeutics. *Cancer Cell.* **2014**, *25*, 282–303. [CrossRef] [PubMed]
77. Yamaoka, T.; Ohba, M.; Ohmori, T. Molecular-Targeted Therapies for Epidermal Growth Factor Receptor and Its Resistance Mechanisms. *Int. J. Mol. Sci.* **2017**, *18*, 2420. [CrossRef]
78. Lau, S.; Chooback, N.; Ho, C.; Melosky, B. Outcome Differences Between First- and Second-generation EGFR Inhibitors in Advanced EGFR Mutated NSCLC in a Large Population-based Cohort. *Clin. Lung Cancer.* **2019**, *20*, e576–e583. [CrossRef]
79. Reardon, D.A.; Nabors, L.B.; Mason, W.P.; Perry, J.R.; Shapiro, W.; Kavan, P.; Mathieu, D.; Phuphanich, S.; Cseh, A.; Fu, Y.; et al. Phase I/randomized phase II study of afatinib, an irreversible ErbBfamily blocker, with or without protracted temozolomide in adults with recurrent glioblastoma. *Neuro Oncol.* **2015**, *17*, 430–439. [CrossRef]
80. Subbiah, V.; Khawaja, M.R.; Hong, D.S.; Amini, B.; Yungfang, J.; Liu, H.; Johnson, A.; Schrock, A.B.; Ali, S.M.; Sun, J.X.; et al. First-in-human trial of multikinase VEGF inhibitor regorafenib and anti-EGFR antibody cetuximab in advanced cancer patients. *JCI Insight* **2017**, *2*, e90380. [CrossRef]
81. Sepúlveda-Sánchez, J.M.; Vaz, M.Á.; Balañá, C.; Gil-Gil, M.; Reynés, G.; Gallego, Ó.; Martínez-García, M.; Vicente, E.; Quindós, M.; Luque, R.; et al. Phase II trial of dacomitinib, a pan-human EGFR tyrosine kinase inhibitor, in recurrent glioblastoma patients with EGFR amplification. *Neuro Oncol.* **2017**, *19*, 1522–1531. [CrossRef]
82. Karpel-Massler, G.; Westhoff, M.A.; Kast, R.E.; Wirtz, C.R.; Halatsch, M.E. Erlotinib in glioblastoma: Lost in translation? *Anticancer Agents Med. Chem.* **2011**, *11*, 748–755. [CrossRef]
83. Rich, J.N.; Reardon, D.A.; Peery, T.; Dowell, J.M.; Quinn, J.A.; Penne, K.L.; Wikstrand, C.J.; Van Duyn, L.B.; Dancey, J.E.; McLendon, R.E.; et al. Phase II trial of gefitinib in recurrent glioblastoma. *J. Clin. Oncol.* **2004**, *22*, 133–142. [CrossRef] [PubMed]
84. Thiessen, B.; Stewart, C.; Tsao, M.; Kamel-Reid, S.; Schaiquevich, P.; Mason, W.; Easaw, J.; Belanger, K.; Forsyth, P.; McIntosh, L.; et al. A phase I/II trial of GW572016 (lapatinib) in recurrent glioblastoma multiforme: Clinical outcomes, pharmacokinetics and molecular correlation. *Cancer Chemother. Pharmacol.* **2010**, *65*, 353–361. [CrossRef] [PubMed]
85. Du, X.J.; Li, X.M.; Cai, L.B.; Sun, J.C.; Wang, S.Y.; Wang, X.C.; Pang, X.L.; Deng, M.; Chen, F.F.; Wang, Z.Q.; et al. Efficacy and safety of nimotuzumab in addition to radiotherapy and temozolomide for cerebral glioblastoma: A phase II multicenter clinical trial. *J. Cancer* **2019**, *10*, 3214–3223. [CrossRef] [PubMed]
86. Makhlin, I.; Salinas, R.D.; Zhang, D.; Jacob, F.; Ming, G.L.; Song, H.; Saxena, D.; Dorsey, J.F.; Nasrallah, M.P.; Morrissette, J.J.; et al. Clinical activity of the EGFR tyrosine kinase inhibitor osimertinib in EGFR-mutant glioblastoma. *CNS Oncol.* **2019**, *8*, CNS43. [CrossRef]
87. Liu, X.; Chen, X.; Shi, L.; Shan, Q.; Cao, Q.; Yue, C.; Li, H.; Li, S.; Wang, J.; Gao, S.; et al. The third-generation EGFR inhibitor AZD9291 overcomes primary resistance by continuously blocking ERK signaling in glioblastoma. *J. Exp. Clin. Cancer Res.* **2019**, *38*, 219. [CrossRef]
88. Chagoya, G.; Kwatra, S.G.; Nanni, C.W.; Roberts, C.M.; Phillips, S.M.; Nullmeyergh, S.; Gilmore, S.P.; Spasojevic, I.; Corcoran, D.L.; Young, C.C.; et al. Efficacy of osimertinib against EGFRvIII+ glioblastoma. *Oncotarget* **2020**, *11*, 2074–2082. [CrossRef]
89. Napier, T.S.; Udayakumar, N.; Jani, A.H.; Hartman, Y.E.; Houson, H.A.; Moore, L.; Amm, H.M.; van den Berg, N.S.; Sorace, A.G. Warram JM. Comparison of Panitumumab-IRDye800CW and 5-Aminolevulinic Acid to Provide Optical Contrast in a Model of Glioblastoma Multiforme. *Mol. Cancer Ther.* **2020**, *19*, 1922–1929. [CrossRef]
90. Reardon, D.A.; Desjardins, A.; Vredenburgh, J.J.; O'Rourke, D.M.; Tran, D.D.; Fink, K.L.; Nabors, L.B.; Li, G.; Bota, D.A.; Lukas, R.V.; et al. Rindopepimut with Bevacizumab for Patients with Relapsed EGFRvIII-Expressing Glioblastoma (ReACT): Results of a Double-Blind Randomized Phase II Trial. *Clin. Cancer Res.* **2020**, *26*, 1586–1594. [CrossRef]
91. Kreisl, T.N.; McNeill, K.A.; Sul, J.; Iwamoto, F.M.; Shih, J.; Fine, H.A. A phase I/II trial of vandetanib for patients with recurrent malignant glioma. *Neuro Oncol.* **2012**, *14*, 1519–1526. [CrossRef]
92. Clinical Trials. Available online: https://clinicaltrials.gov/ct2/show/NCT02844439 (accessed on 7 December 2020).
93. Schaller, T.H.; Foster, M.W.; Thompson, J.W.; Spasojevic, I.; Normantaite, D.; Moseley, M.A.; Sanchez-Perez, L.; Sampson, J.H. Pharmacokinetic Analysis of a Novel Human EGFRvIII:CD3 Bispecific Antibody in Plasma and Whole Blood Using a High-Resolution Targeted Mass Spectrometry Approach. *J. Proteome Res.* **2019**, *18*, 3032–3041. [CrossRef]
94. Orellana, L. Convergence of EGFR glioblastoma mutations: Evolution and allostery rationalizing targeted therapy. *Mol. Cell Oncol.* **2019**, *6*, 5. [CrossRef] [PubMed]
95. Quang, T.S.; Brady, L.W. Radioimmunotherapy as a novel treatment regimen: 125I-labeled monoclonal antibody 425 in the treatment of high-grade brain gliomas. *Int. J. Radiat. Oncol. Biol. Phys.* **2004**, *58*, 972–975. [CrossRef] [PubMed]
96. Jacob, F.; Salinas, R.D.; Zhang, D.Y.; Nguyen, P.T.T.; Schnoll, J.G.; Wong, S.Z.H.; Thokala, R.; Sheikh, S.; Saxena, D.; Prokop, S.; et al. A Patient-Derived Glioblastoma Organoid Model and Biobank Recapitulates Inter- and Intra-tumoral Heterogeneity. *Cell* **2020**, *180*, 188–204. [CrossRef] [PubMed]

97. Petersen, C.T.; Krenciute, G. Next Generation CAR T Cells for the Immunotherapy of High-Grade Glioma. *Front Oncol.* **2019**, *9*, 69. [CrossRef] [PubMed]
98. Ye, C.; Pan, B.; Xu, H.; Zhao, Z.; Shen, J.; Lu, J.; Yu, R.; Liu, H. Co-delivery of GOLPH3 siRNA and gefitinib by cationic lipid-PLGA nanoparticles improves EGFR-targeted therapy for glioma. *J. Mol. Med.* **2019**, *97*, 1575–1588. [CrossRef]
99. Korshunov, A.; Okonechnikov, K.; Sahm, F.; Ryzhova, M.; Stichel, D.; Schrimpf, D.; Ghasemi, D.R.; Pajtler, K.W.; Antonelli, M.; Donofrio, V.; et al. Transcriptional profiling of medulloblastoma with extensive nodularity (MBEN) reveals two clinically relevant tumor subsets with VSNL1 as potent prognostic marker. *Acta Neuropathol.* **2020**, *139*, 583–596. [CrossRef]
100. Taylor, O.G.; Brzozowski, J.S.; Skelding, K.A. Glioblastoma Multiforme: An Overview of Emerging Therapeutic Targets. *Front. Oncol.* **2019**, *9*, 963. [CrossRef]
101. Lin, Y.J.; Chen, K.T.; Huang, C.Y.; Wei, K.C. Review ArticleNon-Invasive Focused Ultrasound-Based Synergistic Treatment of Brain Tumors. *J. Cancer Res. Pract.* **2016**, *3*, 63–68. [CrossRef]
102. Wei, K.C.; Chu, P.C.; Wang, H.Y.; Huang, C.Y.; Chen, P.Y.; Tsai, H.C.; Lu, Y.J.; Lee, P.Y.; Tseng, I.C.; Feng, L.Y.; et al. Focused ultrasound-induced blood-brain barrier opening to enhance temozolomide delivery for glioblastoma treatment: A preclinical study. *PLoS ONE* **2013**, *8*, e58995. [CrossRef]
103. Westphal, M.; Maire, C.L.; Lamszus, K. EGFR as a Target for Glioblastoma Treatment: An Unfulfilled Promise. *CNS Drugs* **2017**, *31*, 723–735. [CrossRef]
104. Freeman, A.C.; Platt, S.R.; Holmes, S.; Kent, M.; Robinson, K.; Howerth, E.; Eagleson, J.; Bouras, A.; Kaluzova, M.; Hadjipanayis, C.G. Convection-enhanced delivery of cetuximab conjugated iron-oxide nanoparticles for treatment of spontaneous canine intracranial gliomas. *J. Neurooncol.* **2018**, *137*, 653–663. [CrossRef] [PubMed]
105. Roskoski, R., Jr. Properties of FDA-approved small molecule protein kinase inhibitors. *Pharmacol. Res.* **2019**, *144*, 19–50. [CrossRef]
106. Uhm, J.H.; Ballman, K.V.; Wu, W.; Giannini, C.; Krauss, J.C.; Buckner, J.C.; James, C.D.; Scheithauer, B.W.; Behrens, R.J.; Flynn, P.J.; et al. Phase II evaluation of gefitinib in patients with newly diagnosed Grade 4 astrocytoma: Mayo/North Central Cancer Treatment Group Study N0074. *Int. J. Radiat. Oncol. Biol. Phys.* **2011**, *80*, 347–353. [CrossRef]
107. Chakravarti, A.; Wang, M.; Robins, H.I.; Lautenschlaeger, T.; Curran, W.J.; Brachman, D.G.; Schultz, C.J.; Choucair, A.; Dolled-Filhart, M.; Christiansen, J.; et al. RTOG 0211: A phase 1/2 study of radiation therapy with concurrent gefitinib for newly diagnosed glioblastoma patients. *Int. J. Radiat. Oncol. Biol. Phys.* **2013**, *85*, 1206–1211. [CrossRef] [PubMed]
108. Shepherd, F.A.; Rodrigues Pereira, J.; Ciuleanu, T.; Tan, E.H.; Hirsh, V.; Thongprasert, S.; Campos, D.; Maoleekoonpiroj, S.; Smylie, M.; Martins, R.; et al. Erlotinib in previously treated non-small-cell lung cancer. *N. Engl. J. Med.* **2005**, *353*, 123–132. [CrossRef] [PubMed]
109. Peereboom, D.M.; Shepard, D.R.; Ahluwalia, M.S.; Brewer, C.J.; Agarwal, N.; Stevens, G.H.; Suh, J.H.; Toms, S.A.; Vogelbaum, M.A.; Weil, R.J.; et al. Phase II trial of erlotinib with temozolomide and radiation in patients with newly diagnosed glioblastoma multiforme. *J. Neurooncol.* **2010**, *98*, 93–99. [CrossRef] [PubMed]
110. Wen, P.Y.; Chang, S.M.; Lamborn, K.R.; Kuhn, J.G.; Norden, A.D.; Cloughesy, T.F.; Robins, H.I.; Lieberman, F.S.; Gilbert, M.R.; Mehta, M.P.; et al. Phase I/II study of erlotinib and temsirolimus for patients with recurrent malignant gliomas: North American Brain Tumor Consortium trial 04-02. *Neuro Oncol.* **2014**, *16*, 567–578. [CrossRef] [PubMed]
111. Raizer, J.J.; Giglio, P.; Hu, J.; Groves, M.; Merrell, R.; Conrad, C.; Phuphanich, S.; Puduvalli, V.K.; Loghin, M.; Paleologos, N.; et al. A phase II study of bevacizumab and erlotinib after radiation and temozolomide in MGMT unmethylated GBM patients. *J. Neurooncol.* **2016**, *126*, 185–192. [CrossRef]
112. Reardon, D.A.; Groves, M.D.; Wen, P.Y.; Nabors, L.; Mikkelsen, T.; Rosenfeld, S.; Raizer, J.; Barriuso, J.; McLendon, R.E.; Suttle, A.B.; et al. A phase I/II trial of pazopanib in combination with lapatinib in adult patients with relapsed malignant glioma. *Clin. Cancer Res.* **2013**, *19*, 900–908. [CrossRef]
113. Liu, X.; Ide, J.L.; Norton, I.; Marchionni, M.A.; Ebling, M.C.; Wang, L.Y.; Davis, E.; Sauvageot, C.M.; Kesari, S.; Kellersberger, K.A.; et al. Molecular imaging of drug transit through the blood-brain barrier with MALDI mass spectrometry imaging. *Sci. Rep.* **2013**, *3*, 2859. [CrossRef]
114. Lassman, A.B.; Rossi, M.R.; Raizer, J.J.; Abrey, L.E.; Lieberman, F.S.; Grefe, C.N.; Lamborn, K.; Pao, W.; Shih, A.H.; Kuhn, J.G.; et al. Holland. Molecular study of malignant gliomas treated with epidermal growth factor receptor inhibitors: Tissue analysis from North American Brain Tumor Consortium Trials 01–03 and 00–01. *Clin. Cancer Res.* **2005**, *11*, 7841–7850. [CrossRef] [PubMed]
115. Sepúlveda, J.M.; Sánchez-Gómez, P.; Vaz Salgado, M.Á.; Gargini, R.; Balañá, C. Dacomitinib: An investigational drug for the treatment of glioblastoma. *Expert Opin. Investig. Drugs* **2018**, *27*, 823–829. [CrossRef]
116. Clinical Trials. Available online: https://clinicaltrials.gov/ct2/results?cond=glioma&term=vandetanib&cntry=&state=&city=&dist= (accessed on 20 June 2020).
117. Wee, P.; Wang, Z. Epidermal growth factor receptor cell proliferation signaling pathways. *Cancers* **2017**, *9*, 52. [CrossRef]
118. Neyns, B.; Sadones, J.; Joosens, E.; Bouttens, F.; Verbeke, L.; Baurain, J.F.; D'Hondt, L.; Strauven, T.; Chaskis, C.; In't Veld, P.; et al. Stratified phase II trial of cetuximab in patients with recurrent high-grade glioma. *Ann. Oncol.* **2009**, *20*, 1596–1603. [CrossRef]
119. Yang, J.; Shi, Z.; Liu, R.; Wu, Y.; Zhang, X. Combined-therapeutic strategies synergistically potentiate glioblastoma multiforme treatment via nanotechnology. *Theranostics* **2020**, *10*, 3223–3239. [CrossRef]
120. Huang, H.C.; Pigula, M.; Fang, Y.; Hasan, T. Immobilization of Photo-Immunoconjugates on Nanoparticles Leads to Enhanced Light-Activated Biological Effects. *Small* **2018**, *14*, e1800236. [CrossRef]

121. Heun, J.; Holen, K.; Paul, P. Treatment with panitumumab after a severe infusion reaction to cetuximab in a patient with metastatic colorectal cancer: A case report. *Clin. Colorectal. Cancer* **2007**, *6*, 529–531. [CrossRef]
122. Clinical Trials. Available online: https://clinicaltrials.gov/ct2/show/NCT03510208 (accessed on 5 December 2020).
123. Choi, B.D.; Kuan, C.T.; Cai, M.; Archer, G.E.; Mitchell, D.A.; Gedeon, P.C.; Sanchez-Perez, L.; Pastan, I.; Bigner, D.D.; Sampson, J.H. Systemic administration of a bispecific antibody targeting EGFRvIII successfully treats intracerebral glioma. *Proc. Natl. Acad. Sci. USA* **2013**, *110*, 270–275. [CrossRef]
124. Orellana, L.; Thorne, A.H.; Lema, R.; Gustavsson, J.; Parisian, A.D.; Hospital, A.; Cordeiro, T.N.; Bernadó, P.; Scott, A.M.; Brun-Heath, I.; et al. Oncogenic mutations at the EGFR ectodomain structurally converge to remove a steric hindrance on a kinase-coupled cryptic epitope. *Proc. Natl. Acad. Sci. USA* **2019**, *116*, 10009–10018. [CrossRef]
125. Johns, T.G.; McKay, M.J.; Cvrljevic, A.N.; Gan, H.K.; Taylor, C.; Xu, H.; Smyth, F.E.; Scott, A.M. MAb 806 enhances the efficacy of ionizing radiation in glioma xenografts expressing the de2-7 epidermal growth factor receptor. *Int. J. Radiat. Oncol. Biol. Phys.* **2010**, *78*, 572–578. [CrossRef]
126. Emrich, J.G.; Brady, L.W.; Quang, T.S.; Class, R.; Miyamoto, C.; Black, P.; Rodeck, U. Radioiodinated (I-125) monoclonal antibody 425 in the treatment of high grade glioma patients: Ten-year synopsis of a novel treatment. *Am. J. Clin. Oncol.* **2002**, *25*, 541–546. [CrossRef] [PubMed]
127. Johnson, L.A.; Scholler, J.; Ohkuri, T.; Kosaka, A.; Patel, P.R.; McGettigan, S.E.; Nace, A.K.; Dentchev, T.; Thekkat, P.; Loew, A.; et al. Rational development and characterization of humanized anti-EGFR variant III chimeric antigen receptor T cells for glioblastoma. *Sci. Transl. Med.* **2015**, *7*, 275ra22. [CrossRef] [PubMed]
128. Sahin, A.; Sanchez, C.; Bullain, S.; Waterman, P.; Weissleder, R.; Carter, B.S. Development of third generation anti-EGFRvIII chimeric T cells and EGFRvIII-expressing artificial antigen presenting cells for adoptive cell therapy for glioma. *PLoS ONE* **2018**, *13*, e0199414. [CrossRef] [PubMed]
129. Clinical Trials. Available online: https://clinicaltrials.gov/ct2/show/results/NCT01454596?term=CAR-T&cond=glioma&draw=3&rank=12 (accessed on 20 September 2020).
130. Binder, D.C.; Ladomersky, E.; Lenzen, A.; Zhai, L.; Lauing, K.L.; Otto-Meyer, S.D.; Lukas, R.V.; Wainwright, D.A. Lessons learned from rindopepimut treatment in patients with EGFRvIII-expressing glioblastoma. *Transl. Cancer Res.* **2018**, *7* (Suppl. S4), S510–S513. [CrossRef]
131. Schuster, J.; Lai, R.K.; Recht, L.D.; Reardon, D.A.; Paleologos, N.A.; Groves, M.D.; Mrugala, M.M.; Jensen, R.; Baehring, J.M.; Sloan, A.; et al. A phase II, multicenter trial of rindopepimut (CDX-110) in newly diagnosed glioblastoma: The ACT III study. *Neuro Oncol.* **2015**, *17*, 854–861. [CrossRef]
132. Weller, M.; Butowski, N.; Tran, D.D.; Recht, L.D.; Lim, M.; Hirte, H.; Ashby, L.; Mechtler, L.; Goldlust, S.A.; Iwamoto, F.; et al. Rindopepimut with temozolomide for patients with newly diagnosed, EGFRvIII-expressing glioblastoma (ACT IV): A randomised, double-blind, international phase 3 trial. *Lancet Oncol.* **2017**, *18*, 1373–1385. [CrossRef]
133. Read, J.; Ingram, A.; Al Saleh, H.A.; Platko, K.; Gabriel, K.; Kapoor, A.; Pinthus, J.; Majeed, F.; Qureshi, T.; Al-Nedawi, K. Nuclear transportation of exogenous epidermal growth factor receptor and androgen receptor via extracellular vesicles. *Eur. J. Cancer.* **2017**, *70*, 62–74. [CrossRef]
134. Ricklefs, F.; Mineo, M.; Rooj, A.K.; Nakano, I.; Charest, A.; Weissleder, R.; Breakefield, X.O.; Chiocca, E.A.; Godlewski, J.; Bronisz, A. Extracellular Vesicles from High-Grade Glioma Exchange Diverse Pro-oncogenic Signals That Maintain Intratumoral Heterogeneity. *Cancer Res.* **2016**, *76*, 2876–2881. [CrossRef]
135. Bronisz, A.; Wang, Y.; Nowicki, M.O.; Peruzzi, P.; Ansari, K.; Ogawa, D.; Balaj, L.; De Rienzo, G.; Mineo, M.; Nakano, I.; et al. Extracellular vesicles modulate the glioblastoma microenvironment via a tumor suppression signaling network directed by miR-1. *Cancer Res.* **2014**, *74*, 738–750. [CrossRef]
136. Liao, H.W.; Hsu, J.M.; Xia, W.; Wang, H.L.; Wang, Y.N.; Chang, W.C.; Arold, S.T.; Chou, C.K.; Tsou, P.H.; Yamaguchi, H.; et al. PRMT1-mediated methylation of the EGF receptor regulates signaling and cetuximab response. *J. Clin. Investig.* **2015**, *125*, 4529–4543. [CrossRef]
137. Karim, R.; Palazzo, C.; Evrard, B.; Piel, G. Nanocarriers for the treatment of glioblastoma multiforme: Current state-of-the-art. *J. Control. Release* **2016**, *10*, 23–37. [CrossRef] [PubMed]
138. Mortensen, J.H.; Jeppesen, M.; Pilgaard, L.; Agger, R.; Duroux, M.; Zachar, V.; Moos, T. Targeted antiepidermal growth factor receptor (cetuximab) immunoliposomes enhance cellular uptake in vitro and exhibit increased accumulation in an intracranial model of glioblastoma multiforme. *J. Drug Deliv.* **2013**, *2013*, 209205. [CrossRef] [PubMed]
139. Clinical Trials. Available online: https://clinicaltrials.gov/ct2/show/NCT02340156?term=nanoparticles&cond=glioma&draw=2&rank=7 (accessed on 15 April 2020).
140. Kim, S.S.; Rait, A.; Kim, E.; Pirollo, K.F.; Chang, E.H. A tumor-targeting p53 nanodelivery system limits chemoresistance to temozolomide prolonging survival in a mouse model of glioblastoma multiforme. *Nanomedicine* **2015**, *11*, 301–311. [CrossRef] [PubMed]
141. Saleem, H.; Kulsoom, A.; Küçükosmanoglu, A.; Houweling, M.; Cornelissen, F.M.G.; Heiland, D.H.; Hegi, M.E.; Kouwenhoven, M.C.M.; Bailey, D.; Würdinger, T.; et al. The TICking clock of EGFR therapy resistance in glioblastoma: Target Independence or target Compensation. *Drug Resist. Updates* **2019**, *43*, 29–37. [CrossRef]

142. Nathanson, D.A.; Gini, B.; Mottahedeh, J.; Visnyei, K.; Koga, T.; Gomez, G.; Eskin, A.; Hwang, K.; Wang, J.; Masui, K.; et al. Targeted therapy resistance mediated by dynamic regulation of extrachromosomal mutant EGFR DNA. *Science* **2014**, *343*, 72–76. [CrossRef]
143. Stein, S.; Zhao, R.; Haeno, H.; Vivanco, I.; Michor, F. Mathematical modeling identifies optimum lapatinib dosing schedules for the treatment of glioblastoma patients. *PLoS Comput Biol.* **2018**, *14*, e1005924. [CrossRef]
144. Yu, H.A.; Ahn, M.J.; Cho, B.C.; Gerber, D.E.; Natale, R.B.; Socinski, M.A.; Giri, N.; Quinn, S.; Sbar, E.; Zhang, H.; et al. Phase 2 study of intermittent pulse dacomitinib in patients with advanced non-small cell lung cancers. *Lung Cancer* **2017**, *112*, 195–199. [CrossRef]
145. Zhao, H.F.; Wang, J.; Shao, W.; Wu, C.P.; Chen, Z.P.; To, S.T.; Li, W.P. Recent advances in the use of PI3K inhibitors for glioblastoma multiforme: Current preclinical and clinical development. *Mol. Cancer* **2017**, *16*, 100. [CrossRef]
146. Schettini, F.; De Santo, I.; Rea, C.G.; De Placido, P.; Formisano, L.; Giuliano, M.; Arpino, G.; De Laurentiis, M.; Puglisi, F.; De Placido, S.; et al. CDK 4/6 Inhibitors as Single Agent in Advanced Solid Tumors. *Front. Oncol.* **2018**, *8*, 608. [CrossRef]
147. Cao, Y.; Li, X.; Kong, S.; Shang, S.; Qi, Y. CDK4/6 inhibition suppresses tumour growth and enhances the effect of temozolomide in glioma cells. *J. Cell. Mol. Med.* **2020**, *24*, 5135–5145. [CrossRef]
148. Sharma, P.; Roberts, C.; Herpai, D.; Fokt, I.D.; Priebe, W.; Debinski, W. Drug Conjugates for Targeting Eph Receptors in Glioblastoma. *Pharmaceuticals* **2020**, *13*, 77. [CrossRef] [PubMed]
149. Lowenstein, P.R.; Castro, M.G. Multiple Expressed Endogenous Glioma Epitopes as Novel Vaccines for Gliomas. *Clin. Cancer Res.* **2016**, *22*, 4760–4762. [CrossRef] [PubMed]
150. Valentini, D.; Rao, M.; Meng, Q.; von Landenberg, A.; Bartek, J., Jr.; Sinclair, G.; Paraschoudi, G.; Jäger, E.; Harvey-Peredo, I.; Dodoo, E.; et al. Identification of neoepitopes recognized by tumor-infiltrating lymphocytes (TILs) from patients with glioma. *Oncotarget* **2018**, *9*, 19469–19480. [CrossRef] [PubMed]
151. Alayo, Q.A.; Ito, H.; Passaro, C.; Zdioruk, M.; Mahmoud, A.B.; Grauwet, K.; Zhang, X.; Lawler, S.E.; Reardon, D.A.; Goins, W.F.; et al. Glioblastoma infiltration of both tumor- and virus-antigen specific cytotoxic T cells correlates with experimental virotherapy responses. *Sci. Rep.* **2020**, *10*, 5095. [CrossRef] [PubMed]
152. Meng, X.; Zhao, Y.; Han, B.; Zha, C.; Zhang, Y.; Li, Z.; Wu, P.; Qi, T.; Jiang, C.; Liu, Y.; et al. Dual functionalized brain-targeting nanoinhibitors restrain temozolomide-resistant glioma via attenuating EGFR and MET signaling pathways. *Nat. Commun.* **2020**, *11*, 594. [CrossRef]
153. Jones, S.; King, P.J.; Antonescu, C.N.; Sugiyama, M.G.; Bhamra, A.; Surinova, S.; Angelopoulos, N.; Kragh, M.; Pedersen, M.W.; Hartley, J.A.; et al. Targeting of EGFR by a combination of antibodies mediates unconventional EGFR trafficking and degradation. *Sci. Rep.* **2020**, *10*, 663. [CrossRef]
154. Guo, G.; Narayan, N.R.; Horton, L.; Patel, T.; Habib, A. The Role of EGFR-Met Interactions in the Pathogenesis of Glioblastoma and Resistance to Treatment. *Curr. Cancer Drug Targets* **2017**, *17*, 297–302. [CrossRef]

International Journal of
Molecular Sciences

MDPI

Review

PI3K/AKT/mTOR Signaling Pathway in Breast Cancer: From Molecular Landscape to Clinical Aspects

Daniela Miricescu [1], Alexandra Totan [1], Iulia-Ioana Stanescu-Spinu [1,*], Silviu Constantin Badoiu [2,*], Constantin Stefani [3] and Maria Greabu [1]

[1] Department of Biochemistry, Faculty of Dental Medicine, Carol Davila University of Medicine and Pharmacy, 8 Eroii Sanitari Blvd, 050474 Bucharest, Romania; daniela.miricescu@umfcd.ro (D.M.); alexandra.totan@umfcd.ro (A.T.); maria.greabu@umfcd.ro (M.G.)

[2] Department of Anatomy and Embryology, Faculty of Medicine, Carol Davila University of Medicine and Pharmacy, 8 Eroii Sanitari Blvd, 050474 Bucharest, Romania

[3] Department of Family Medicine and Clinical Base, Dr. Carol Davila Central Military Emergency University Hospital, 134 Calea Plevnei, 010825 Bucharest, Romania; constantin.stefani@umfcd.ro

* Correspondence: iulia.stanescu@umfcd.ro (I.-I.S.-S.); silviu.badoiu@umfcd.ro (S.C.B.)

Abstract: Breast cancer is a serious health problem worldwide, representing the second cause of death through malignancies among women in developed countries. Population, endogenous and exogenous hormones, and physiological, genetic and breast-related factors are involved in breast cancer pathogenesis. The phosphatidylinositol 3-kinase (PI3K)/protein kinase B (AKT)/mammalian target of rapamycin (mTOR) is a signaling pathway involved in cell proliferation, survival, invasion, migration, apoptosis, glucose metabolism and DNA repair. In breast tumors, PIK3CA somatic mutations have been reported, located in exon 9 and exon 20. Up to 40% of PIK3CA mutations are estrogen receptor (ER) positive and human epidermal growth factor receptor 2 (HER2) -negative in primary and metastatic breast cancer. HER2 is overexpressed in 20–30% of breast cancers. HER1, HER2, HER3 and HER4 are membrane receptor tyrosine kinases involved in HER signaling to which various ligands can be attached, leading to PI3K/AKT activation. Currently, clinical studies evaluate inhibitors of the PI3K/AKT/mTOR axis. The main purpose of this review is to present general aspects of breast cancer, the components of the AKT signaling pathway, the factors that activate this protein kinase B, PI3K/AKT-breast cancer mutations, PI3K/AKT/mTOR-inhibitors, and the relationship between everolimus, temsirolimus and endocrine therapy.

Keywords: breast cancer; estrogen receptor-positive; HER2; PI3K/AKT/mTOR pathway; endocrine resistance

Citation: Miricescu, D.; Totan, A.; Stanescu-Spinu, I.-I.; Badoiu, S.C.; Stefani, C.; Greabu, M. PI3K/AKT/mTOR Signaling Pathway in Breast Cancer: From Molecular Landscape to Clinical Aspects. *Int. J. Mol. Sci.* **2021**, 22, 173. https://dx.doi.org/10.3390/ijms22010173

Received: 29 November 2020
Accepted: 23 December 2020
Published: 26 December 2020

1. Breast Cancer–Incidence and Risk Factors

Breast cancer is a serious medical condition which especially affects women, but also men, in the Unites States of America (USA) 1% of the total cases of breast malignant tumors being diagnosed in male patients [1]. Breast cancer is the main type of carcinoma in women [2,3] and the second most common type of neoplasia in the general population, in the world [4]. Almost 270,000 women are newly diagnosed with breast cancer every year in the USA [5–7]. In developed countries, it is the second cause of death through malignancies in women after lung cancer [5–7]. About 42,000 women died in the USA as a consequence of breast cancer in 2019 [5–7]. The lifetime risk of dying of breast cancer is approximately 2.6% [7]. The lifetime risk of developing breast cancer in the United States of America is 12.4% [8]. The incidence of this pathology varies with race and ethnicity [9]. Breast cancer incidence also presents variations related to the geographic zone, ranging from 27/100,000 (Central-East Asia and Africa) to 85–94/100,000 (Australia, North America and Western Europe) [4,10]. The mortality rate varies from six cases per 100,000 people in East Asia to 20 cases per 100,000 people in Western Africa [4]. Although the incidence of breast cancer is higher in developed countries, the mortality is higher in low-income countries because of late diagnosis, lack of screening and poor access to treatment [11]. Many factors influencing breast cancer have been studied. Age is the most prominent risk factor for breast cancer, after gender [12]. The incidence rate of breast cancer increases with age and reaches a peak around the age of 50. Age of menarche and age of menopause can also influence breast cancer development, given the effects of the ovarian hormones on the mammary gland from puberty till menopause. Early menarche, time of menopause onset and oophorectomy can play an important role in breast cancer development as well [12–15].

Regarding gender prevalence, it is common knowledge that breast cancer is much more frequent in women than men [16]. Because it is so rare in men, breast cancer is underdiagnosed and, in many cases, the diagnostic is obtained in more advanced stages than in women. A controversial risk factor for breast cancer is represented by the blood group; women with blood group A and Rh positive present a higher risk for breast cancer than women with blood group AB and Rh negative [17]. In addition, studies have shown that black women have a higher incidence of breast cancer than white women before the age of 40. After 40 years of age, the situation reverses [18]. Asian, Hispanic and Indian American women have a lower incidence rate of breast cancer than white and black American women [19].

Full-term pregnancy is a protective factor against breast cancer [20–22], with nulliparous women having an increased risk for breast cancer [23]. On the other hand, during the first pregnancy, important and permanent changes in the mammary gland cells occur: the glandular cells have a longer cell cycle and a prolonged G1 phase (Gap 1 phase)—it is the phase that allows DNA to repair. Consequently, the risk of the appearance of DNA alterations (that will be transmitted with the proliferation of breast gland cells during pregnancy) increases with the age of the women at first full term pregnancy [24–26]. Additionally, more than one birth and more closely spaced births are considered to be protective factors against breast cancer [24].

Breastfeeding is a protective factor against breast cancer, the longer the breastfeeding, the greater the protective effect [27,28]. At the same time, the association of two protective factors (two or more childbirths and lactation for more than 13 months) could reduce the risk of developing breast cancer by up to 60% [29,30].

In postmenopausal women, the risk of breast cancer increases proportionally with estrogen levels, with the use of antiestrogens (tamoxifen, raloxifene) proving to be effective in preventing the development of breast cancer [31,32]. Testosterone can also promote breast cancer development in postmenopausal women [31], acting by its conversion to estrogen and by its anabolic effect upon breast cancer cells (apparently for receptor-positive breast cancers) [31,33].

Although clinical and epidemiological studies have controversial results, prolactin, insulin-like growth factor (IGF-1) and oral contraceptives used longer than 10 years have all been incriminated as risk factors for breast cancer [34–40]. Additionally, hormone replacement therapy can enhance the risk of developing breast cancer and amplify the rate of mortality. The risk decreases after a minimum of 5 years of discontinuation of the treatment [41,42]. Moreover, the increased risk of developing breast cancer varies, pending to the hormones used, while the association of progesterone determines a greater risk for breast cancer [43–45]. The influence of ovulation stimulating drugs in breast cancer pathogenesis is still controversial, with studies reporting different results [46,47].

Family history of breast cancer and mutations of the genes codifying the synthesis of enzymes (matrix metalloproteinases, hormone-metabolizing enzymes), interferon alpha, estrogen and progesterone receptors or the genes involved in DNA repair represent major risk factors as well [48–53].

Diabetes mellitus (mainly type II) correlates with an increased risk of breast cancer, especially when associated with an elevated BMI (Body Mass Index) and with high levels of IGF-1 [54–56]. The use of metformin increases the survival rate in type II diabetic patients with breast cancer [57]. An increased BMI correlates with an increased risk of breast cancer [58,59], with obesity after menopause representing a proven risk factor for breast carcinoma [60,61]. Increased intake of meat and saturated fats [62,63] and the consumption of milk before menopause determine increased risk of breast cancer [64], while decreased serum levels of vitamin D are associated with a higher risk of breast cancer [65]. The administration of vitamin D supplements reduces the risk of breast cancer [66], but alcohol consumption constitutes a risk factor [67,68], as well as active and/or passive smoking [69,70], although there are studies that show different results [71].

On the other hand, physical activity reduces the risk of breast cancer [72,73]. However, working overnight leads to exposure to artificial light, increasing the level of estrogen and, consequently, augmenting the risk of breast cancer [74–76].

Women with higher social and economic status have an increased risk of breast cancer (older age at the first childbirth, older age at menopause, sedentary life, and unhealthy diet). However, they have more frequent medical examinations, hence can benefit from earlier diagnosis [77,78]. Meanwhile, women with lower social and economic status are later diagnosed and have a poorer prognosis [79].

Non-proliferative breast diseases, such as mild hyperplasia, cysts and apocrine metaplasia, do not increase the risk of breast cancer, and neither do breast implants [80,81]. Meanwhile, proliferative breast diseases without atypia (moderate hyperplasia, intraductal papilloma, sclerosing adenoma) increase the risk of breast cancer up to two-fold. Moreover, proliferative breast diseases with atypia (ductal hyperplasia with atypia and lobular hyperplasia with atypia) enhance the risk of developing this type of carcinoma up to six-fold [80]. Increased mammographic density is the most important risk factor for breast carcinoma after family history of breast cancer [82], with breast density being an independent risk factor [83,84]. Breast exposure to radiation by accident, therapeutically or for screening, can also increase the risk of malignancy [85–87].

2. Molecular Types of Breast Cancer

Using gene expression profiling, breast cancer molecular types were described [88]. In order to define the types of breast cancer, the expression of three tumor markers was studied:

1. ER-estrogen receptor status.
2. PR-progesterone receptor status.

Note that HR-represents the joint assessment of ER and PR status.

3. HER2-human epidermal growth factor receptor 2 status.

This is how the main molecular subtypes of breast cancer were defined:

1. HR+/HER2– corresponding to Luminal A subtype.
2. HR+/HER2+ corresponding to Luminal B subtype.
3. HR−/HER2+ corresponding to HER2 enriched subtype.
4. HR−/HER2– corresponding to triple negative subtype.

The most frequent is HR+/HER−, accounting for about 70% of breast cancers [89]. The prognosis is different for each molecular type. Luminal A breast cancer is hormone-receptor-positive (estrogen-receptor and/or progesterone-receptor-positive), HER2 negative, and has low levels of the protein Ki-67, which helps control how fast cancer cells grow. Luminal A cancers are low-grade, tend to grow slowly and have the best prognosis: 80–85% 5-year survival [90]. Luminal B breast cancer is hormone-receptor-positive (estrogen-receptor and/or progesterone-receptor-positive), and either HER2 positive or HER2 negative with high levels of Ki-67. The expression status of proliferation linked genes is one of the most important factors of the difference between luminal A and luminal B breast cancers. Luminal B cancers generally grow slightly faster than luminal A cancers and their prognosis is worse [90].

HER2-enriched breast cancer is hormone-receptor negative (estrogen-receptor and progesterone-receptor negative) and HER2 positive. HER2-enriched cancers tend to grow faster than luminal cancers and can have a worse prognosis: approximately 50–60% 5-year survival. They are often successfully treated with targeted therapies aimed at the

HER2 protein [90]. Triple-negative/basal-like breast cancer is hormone-receptor negative (estrogen-receptor and progesterone-receptor negative) and HER2 negative, being more common in women with BRCA1 gene mutations. There is not a perfect match between basal-like breast cancer and triple-negative breast cancer [91]. Recently, seven different subtypes have been described for the triple-negative breast cancer based on analysis of gene expression profiles: basal-like 1 (BL1), basal-like 2 (BL2), immunomodulatory (IM), mesenchymal-like (M), mesenchymal stem–like (MSL), luminal androgen receptor (LAR), and unstable (UNS) [92]. The basal-like and HER2+ subtypes are more aggressive, having a higher proportion of major gene expression signatures [91]. Normal-like breast cancer is similar to luminal A disease: hormone-receptor-positive (estrogen-receptor and/or progesterone-receptor-positive), HER2 negative, and has low levels of the protein Ki-67, which helps control how fast cancer cells grow. Still, while normal-like breast cancer has a good prognosis, its outlook is slightly worse compared to luminal A cancer. Ki-67 is a nuclear antigen present in some phases of the cell cycle (mid G1, S, G2, and the entire M) that was found to be overexpressed in women with a shorter metastases-free survival period [91,93].

3. PI3K/AKT/mTOR Signaling Pathway

Cells intercommunicate in a process called extracellular signaling. They produce specific molecules that bind to specific receptors of other cells and activate intracellular signaling pathways. This is how cells respond to changes and adapt [94].

The PI3K/AKT/mTOR complex is a signaling pathway with a major role in essential cellular activities, such as: cell metabolism, cell growth, cell proliferation, apoptosis, and angiogenesis [94]. A ligand (for instance insulin or an insulin-like growth factor) binds to a cell-membrane receptor (such as receptors for tyrosine kinases or G-protein-coupled-receptors-GPCR). The specific receptor, activated by the extracellular ligand, activates PI3K (phosphatidylinositol (3,4,5)-trisphosphate kinase). The activated PI3K catalyzes phosphorylation of PIP2 at the 3 position of the inositol ring to generate PIP3, which recruits two protein kinases to the plasma membrane via their pleckstrin homology interaction domains (PH domains): AKT (also called protein kinase B, or PKB) and PDK1 (phosphoinositide-dependent protein kinase 1). Once recruited to the cell membrane, the AKT is phosphorylated by mTORC2 (mTOR complex 2) on Ser473, changing the conformation of the AKT and allowing its phosphorylation on Thr308 by PDK1. The activated AKT phosphorylates target proteins from the cell membrane, then loses its connection with the cell membrane and phosphorylates other target proteins in the cytosol and cell nucleus. The phosphorylation of target proteins results in the stimulation of cell survival, growth, and proliferation [94].

3.1. *PI3K/AKT/mTOR Signaling Pathway Members*

3.1.1. PI3K-Phosphatidylinositol 3-Kinase-(Phosphoinositide 3-Kinase)

PI3K is a plasma-membrane-bound enzyme activated by RTKs (receptor tyrosine kinases) and by GPCRs (G protein-coupled receptors). GPCRs are the largest class of cellular surface receptors, with a generic structure; each GPCR is a transmembrane single polypeptide chain that uses G proteins to transmit the signal into the cytosol [94–97]. RTKs are a large family of plasma membrane receptors, too, with intrinsic protein kinase activity [95–97]. PI3Ks phosphorylate the 3′ position of the inositol head group of phosphatidylinositol (PIP2 and PIP3) lipids. PIP3 is the effector of multiple downstream targets of the phosphoinositide 3 kinase (PI3K) pathway [95–97]. There are many phosphatidylinositol 3-kinases-PI3Ks, divided into three groups or classes: PI3Ks class I, PI3Ks class II and PI3Ks class III [98].

PI3Ks class I is subdivided into PI3Ks class IA, PI3Ks class IB and PI3Ks class IC. IA-PI3Ks are heterodimers, consisting of a regulatory unit (p85α, p85β, p85γ) that activates the catalytic unit (p110α, p110β, p110δ, p110γ [98]). The IA-PI3Ks are activated directly by cell surface receptors: G protein-coupled receptors, RTKs and the small G protein RAS. Small GTPases form a superfamily within the larger class of regulatory GTP hydrolases, while RAS proteins are small GTPases that regulate cell growth, proliferation and differentiation [94–97]. IA-PI3Ks are present in many types of tissues and are activated by G protein- coupled receptors [98,99]. IB-PI3Ks are heterodimers containing the p101 regulatory subunit, which activates the p110γ catalytic subunit [99–101].

Class II PI3Ks has three isoforms: PI3KC2α and PI3KC2β are expressed in most of the tissues and organs, while PI3KC2γ is expressed only in the liver [102]. They regulate intracellular membrane dynamics and membrane traffic [102,103]. Class III PI3Ks has only one member identified: VPS34, which is connected to regulation of phagocytosis, pinocytosis, endosomal sorting and autophagy [104].

3.1.2. AKT

The serine/threonine protein kinase AKT is the principal downstream molecule of the PI3K signaling pathway. There are three subtypes (isoforms) of AKT [105,106]: AKT1 (expressed in the majority of tissues), AKT2 (expressed mainly in tissues with high sensitivity to insulin: liver, pancreas, muscles), and AKT3 (expressed in the brain and testicles). AKT is activated by PIP2-driven and PIP3-driven recruitment to the plasma membrane. Here, the phosphorylation of Thr308 and of Ser473 determines the activation of AKT [107,108].

Activated AKT mediates the regulation of the cell cycle, growth, proliferation, and energy metabolism [109]. AKT has over 100 substrates [105], including: transcription factors, inhibitors of cell cycle progression, protein kinases, GTPase-activation proteins, and apoptosis inducers [110,111].

Glycogen synthase kinase-3 (GSK-3), one of the main AKT protein substrates, is a protein kinase that phosphorylates and inhibits the glycogen synthase. GSK-3 lies downstream of multiple cellular signaling pathways, such as: the phosphatidylinositol-3–kinase-dependent pathway that is stimulated by insulin and growth factors, and the Wnt signaling pathway that is required for embryonic development [112]. GSK-3 is primarily regulated by inhibition [113]. There are two isoforms, GSK-3α and GSK-3β, generated from distinct genes, but with great structural homology (almost 97%) and similar roles, being encountered in many tissues [112,113] and especially in the brain. AKT phosphorylates GSK-3 and inactivates it; consequently, there is an increase in the cellular uptake of glucose and glycogen synthesis. This determines a decrease of blood sugar levels [114].

The inhibition of GSK-3 triggered by growth factors, through AKT activation, has anti-apoptotic effects [113]. GSK-3 has a broad range of substrates (more than 100) including signaling proteins, structural proteins, and transcription factors involved in metabolism [115].

3.1.3. mTOR

TOR is a large protein-kinase inactivated by a bacterial toxin called Rapamycin—hence the name Target of Rapamycin. It was identified in yeasts, but it also exists in mammalian cells, being named mTOR (mammalian Target of Rapamycin) [94,116]. In cells, it exists as two distinct multiprotein complexes: mTORC1 and mTORC2. mTORC1 contains mTOR, protein Raptor and mLST8 (the acronym for mammalian Lethal with SEC13 protein 8). mLST8 interacts directly with mTOR and enhances its kinase activity, with this protein being found in human colon and prostate cancer cells [117]. mTORC1 is sensitive to Rapamycin and promotes cell growth and survival by stimulating nutrient uptake and metabolism [94,116]. It also stimulates cell growth by promoting ribosome production and protein synthesis and by inhibiting protein degradation [94,116]. mTORC1 may be activated through different pathways, but mainly through the PI3P/AKT pathway, which is activated by extracellular growth factors and nutrients. Activated AKT phosphorylates the Tuberous Sclerosis protein 2 (TSC2), which becomes inactive. Thus, TSC2 cannot keep Rheb (a Ras-related GTPase) in its inactive form. Consequently, Rheb-GDP (inactive) becomes Rheb-GTP (active), contributing to the activation of mTORC1 [94,116,118]. Other targets of the mTORC1 are: S6K (a protein kinase that phosphorylates the ribosomal protein S6) and 4E-BP (an inhibitor of the translation initiation factor eIF4E); the consequences are increased production of ribosomes and increased protein synthesis [118].

mTORC2 consists of mTOR, protein Rictor, Sin1 and mLST1 and is not sensitive to Rapamycin. mTORC2 promotes AKT activation by directly phosphorylating its hydrophobic motif (Ser473). This permits further phosphorylation of AKT, at Thr308, by PDK1 and so AKT becomes fully active [94,116]. Sin1 contains a phospholipid-binding pleckstrin homology (PH) domain that facilitates the association of mTORC2 with membranes [119]. Phosphorylation of Sin1 at Thr86 and Thr398 (by S6K or AKT) dissociates Sin1 from mTORC2, thus resulting mTORC2 inhibition [120]. Acting on S6K, mTORC1 directly regulates mTORC2 [120]. mTORC2 is mainly involved in the reconstruction of the cytoskeleton (through the Rho family GTPases) and cell survival [94].

3.1.4. FoxO1

Forkhead box other 1 is a member of the Forkhead transcription factor family. The family is divided into 17 subfamilies named FoxA to FoxQ [121]. There is a common feature of the Forkhead family, namely a conserved DNA-binding domain called Fox [122].

FoxO1 is important for the glucose and lipids' metabolism. It enhances the synthesis of enzymes involved in gluconeogenesis, has a suppressive effect on the synthesis of enzymes of glycolysis, inhibits the pentose phosphate pathway, and diminishes the triacylglycerol synthesis. Insulin activates the RTKs and initiates the PI3P/AKT pathway. The activated AKT phosphorylates the FoxO1 existing in the cytosol. The phosphorylated FoxO1 is tagged by the attachment of ubiquitin and is then degraded by proteasomes. The unphosphorylated FoxO1 remains active, passes from the cytosol into the nucleus, binds to a response element, and triggers the transcription of its associated genes, such as PEP-carboxykinase, glucose 6-phosphatase, etc. FoxO family members have an important role in oxidative stress resistance, cell proliferation, apoptosis, and differentiation [123,124].

3.1.5. PTEN

PTEN (phosphatase and tensin homolog) is a PIP3 specific phosphatase that dephosphorylates the PIP3 molecules, resulting in PIP2 molecules (Figure 1). PIP2 is not a binding dock for AKT, so AKT cannot be recruited to the cell membrane. As a consequence, AKT cannot be phosphorylated by mTORC2 on Ser473; therefore, the conformation of the AKT does not change anymore and the phosphorylation on Thr308 by PDK1 is not permitted. The result is that AKT cannot be activated and the PI3K/AKT/mTOR signaling pathway is suppressed [125,126]. That is why PTEN acts as a tumor suppressor, by inhibiting cell proliferation [125,126]. In many malignant tumors, the PTEN gene has suffered mutations, resulting in abnormal PTEN, which cannot exert its inhibitory effect on the PIP3/AKT/mTOR pathway [127]. The plasmatic levels of PIP3 rise and the activity of AKT is continuously stimulated [128]. By modulating the PIP3/AKT/mTOR pathway, PTEN is linked to glucose homeostasis [129].

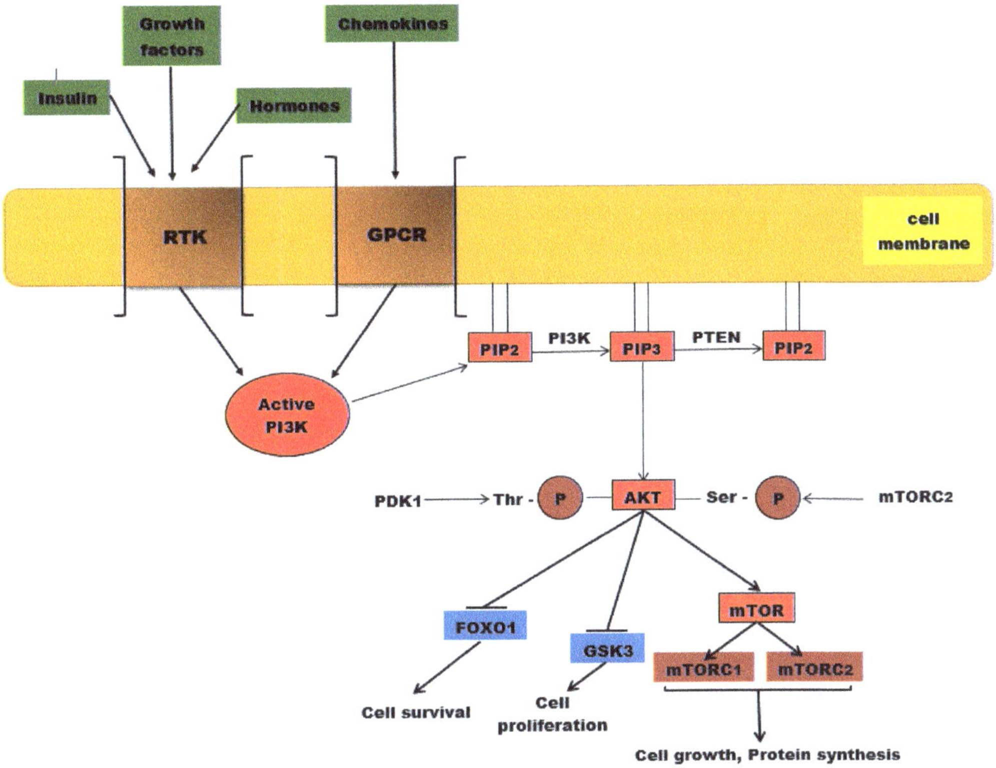

Figure 1. The PI3K/AKT/mTOR signaling pathway. PI3K is activated by the binding of ligands (insulin, growth factors, hormones) to RTKs, but also to GPCR (chemokines). Once activated, this protein kinase will catalyze the phosphorylation of PIP2 to PIP3. AKT is recruited to the plasma membrane where it undergoes two phosphorylation processes, one catalyzed by PDK1 at the level of threonine residue and the second reaction being catalyzed by mTORC2. Once activated by phosphorylation, AKT will phosphorylate other substances such as the mTOR complex, which will be associated in the end with protein synthesis and cell growth. Other phosphorylated substrates, such as GSK-3 and Fox01, will be inhibited, associated with cell proliferation and survival. PTEN is the major negative regulator of this signaling pathway involved in PIP3 dephosphorylation.

4. PI3K/AKT/mTOR Mutations in Breast Cancer

The PI3K pathway undergoes many changes in breast cancer caused by mutations or amplifications of genes which encode the catalytic subunits p110α (PIK3CA) and p110 β (PIK3CB), but also the regulatory subunit PI3K, p85α (PIK3R1) [130]. In human neoplasms, PIK3CA is the frequently mutated gene that encodes the p110α catalytic subunit of the PI3K pathway, and was found amplified in head and neck, cervical, gastric, lung and breast cancers. In prostate, breast, endometrium and colon cancers, the highest incidence of PIK3CA mutations has been detected [131].

Approximately 30–40% of patients with breast cancer present PIK3CA mutations, which will induce hyperactivation of the α isoform (p110α) of PI3K. Recently, the FDA (Food and Drug Administration) approved testing of breast cancer patients with PIK3CA mutations using breast tumor tissue and/or circulating tumor DNA, isolated from plasma specimens. The results reported 11 PIK3CA hotspots mutations, located mainly in exons 9 and 20. Gene PIK3CA mutations have been detected using a PCR test, with the results revealing the following exon mutations—exon 9: E542K, E545A, E545D, E545G, E545K, Q546E, and Q546R; and exon 20: H1047L, H1047R, and H1047Y [132]. PI3Kα is activated both by binding insulin or growth factors to RTKs and by oncogenic mutations [133]. In breast cancer, the PI3K/AKT pathway is activated through PIK3CA or AKT1 mutations and PTEN loss [134].

In 2004, Samuels Y and co-workers reported somatic mutations of PIK3CA coding p110α in various solid malignancies for the first time [135]. Samuels Y et al., observed that the majority of PIK3CA somatic mutations are located at the level of exon 9 (E542K or E545K) and exon 20 (H1047R or H1047L). In the helical domain of p110α, there are exon 9 mutations that are considered to enable p110α to escape the inhibitory effect of p85 via the Src-homology 2 (SH2) domain. Near the activation loop of the kinase domain, mutations of exon 20 are located. The study reported 10% frequency of PIK3CA somatic mutations in breast cancer, but later studies reported ~30%. [136]. Karakas B et al., also reported that the catalytic subunit of the PI3K gene called PIK3CA or p110α is frequently mutated in breast cancer [137].

In the coding sequence, PIK3CA mutations are concentrated in three hotspots, with two located in the helical domain of p110α and the last situated in the catalytic domain. The hotspot mutations represent the single nucleotide substitutions that will determine amino acid substitutions, E542 K, E545 K and H1047R. Unfortunately, these hotspot mutations induce a gain-of-function and prompt transformation and tumorigenicity. The results from 6338 tumors revealed that 2261 patients presented PIK3CA mutations (35.7%) [138].

A total of 73% of all PIK3CA mutations are: H1047R (35%), E545K (17%), E542K (11%), N345K (6%), and H1047L (4%). In patients with triple negative breast cancer, PIK3CA mutation rates were decreased (16%) compared to HR+/HER2 (42%) and HER2+ (31%) breast cancer subtypes. Moreover, in patients with advanced HR+/HER2−breast cancer, 28% of PIK3CA mutations were identified in circulating tumor DNA [132].

Tumor sequencing studies have reported that these somatic mutations of PI3CA, concentrated in certain hotspots, will lead to tumor progression by gaining a function for PI3CA [139]. Moreover, PIK3CA mutations in human breast cancers, at E545K in exon 9 and H1047R in exon 20, have been reported even by studies using cell lines such as MCF10A immortalized breast epithelial cells. PIK3CA was the most frequent mutation observed, associated with an increased kinase activity of the PI3K pathway. Mutant PIK3CA promotes cell growth and invasion of human cancer cells [136].

Bachman KE and co-workers reported an incidence of 25% PIKCA mutation in human breast cancer. The study did not reveal any correlation between PIK3CA and the presence or absence of ER/PR labelling, or even with Her-2/neu. PI3CA mutations affect the PI3K/AKT/mTOR signaling pathway independent of ER/PR and Her-2/neu. Analyzing the fifty-three samples, the study reported three mutations in exon 9, 8 uncovered mutations in exon 20, and novel somatic mutations were detected—two in exon 1 and one in exon 2 [140]. Stemke-Hale K and colleagues analyzed 547 breast tumor samples and 41 cell lines using mass spectrometry sequencing and reverse-phase protein arrays to detect mutations in PI3KCA, AKT and PTEN. The study revealed that the most common PIK3CA mutations were found in hormone receptor-positive forms (34.5%) followed by HER2- positive cases with an incidence of 22.7%, compared with basal-like tumors (8.3%). Moreover, in hormone receptor-positive cancers, mutations on AKT1 represented 1.4% and PTEN 2.3%, respectively [141]. Using cell cultures, the study reported that AKT1 mutations were absent, while PIK3CA and PTEN mutations appeared in 39% and 20% of the cases, respectively. In tumors and cell lines, PIK3CA mutations compared with the loss of PTEN and AKT1 mutations were associated with less activation of AKT. The most frequent modifications on the PI3K/AKT/mTOR signaling pathway were PTEN loss and PIK3CA mutation [141].

Li SY. et al., analyzed 250 primary human breast tumors and detected that 35% of PIK3CA mutations were located in C2 helical and kinase domains. The PIK3CA mutations were associated with larger tumors and significantly worse survival rate, especially in positive estrogen receptor status or non-amplified ERBB2 [142]. Moreover, PIK3CA mutations may sometimes harbor PTEN loss or HER2 overexpression in breast tumors [130].

p110α, the catalytic subunit of the phosphoinositide 3-kinase alpha (PI3Kα) complex, which is necessary for normal growth and proliferation, [133] is essential for signaling and the growth of tumors driven by PIK3CA mutations or RTKs [130]. It has been shown that p110β mediates tumorigenesis in PTEN-deficient cells [139]. Breast cancers show poor disease outcome if they are associated with increased levels of AKT phosphorylation/activation and PTEN loss. Moreover, the loss of PTEN activity and activation of the PI3K signaling pathway are associated with resistance to endocrine therapy [134]. Endometrial, prostate, breast, thyroid and kidney tumors present somatic PTEN alterations, leading to uncontrolled PI3K activation [143]. PTEN, the most important regulator of the PI3K/AKT/mTOR signaling pathway, is involved in cell growth and survival, cellular migration and genomic stability. In 1997, it was discovered that PTEN acts as a key tumor suppressor gene for various tumor types, being involved in cell cycle progression, cell growth and survival. Moreover, PTEN is implicated in DNA repair and genome stability. In response to DNA damage, PTEN is phosphorylated (Tyr) and binds to chromatin, promoting DNA repair [144].

The somatic mutations (missense and nonsense mutations, monoallelic or biallelic deletion on the PTEN gene), epigenetic alterations (methylation promotor), PTEN protein degradation and the post-translational modification of PTEN protein will conduce to PTEN inactivation. In breast tumors, the loss of heterozygosity at the PTEN locus was detected in 40–50% cases. The loss of PTEN function due to PTEN mutations is found in 5–10% of breast cancers [144].

In luminal breast cancers, the PI3K pathway is one of the most altered pathways, correlated with PIK3CA mutations, loss of PTEN, or downstream protein phosphorylation [145]. Zardavas D et al., reported the results obtained from 10,319 patients included in 19 studies where PIK3CA mutations were present in 32% of patients. PIK3CA mutations were associated with ER positivity, and were increased with age, lower grade, and smaller size. In breast cancer subtypes-ER-negative/HER2-negative, HER2-positive, and ER-positive/HER2-negative, the prevalence of PIK3CA mutations was 18%, 22%, and 37%, respectively [146]. Ling D et al., conducted a study in which tumors from 507 breast cancer patients were collected from the West China Hospital between 2008 and 2013. The study's results revealed 3.% AKT1 mutations with ER+/PR+/HER2. The incidence of the PIK3CA mutations was reported at 46.5%. These mutations were associated with ER+/PR+/HER2- status, and it was observed that 35 patients carried two or three variants of the PIK3CA gene [147]. PIK3CA mutations, associated with many distinct cancers, include hotspot single–amino acid substitutions in the helical (E542K and E545K) or kinase (H1047R) domains. In multiple cancer types, including breast cancer, PIK3CA is considered oncogenic, mutations of the alpha catalytic subunit of PI3K having an incidence of 40% in ER+/HER2− primary and metastatic tumors. Therefore, PIK3CA is a target for cancer therapy [134]. Anderson EJ et al., reported 36% PIK3CA mutation in HR+/HER2- metastatic breast cancer [148].

Mutations also occur in RTKs such as HER2 (ERBB2) and fibroblast growth factor receptor (FGFR)1, in AKT1, AKT2, PDK1, and loss of PTEN and INPP4B (inositol polyphosphate-4-phosphatase type II). The activation of PI3K occurs through the binding of growth factors to RTKs and GPCR. Moreover, PIK3CA mutations appear in breast tumors associated with PTEN loss or HER overexpression [130].

Lehmann BD and co-workers detected highly clonal PIK3CA mutations in the triple negative breast cancer subtype that present a luminal phenotype and express androgen receptors (40%) versus triple negative breast cancer without androgen receptors (4%) [149]. A total of 15–20% of breast cancer cases present an overexpression of human epidermal growth factor receptor-2 (HER2), associated with an aggressive clinical behavior [150]. Luminal A tumors are associated with PIK3CA mutations in 45% of the cases, while AKT1 and PTEN mutations both appear in 4% of the patients. At the same time, PIK3CA genes are mutated in 29% of the cases with subtype luminal B, in 39% of HER2-enriched breast cancers and only in 7% of basal-like tumors [151].

HER Receptors and Breast Cancers

EGFR (epidermal growth factor receptor, also known as ERBB1/HER1), ERBB2 (HER2), ERBB3 (HER3), and ERBB4 (HER4) represent the ERBB family of RTKs, which are cytoplasmic membrane-anchored proteins. All four receptors display similarities in structure and sequence, contain an extracellular ligand-binding domain, a transmembrane domain, and an intracellular tyrosine kinase domain [152].

Cell growth, survival, and differentiation are regulated by HER receptors via various signaling pathways and even participate in cellular proliferation and differentiation. The HER2 gene encodes a 185-kDa transmembrane protein, being located on the long arm of chromosome 17 [153]. When HER2 is overexpressed or amplified, it stimulates tumor growth, invasiveness, and survival via the activation of several signaling cascades, such as PI3K/AKT pathways. HER2 phosphorylation may lead to PI3K/AKT/mTOR pathway activation [154].

The formation of HER2-EGFR dimers, HER2 homodimers and even HER2-HER3 dimers will promote tumor development by increasing tumor cell metabolic functions, cell survival, proliferation and invasiveness [155]. In breast cancer, overactivation of HER receptors is caused by several factors such as gene amplification, truncation of the extracellular domain, mutations in the kinase domain, and co-expression of HER receptor ligands [156]. HER2 overexpression is associated with poor clinical outcome and disease progression [152].

In primary invasive breast cancer, approximately 18–20% of cases present an amplification or overexpression of the HER2 oncogene [154]. HER 2 (c-erbB-2) is a cell membrane surface-bound RTK, while HER2/neu, its extracellular domain, is normally implicated in the signal transduction pathways that will conduce to cell growth and differentiation. In approximately 15–20% of breast cancer cases, HER2-overexpression was observed [157]. HER2 overexpression and PIK3CA mutations have been observed in both invasive breast cancers and ductal carcinoma in situ. In intraepithelial neoplastic lesions, PIK3CA mutations have a decreased frequency, so these mutations can enhance PI3K pathway activation by HER 2 (ERBB2) [130].

EGFR, HER3, and HER4 are amplified and overexpressed in more than 20% of breast cancers. Moreover, HER2 is the oncogenic driver of these pathologies, involved in the genesis and progression of these tumors [158]. EGFR and HER4 can activate PI3K after their binding to RTKs, especially by transphosphorylation of HER3, which can act as a critical partner for HER2 in the genesis and progression of the tumor [158]. After phosphorylation of tyrosine residues within the cytoplasmic domain, dimerization of the receptor takes place and various signaling pathways are activated, which are further involved in cellular proliferation, transcription, motility, and inhibition of apoptosis [154].

Yang Z et al., conducted a study that included 142 patients with metastatic breast cancer, detecting alterations in estrogen receptor (ER), progesterone receptor (PR), and HER2 status as follows: 20.70%, 37.78%, and 11.48%, respectively [159].

5. Mechanisms of Endocrine Resistance in Breast Cancer

Breast cancer, the most common form of cancer in women, has a very high mortality rate, causing the death of a woman every 13 minutes. One of the most important factors involved in breast cancer development and progression is represented by the expression of proteins for hormone receptors, with estrogen positive breast cancers representing seventy percent of the total cases. Therefore, endocrine therapy plays a crucial role in breast cancer therapy [160–165].

HER2+ breast cancers account for 15% to 20% of all cases and are treated primarily with drugs that target HER2 (trastuzumab, pertuzumab). In addition, half of these cases are also ER positive; therefore, these women are perfect candidates for endocrine therapy, despite the shorter and lower response rate compared to HER2 negative breast cancers [166].

However, de novo and acquired resistance appear in all cases of metastatic breast cancer and in approximately 25% of ER positive breast cancer patients, limiting the efficiency of the treatment used to target the estrogen receptor [162,166]. Although the exact mechanisms that lead to endocrine resistance have not been identified, there are several theories revealing cell cycle changes and alterations of the ER pathway as causes for endocrine resistance [166]. Moreover, studies have shown that growth factor receptor signaling pathways are involved in the development of this aggressive pathology. In addition, PI3K mutations or loss of heterozygosity, methylation of PTEN, and AKT activation promote hormonal therapy resistance, thus the new treatment protocols are based on the use of medicines targeting not only the estrogen receptor but also these signaling pathways [162,167].

Previous studies have shown that ligand-independent estrogen receptor activity caused by mutations in the encoding gene for ER (ESR1) can lead to endocrine resistance through an increased number of mutant clones [166].

Estrogen positive breast cancers are currently treated with three types of agents: selective estrogen receptor modulators (tamoxifen), estrogen synthesis inhibitors (aromatase inhibitors) and selective estrogen receptor down-regulators (fulvestrant) [167].

The main goal of endocrine targeted therapy is to remove the endogenous activating ligands of the estrogen receptors. Tamoxifen and other antiestrogens (such as fulvestrant) fulfil their role through competitive inhibition, while aromatase inhibitors (letrozole, anastrozole) block estrogen synthesis [166,168].

Tamoxifen was the first therapeutic agent targeting cancer on a molecular level, showing great results in women with breast cancer, especially in estrogen receptor-positive premenopausal women. Although tamoxifen has proven to be a very efficient drug in preventing recurrence, the estrogen receptor-positive subtype remains the most aggressive type of breast cancer [160]. Tamoxifen is an antiestrogen that performs as a partial agonist and has been the standard of care for premenopausal women for many decades. However, aromatase inhibitors have been proven to increase the survival rate for postmenopausal women and have replaced tamoxifen as the main therapy [166].

Previous studies indicate that estrogen receptor breast cancers have a low recurrence rate, but the risk increases over 3 to 5 years after the initial treatment. This late recurrence, called dormancy, is often associated with ER positive breast cancers and could be determined by the therapeutic agents used to treat the disease [166].

HER2 has been incriminated in many important pathways involved in tamoxifen resistance, with an increased expression of HER2 being associated with resistance to hormonal therapy [169]. In this case, tamoxifen can perform both as an agonist or an antagonist, depending on the recruitment of coactivators or repressors of the estrogen receptor α transcription complex. In the presence of HER2, the augmented expression of AIB1 (amplified in breast cancer 1 protein), a regulator of the estrogen receptor α, leads to tamoxifen resistance [170].

Moreover, growth factor receptors such as IGF1R (insulin-like growth factor receptor 1) and EGFR (epidermal growth factor receptor) can cause a lack of response to tamoxifen by activating the MAPK (mitogen-activated protein kinase) and the PI3K signaling pathway [170,171].

The cross-talk between these receptors and ER is very complex. MAPK leads to estrogen-independent phosphorylation, with AKT playing an essential role for ERα [171]. Activation of the PI3K signaling pathway and AKT phosphorylation promotes estrogen-independent growth in tumor cells and resistance to anti-estrogens. The overexpression of HER2, FGFR1 or loss of INPP4B (inositol polyphosphate-4-phosphatase type II) was also observed in tamoxifen-resistant cells. PI3KCA (the alpha catalytic subunit of PI3K) is more often affected in estrogen positive breast cancers, while ER negative breast cancers are characterized by PTEN loss [161].

Additionally, it has been demonstrated that tamoxifen dysregulated metabolism (caused by cytochrome P450 proteins' polymorphism), cellular accumulation of the drug, hypermethylation of CpG islands and expression of P-glycoprotein, and the activity of the histone deacetylase promote tamoxifen resistance as well. Mechanisms of aromatase inhibitor resistance implicate the PI3K pathway, MAPK, HER2 and the estrogen receptor [170].

Recent studies have shown that microRNA can promote unresponsiveness to endocrine therapy, hence new therapies could be developed to target these small RNA molecules [170]. Hoppe R et al., have discovered that miR-126 and miR-10a are overexpressed in estrogen receptor-positive breast cancers. Moreover, they found a correlation between this overexpression and the amount of time without recurrences. However, mi-R221 and miR-222 are associated with tamoxifen resistance through the decrease in estrogen receptor protein expression in tumor cells [172]. Moreover, miR-451, a microRNA that acts on PI3K/AKT and controls P-glycoprotein, presents a decreased expression in MCF-7 tamoxifen-resistant and doxorubicin-resistant cells [173].

Moreover, epithelial to mesenchymal transition is coded by genes that are active during embryogenesis tissue formation and wound healing, but also during carcinoma progression. Transcription factors (FOXC2, ZEB1/TCF8, E12/E47 and many others) determine abnormal survival via PDGFR (platelet derived growth factor), FGFR (fibroblast growth factor receptor) and EGFR, but also through PI3K, AKT and mTOR [174].

In order to improve breast cancer prognosis, new therapeutic strategies have to use combinations of drugs targeting ER and HER2, but also the downstream signaling pathways [167].

The combination of tamoxifen with PI3K inhibitors augments the effects of the antiestrogens, thus reflecting the influence of the PI3K/AKT pathway in acquired endocrine resistance [161]. In addition, studies have shown that the association of PI3K/mTOR inhibitors (BEZ235) improved the outcome of the treatment, showing better results than when the drugs were administered individually [175]. Moreover, Cavazzoni A et al., found that the addition of everolimus, an mTORC1 inhibitor, improved the effects of letrozole [176]. On the other hand, PI3K inhibitors are in the initial phase of development [177].

6. mTOR Inhibitors: Everolimus and Temsirolimus

The PAM (phosphoinositide 3 kinase (PI3K)/AKT/mammalian target of rapamycin (mTOR)) pathway is often altered in cancers, being involved in more than 70% of breast cancer cases. Studies show that PIK3CA2 is the most commonly mutated oncogene in ER positive breast cancers (in 35% of the clinical cases) and is frequently involved in HER positive forms of breast cancer. In addition, several studies reflect a correlation between the activation of the PAM pathway and resistance to endocrine therapy. Therefore, an increased number of clinical trials have focused on inhibitors of this key signaling pathway that is involved in essential cellular processes, such as proliferation and metabolism, being imperative for cellular survival. Although the studies have shown great perspectives for a series of PI3K and AKT inhibitors, the adverse effects of these compounds have led to the limitation of clinical trials to only one mTOR inhibitor: everolimus [178].

Everolimus is an oral rapalog (rapamycin analog) approved by the FDA as an antitumor agent in ER positive/HER negative breast cancer [178,179]. RAD-001 (40-O-(2-hydroxyethyl)–rapamycin), better known as everolimus, exerts its effects on mTOR by binding cyclophilin FKBP-12 associated with raptor and mLST8, thus acting as an inhibitor for downstream signaling [179,180].

mTOR is downstream of PI3K/AKT and consists of 2 complexes (mTORC1 and mTORC2) that function differently despite their similar structure. mTORC1 promotes mRNA translocation, protein synthesis and lipid synthesis, thus stimulating cell growth, while mTORC2 is involved in AKT phosphorylation and cellular organization. Rapalogs, just as rapamycin, target mTORC1 and can produce the phosphorylation of the activation function domain 1 of the estrogen receptor via substrate ribosomal S6K1 (S6 kinase-1), hence leading to the activation of the ligand-independent receptor [179].

In addition, mTORC1 helps mRNA translation by producing the dissociation of 4E-BP1 from eIF4E through phosphorylation and causing the formation of a pre-initiation translation complex by associating eIF4E with eIF4G (a scaffolding protein) and initiation factors. Thus, in stress conditions, mTORC1 represents a restriction point in the cells and is a key target for cancer therapy [178].

Interestingly, AKT can be hyperphosphorylated by rapalogs. mTORC1 inhibition and inadequate inhibition produce activation of AKT and cause cell proliferation [178,179]. The pharmaceutical properties of everolimus are superior to rapamycin. This hydroxyethyl ether derivative of rapamycin does not inhibit mTOR at pharmacologically achievable drug concentrations [179]. Rapalogs' influence on mTORC2 is very controversial. While some specialists consider that everolimus can exert its effects on both mTOR complexes, most authors consider that this agent can only act on mTORC1 (Figure 2) [181].

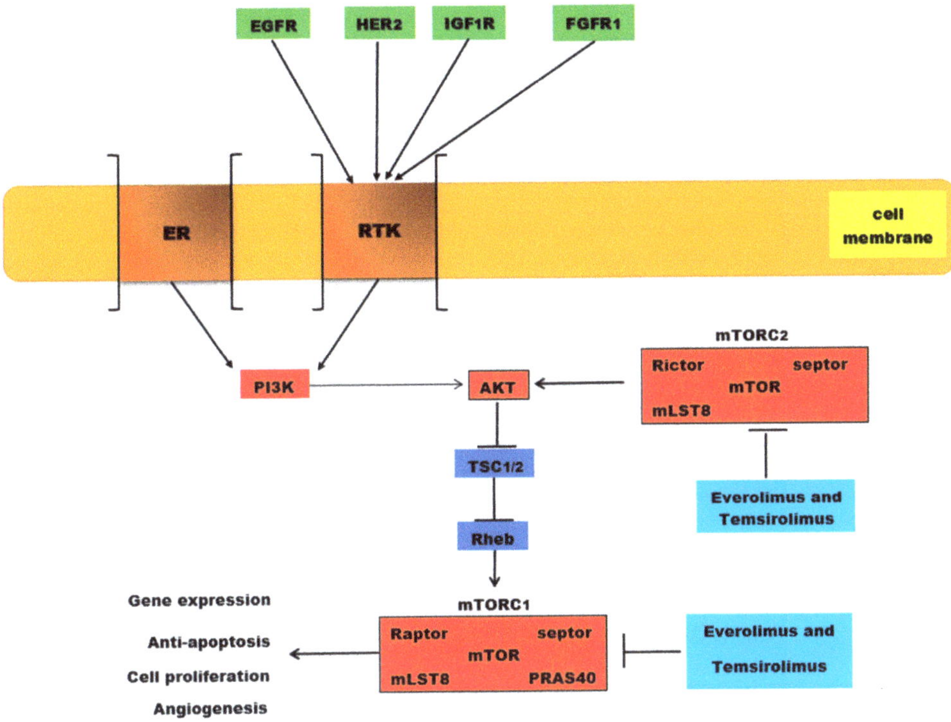

Figure 2. The PI3K/AKT/mTOR signaling pathway and breast cancer. The PI3K/AKT/mTOR signaling pathway is activated by ER, but also by EGFR, HER 2, IGFR1R, and FGFR1 at RTKs level. Once activated, protein kinase B or AKT inhibits TSC $\frac{1}{2}$ by phosphorylation, further leading to the inhibition of Rheb and activation of mTORC1. This activation is associated with anti-apoptotic effects, increased gene expression, cell proliferation and angiogenesis. Everolimus and temsirolimus are two analogues of rapamycin that inhibit the activity of mTOR, especially mTORC1, but also mTORC2.

Bachelot et al., analyzing the effects of tamoxifen and of tamoxifen combined with everolimus, showed that the clinical benefit rate and time to progression were significantly improved by the rapalog [182].

Balsega J. et al., evaluated 724 patients in the BOLERO (Breast Cancer Trial of Oral Everolimus)-2 study, assessing the effects of everolimus and exemestane (an aromatase inhibitor) in postmenopausal women with advanced stages of hormone positive receptor breast cancer. Their results showed that everolimus improved the progression free survival rate. However, the combination of the two therapeutic agents also showed adverse effects, such as anaemia, pneumonitis or dyspnea, causing the withdrawal of everolimus [180].

The BOLERO-3 study evaluated the effect of everolimus in HER2 positive, trastuzumab-resistant breast cancer. This randomized study Phase III trial included 569 women with advanced breast cancer and resistance to HER-targeted drugs. The results showed that the addition of everolimus to a combination of trastuzumab and vinorelbine (a chemotherapy medication) significantly improved the progression free survival of the patients [183]. Additionally, everolimus showed great promise in association with trastuzumab and paclitaxel in HER2-positive advanced breast cancers [184].

The MANTA trial (phase 2 randomized clinical trial) is a study that included 333 women with ER positive breast cancer, priorly treated with aromatase inhibitors. The participants were divided into 4 categories depending on the treatment they received: group 1 comprised of 67 women that received fulvestrant, the second group was represented by 103 patients that received fulvestrant and vistusertib (a dual mTORC1 and mTORC2 inhibitor) every day, the third group comprised of 98 patients that intermittently received the combination of fulvestrant and vistusertib, while the fourth group was formed by 65 patients that received fulvestrant and everolimus. This study was conducted in 9 countries and it demonstrated that everolimus in combination with fulvestrant significantly improved progression free survival, thus also reflecting the superior therapeutic effect of rapalogs over mTOR dual inhibitors [181].

Moreover, studies have shown that everolimus enhances letrozole effects, blocking the breast cell cycle and stimulating apoptosis [185]. On the other hand, another study that included 120 women with hormone receptor-positive metastatic breast cancer treated with endocrine therapy and more chemotherapy agents revealed that no significant benefits were obtained by adding everolimus to the therapeutic strategy [186].

Temsirolimus (CCI-779), another rapamycin analog, is converted into rapamycin in vivo [133]. Yu et al., highlighted the sensitivity of inhibitor CCI-779 to MCF-7 cells, determined by the amplification of a kinase (p70S6) downstream of AKT regulated by mTOR [187]. The HORIZON trial, a randomized study, showed that temsirolimus in combination with letrozole did not improve the progression free survival of patients with metastatic hormone receptor-positive breast cancer compared to letrozole alone [188]. Similar results were observed by Fleming et al., who studied the effects of temsirolimus in women with advanced breast cancer and found no significant improvement in the progression free survival rate [189] (Table 1). However, a study conducted by Sadler showed that when used in combination with ERA−923, an ER antagonist, temsirolimus displays promising results in ER positive breast cancers [190].

Dual mTOR inhibitors manifest antitumoral activity through the inhibition of mTORC1 and mTORC2, thus representing a promising strategy for breast cancer treatment. However, while the first generation of inhibitors acting on mTORC1 has been approved in clinical trials, dual inhibitors are still tested on cell cultures and animal models.

Vistusertib (AZD2014) is a dual inhibitor of both mTORC1 and mTORC2 that, compared with everolimus, has demonstrated more complete growth inhibition and cell death both in vitro and in vivo, based on a greater inhibitory function against mTORC1 and additional inhibition of mTORC2, especially in ER-positive breast cancer models. In preclinical models, vistusertib induces rapid tumor regression [181]. Bhattacharyya GS et al., conducted a Phase I/II trial that included 400 patients divided into two groups, diagnosed with hormone receptor-positive and HER2-negative breast cancer, who received tamoxifen and sirolimus 2 mg daily. The results of the study revealed that this combination of sirolimus and tamoxifen was effective and well tolerated by breast cancer patients [191]. In 2015, Seiler M and co-workers published the results of the phase IIb trial that consisted of patients with human epidermal growth factor receptor 2–positive (HER2+) trastuzumab-refractory metastatic breast cancer, who received a daily dose of ridaforolimus and trastuzumab. The study observed that the combination ridaforolimus–trastuzumab was well tolerated. Moreover, in trastuzumab-resistant HER2 positive metastatic breast cancer patients, this combination has antitumor activity [192].

Table 1. The main mTOR inhibitors, used in various clinical trials in patients with different breast cancer types.

mTOR Inhibitors	Type of Breast Cancer	Type of Study	References
Everolimus + exemestane	hormone-receptor-positive advanced breast cancer	Phase 3, randomized trial	[180]
Everolimus + fulvestrant	estrogen receptor-positive breast cancer	Phase 2 Manta trial	[181]
Everolimus + tamoxifen	metastatic breast cancer	Phase II Randomized trial	[182]
Everolimus + plustrastuzumab + vinorelbine	HER2-positive breast cancer	Phase 3 trial (Bolero-3)	[183]
Everolimus + trastuzumab + paclitaxel	HER2-positive advanced breast cancer	Phase 2 multicenter study	[184]
Everolimus	metastatic breast cancer	Retrospective study	[186]
Temsirolimus + letrozole	hormone receptor-positive metastatic breast cancer	Phase III randomized trial	[188]
Temsirolimus	metastatic breast cancer	Phase II trial	[189]
Sirolimus + Tamoxifen	hormone receptor-positive and HER2-negative breast cancer	Phase I/II trial	[191]
Ridaforolimus + trastuzumab	Human epidermal growth factor receptor 2–positive (HER2+) trastuzumab-refractory metastatic breast cancer	Phase IIb trail	[192]

Furthermore, the second-generation mTOR inhibitors are able to inhibit the kinase activity of both mTORC1 and mTORC2, having a more potent anticancer activity compared to rapalogs. In vivo and in vitro experiments revealed that AZD8055 is an mTOR kinase inhibitor with antitumor activity. Moreover, AZD8055 can treat breast cancer resistance to endocrine therapy agents, such as tamoxifen and fulvestrant. Shi JJ et al., detected in breast cancer cells that AZD8055 may overcome tamoxifen resistance [193]. Jordan NJ et al., used in vitro breast cancer cells (MCF7-X) that were treated with everolimus (RAD001) or AZD8055 alone or combined with anti-hormone fulvestrant. RAD001 presented a poor growth inhibitory effect on cells, rapidly inhibiting mTORC1 but not mTORC2. In contrast, AZD8055 rapidly inhibited both mTORC1 and mTORC2, and displayed a powerful inhibitory effect on cells' growth [194]. So far, another dual mTOR inhibitor, MLN0128 is used only in vitro on cell models against everolimus-resistant breast cancer, and it inhibits the AKT phosphorylation. MLN0128 may suppress the proliferation of this kind of cell [195]. Bostner J and co-workers detected that raptor protein expression in the nucleus was increased in ER/PgR-positive and HER2-negative tumors with low grade, further associated with the luminal A subtype. Moreover, raptor seems to stimulate the growth of the luminal A subtype and may be a possible target along with endocrine treatment [196]. Zhu L et al., treated human breast cancer cell lines (MCF-7 and ZR-75–1) with tamoxifen or rapamycin, to observe if ER positive breast cancer cell growth is inhibited. It was observed that rapamycin enhanced the effects of endocrine therapy with tamoxifen. In vivo treatment of cells with rapamycin plus tamoxifen significantly inhibited tumor growth [197]. mTOR is involved in PI3K/AKT signaling pathways, being associated with cell survival, proliferation, metabolism, and angiogenesis, and being abnormally activated in breast cancer. mTOR inhibitors have been developed to enhance the antitumor activity through complete mTORC1 inhibition and mTORC2, which promotes AKT activation by phosphorylation.

Corroborated, these results exhibit the key role played by mTOR inhibitors, especially by everolimus, in the treatment of ER positive/HER2 positive breast cancers with endocrine resistance.

7. Conclusions

Unfortunately, breast cancer has a high risk of mortality among women, with seventy percent of estrogen positive cases; therefore, endocrine therapy is crucial for this kind of neoplasia. Many factors are involved in breast cancer pathologies—some of them are risk factors, others are considered to be protective factors or can be without any influence on breast cancer.

The PI3K/AKT/mTOR signaling pathway is activated by enzyme-linked receptors and is of paramount importance in cell differentiation, proliferation, energetic and glucose metabolism, apoptosis, cellular response to oxidative stress, and angiogenesis. In breast cancer, the PI3K pathway presents mutations of genes which encode the catalytic and the regulatory subunits. The most frequent mutations are located in exon 9 and 20, identified from tumor tissue and/or circulating DNA in all breast cancer subtypes. Breast cancer mutations also appear in RTKs receptors such as HER 2, and phosphorylation of this receptor leads to PI3K/AKT/mTOR activation.

PI3K mutations, PTEN methylation and AKT activation will result in hormonal therapy resistance. Everolimus is a rapamycin analog approved by the FDA which inhibits the mTOR complex, involved in mRNA translocation, protein and lipid synthesis, promoting cell growth and cellular organization. Regarding another rapamycin analog, temsirolimus, the clinical studies conducted so far have reported promising results in ER positive breast cancers.

In conclusion, the investigation of mutations that occur in the PI3K/AKT/mTOR signaling pathway, but also of its inhibitors, may be a real benefit for patients diagnosed with breast cancer. By inhibiting cell growth and proliferation, these drugs could play an essential role in malignant cells' death. Therefore, future studies should focus on reducing the side effects of dual inhibitors in order to optimize the efficiency of these drugs. Moreover, new combinations of the inhibitors of these crucial pathways involved in the development of breast carcinoma could provide a new perspective for the management of breast cancers, especially for the cases with resistance to endocrine therapy. Furthermore, genetic profiling of the patients could lead to a better case selection in which PAM inhibitors would prove to be the key therapeutic agents, particularly for patients with poor prognosis.

Funding: This research received no external funding.

Conflicts of Interest: The authors declare no conflict of interest.

References

1. Breast Cancer in Men—CDC Report 11 August 2020. Available online: www.cdc.gov/cancer/men (accessed on 6 October 2020).
2. Sancho-Garnier, H.; Colonna, M. Épidémiologie des cancers du sein: Breast cancer epidemiology. *Presse Med.* **2019**, *48*, 1076–1084. [CrossRef]
3. Graham, A.C. Breast Cancer Epidemiology and Risk Factors. Medscape Report 26 December 2019. Available online: https://emedicine.medscape.com/article/1697353-overview (accessed on 6 October 2020).
4. Ferley, J.; Soerjomatarami, I.; Ervik, M.; Dikshit, R.; Eser, S. *Cancer Incidence and Mortality Worldwide*; IARC: Lyon, France, 2013.
5. National Cancer Institute Surveillance. Epidemiology and End Results Programme (SEER)—Cancer Stat Facts: Female Breast Cancer. 2020. Available online: http://seer.cancer.gov/statfacts/html/breast.html (accessed on 7 October 2020).
6. DeSantis, C.E.; Miller, K.D.; Sauer, A.G.; Siegel, R.L. Cancer Statistics for African Americans, 2019. *CA Cancer J. Clin.* **2017**, *69*, 211–233. [CrossRef]
7. Surveillance, Epidemiology, and End Results (SEER) Program. *SEER*Stat Database: Mortality-All COD, Aggregated with State, Total U.S. (1990–2017) <Early release with Vintage 2017 Katrina/Rita Population Adjustment>*; National Cancer Institute, Division of Cancer Control and Population Sciences, Surveillance Research Program: North Bethesda, MD, USA, 2019; Underlying mortality data provided the by National Center for Health Statistics.
8. DeSantis, C.E.; Ma, J.; Goding, S.A.; Newman, L.A.; Jemal, A. Breast cancer statistics, 2017: Racial disparity in mortality by state. *CA Cancer. J. Clin.* **2017**, *67*, 439–448. [CrossRef] [PubMed]
9. DeSantis, C.E.; Ma, J.; Bryan, L.; Jemal, A. Breast cancer statistics, 2013. *CA Cancer. J. Clin.* **2014**, *64*, 52–62. [CrossRef] [PubMed]
10. Ghoncheh, M.; Pournamdar, Z.; Salehiniya, H. Incidence and mortality and epidemiology of breast cancer in the world. *Asian Pac. J. Cancer Prev.* **2016**, *17*, 43–46. [CrossRef] [PubMed]
11. Ghoncheh, M.; Mohammadian-Hafshejani, A.; Salehiniya, H. Incidence and mortality of breast cancer and their relationship to development in Asia. *Asian Pac. J. Cancer Prev.* **2015**, *16*, 6081–6087. [CrossRef]

12. Bernstein, L.; Ross, R.K. Endogenous hormones and breast cancer risk. *Epidemiol. Rev.* **1993**, *15*, 48–65. [CrossRef]
13. Wu, A.H.; Stanczyk, F.Z.; Seow, A.; Lee, H.P.; Yu, M.C. Soy intake and other lifestyle determinants of serum estrogen levels among postmenopausal Chinese women in Singapore. *Cancer Epidemiol. Biomark. Prev.* **2002**, *11*, 844–851.
14. Thakur, P.; Seam, R.K.; Gupta, M.K.; Gupta, M.; Sharma, M.; Fotedar, V. Breast cancer risk factor evaluation in a Western Himalayan state: A case-control study and comparison with the Western World. *South Asian J. Cancer* **2017**, *6*, 106–109.
15. Colditz, G.A.; Rosner, B. Cumulative risk of breast cancer to age 70 years according to risk factor status: Data from the Nurses' Health Study. *Am. J. Epidemiol.* **2000**, *152*, 950–964. [CrossRef]
16. Giordano, S.H.; Buzdar, A.U.; Hortobagyi, G.N. Breast Cancer in Men. *Ann. Intern. Med.* **2002**, *137*, 678–687. [CrossRef] [PubMed]
17. Meo, S.A.; Suraya, F.; Jamil, B.; Al Rouq, F.; Meo, A.S.; Sattar, K.; Javed, M.; Alasiri, S.A. Association of ABO and Rh blood groups with breast cancer. *Saudi J. Biol. Sci.* **2017**, *24*, 1609–1613. [CrossRef] [PubMed]
18. National Center for Health Statistics. *SEER Cancer Statistics Review, 1973–1999*; National Cancer Institute: Bethesda, MD, USA, 1998.
19. DeSantis, C.E.; Ma, J.; Gaudet, M.M.; Newman, L.A.; Miller, K.D.; Goding, S.A.; Jemal, A.; Siegel, R.L. Breast cancer statistics, 2019. *CA Cancer J. Clin.* **2019**, *69*, 438–451. [CrossRef] [PubMed]
20. Lilienfeld, A.M. The relationship of cancer of the female breast to artificial menopause and marital status. *Cancer* **1956**, *9*, 927–934. [CrossRef]
21. Dai, Q.; Liu, B.; Du, Y. Meta-analysis of the risk factors of breast cancer concerning reproductive factors and oral contraceptive use. *Front. Med. China* **2009**, *3*, 452–458. [CrossRef]
22. Ma, H.; Henderson, K.D.; Sullivan-Halley, J.; Duan, L.; Marshall, S.F.; Ursin, G.; Horn-Ross, P.L.; Largent, J.; Deapen, D.M.; Lacey, J.V., Jr. Pregnancy-related factors and the risk of breast carcinoma in situ and invasive breast cancer among postmenopausal women in the California Teachers Study cohort. *Breast Cancer Res.* **2010**, *12*, R35. [CrossRef]
23. Balekouzou, A.; Yin, P.; Pamatika, C.M.; Bekolo, C.E.; Nambei, S.W.; Djeintote, M.; Kota, K.; Mossoro-Kpinde, C.D.; Shu, C.; Yin, M.; et al. Reproductive risk factors associated with breast cancer in women in Bangui: A case-control study. *BMC. Women's Health* **2017**, *17*, 14. [CrossRef]
24. Rosner, B.; Colditz, G.A.; Willett, W.C. Reproductive risk factors in a prospective study of breast cancer: The Nurses' Health Study. *Am. J. Epidemiol.* **1994**, *139*, 819–835. [CrossRef]
25. Rosner, B.; Colditz, G.A.; Martínez, M.E.; Giovannucci, E.L.; Stampfer, M.J.; Hunter, D.J.; Speizer, F.E.; Wing, A.; Willett, W.C. Nurses' health study: Log-incidence mathematical model of breast cancer incidence. *J. Natl. Cancer Inst.* **1996**, *88*, 359–364. [CrossRef]
26. Mahouri, K.; Zahedani, M.D.; Zare, S. Breast cancer risk factors in south of Islamic Republic of Iran: A case-control study. *EMHJ—East. Mediterr. Health J.* **2007**, *13*, 1265–1273. [CrossRef]
27. Kim, Y.; Yoo, K.Y.; Goodman, M.T. Differences in Incidence, Mortality and Survival of Breast Cancer by Regions and Countries in Asia and Contributing Factors. *Asian Pac. J. Cancer Prev.* **2015**, *16*, 2857–2870. [CrossRef] [PubMed]
28. Freund, C.; Mirabel, L.; Annane, K.; Mathelin, C. Breastfeeding and breast cancer. *Gynecol. Obstet. Fertil.* **2005**, *33*, 739–744. [CrossRef] [PubMed]
29. Jeong, S.H.; An, Y.S.; Choi, J.Y.; Park, B.; Kang, D.; Lee, M.H.; Han, W.; Noh, D.Y.; Yoo, K.Y.; Park, S.K. Risk reduction of breast cancer by childbirth, breastfeeding, and their interaction in korean women: Heterogeneous effects across menopausal status, hormone receptor status, and pathological subtypes. *J. Prev. Med. Public Health* **2017**, *50*, 401–410. [CrossRef] [PubMed]
30. Deng, Y.; Xu, H.; Zeng, X. Induced abortion and breast cancer: An updated meta-analysis. *Medicine* **2018**, *97*, e9613. [CrossRef] [PubMed]
31. Key, T.; Appleby, P.; Barnes, I.; Reeves, G. Endogenous sex hormones and breast cancer in postmenopausal women: Reanalysis of nine prospective studies. *J. Natl. Cancer Inst.* **2002**, *94*, 606–616.
32. Fisher, B.; Costantino, J.P.; Wickerham, D.L.; Redmond, C.K.; Kavanah, M.; Cronin, W.M.; Vogel, V.; Robidoux, A.; Dimitrov, N.; Atkins, J.; et al. Tamoxifen for prevention of breast cancer: Report of the National Surgical Adjuvant Breast and Bowel Project P-1 Study. *J. Natl. Cancer Inst.* **1998**, *90*, 1371–1388. [CrossRef]
33. Eliassen, A.H.; Missmer, S.A.; Tworoger, S.S.; Spiegelman, D.; Barbieri, R.L.; Dowsett, M.; Hankinson, S.E. Endogenous steroid hormone concentrations and risk of breast cancer among premenopausal women. *J. Natl. Cancer Inst.* **2006**, *98*, 1406–1415. [CrossRef]
34. Tworoger, S.S.; Eliassen, A.H.; Rosner, B.; Sluss, P.; Hankinson, S.E. Plasma prolactin concentrations and risk of postmenopausal breast cancer. *Cancer Res.* **2004**, *64*, 6814–6819. [CrossRef]
35. Toniolo, P.; Bruning, P.F.; Akhmedkhanov, A.; Bonfrer, J.M.; Koenig, K.L.; Lukanova, A.; Shore, R.E.; Zeleniuch-Jacquotte, A. Serum insulin-like growth factor-I and breast cancer. *Int. J. Cancer* **2000**, *88*, 828–832. [CrossRef]
36. Bhadoria, A.; Kapil, U.; Sareen, N.; Singh, P. Reproductive factors and breast cancer: A case-control study in tertiary care hospital of North India. *Indian J. Cancer* **2013**, *50*, 316–321.
37. Fioretti, F.; Tavani, A.; Bosetti, C.; La Vecchia, C.; Negri, E.; Barbone, F.; Talamini, R.; Franceschi, S. Risk factors for breast cancer in nulliparous women. *Br. J. Cancer* **1999**, *79*, 1923–1928. [CrossRef] [PubMed]
38. Marchbanks, P.A.; McDonald, J.A.; Wilson, H.G.; Folger, S.G.; Mandel, M.G.; Daling, J.R.; Bernstein, L.; Malone, K.E.; Ursin, G.; Storm, B.L.; et al. Oral contraceptives and the risk of breast cancer. *N. Engl. J. Med.* **2002**, *346*, 2025–2032. [CrossRef] [PubMed]
39. Zolfaroli, I.; Tarín, J.J.; Cano, A. Hormonal contraceptives and breast cancer: Clinical data. *Eur. J. Obstet. Gynecol. Reprod. Biol.* **2018**, *230*, 212–216. [CrossRef] [PubMed]

40. Collaborative Group of Hormonal Factors in Breast Cancer. Breast cancer and hormonal contraceptives: Collaborative reanalysis of individual data on 53,297 women with breast cancer and 100,239 women without breast cancer from 54 epidemiological studies. *Lancet* **1996**, *347*, 1713–1727. [CrossRef]
41. Beral, V.; Million Women Study Collaborators. Breast cancer and hormone-replacement therapy in the Million Women Study. *Lancet* **2003**, *362*, 419–427. [CrossRef]
42. Beral, V.; Bull, D.; Doll, R.; Key, T.; Peto, R.; Reeves, G. Breast cancer and hormone replacement therapy: Collaborative reanalysis of data from 51 epidemiological studies of 52,705 women with breast cancer and 108,411 women without breast cancer. *Lancet* **1997**, *350*, 1047–1059. [CrossRef]
43. Ross, R.K.; Paganini-Hill, A.; Wan, P.C.; Pike, M.C. Effect of hormone replacement therapy on breast cancer risk: Estrogen versus estrogen plus progestin. *J. Natl. Cancer Inst.* **2000**, *92*, 328–332. [CrossRef]
44. Magnusson, C.; Persson, I.; Adami, H.O. More about: Effect of hormone replacement therapy on breast cancer risk: Estrogen versus estrogen plus progestin. *J. Natl. Cancer Inst.* **2000**, *92*, 1183–1184. [CrossRef]
45. Colditz, G.A. Estrogen, estrogen plus progestin therapy, and risk of breast cancer. *Clin. Cancer Res.* **2005**, *11*, 909s–917s.
46. Taheripanah, R.; Balash, F.; Anbiaee, R.; Mahmoodi, M.; Akbari, S.A. Breast Cancer and Ovulation Induction Treatments. *Clin. Breast Cancer* **2018**, *18*, 395–399. [CrossRef]
47. Brinton, L.A.; Scoccia, B.; Moghissi, K.S.; Westhoff, C.L.; Althuis, M.D.; Mabie, J.E.; Lamb, E.J. Breast cancer risk associated with ovulation-stimulating drugs. *Hum. Reprod.* **2004**, *19*, 2005–2013. [CrossRef] [PubMed]
48. Rojas, K.; Stuckey, A. Breast Cancer Epidemiology and Risk Factors. *Clin. Obstet. Gynecol.* **2016**, *59*, 651–672. [CrossRef] [PubMed]
49. Metcalfe, K.A.; Finch, A.; Poll, A.; Horsman, D.; Kim-Sing, C.; Scott, J.; Royer, R.; Sun, P.; Narod, S.A. Breast cancer risks in women with a family history of breast or ovarian cancer who have tested negative for a BRCA1 or BRCA2 mutation. *Br. J. Cancer* **2009**, *100*, 421–425. [CrossRef] [PubMed]
50. Cobain, E.F.; Milliron, K.J.; Merajver, S.D. Updates on breast cancer genetics: Clinical implications of detecting syndromes of inherited increased susceptibility to breast cancer. *Semin. Oncol.* **2016**, *43*, 528–535. [CrossRef] [PubMed]
51. Godet, I.; Gilkes, D.M. BRCA1 and BRCA2 mutations and treatment strategies for breast cancer. *Integr. Cancer Sci. Ther.* **2017**, *4*, 1–17. [CrossRef] [PubMed]
52. Ahmed, F.; Mahmood, N.; Shahid, S.; Hussain, Z.; Ahmed, I.; Jalal, A.; Ijaz, B.; Shahid, A.; Mujtaba, G.; Mustafa, T. Mutations in human interferon *α2b* gene and potential as risk factor associated with female breast cancer. *Cancer Biother. Radiopharm.* **2016**, *31*, 199–208. [CrossRef]
53. Yari, K.; Rahimi, Z.; Moradi, M.T.; Rahimi, Z. The MMP-2-735 C Allele is a risk factor for susceptibility to breast cancer. *Asian Pac. J. Cancer Prev.* **2014**, *15*, 6199–6203. [CrossRef]
54. Gunter, M.J.; Hoover, D.R.; Yu, H.; Wassertheil-Smoller, S.; Rohan, T.E.; Manson, J.E.; Li, J.; Ho, G.Y.; Xue, X.; Anderson, G.L.; et al. Insulin, insulin-like growth factor-I, and risk of breast cancer in postmenopausal women. *J. Natl. Cancer Inst.* **2009**, *101*, 48–60. [CrossRef]
55. Tabassum, I.; Mahmood, H.; Faheem, M. Type 2 Diabetes Mellitus as a risk factor for female breast cancer in the population of northern Pakistan. *Asian Pac. J. Cancer Prev.* **2016**, *17*, 3255–3258.
56. Larsson, S.C.; Mantzoros, C.S.; Wolk, A. Diabetes mellitus and risk of breast cancer: A meta-analysis. *Int. J. Cancer* **2007**, *121*, 856–862. [CrossRef]
57. Tang, G.H.; Satkunam, M.; Pond, G.R.; Steinberg, G.R.; Blandino, G.; Schünemann, H.J.; Muti, P. Association of metformin with breast cancer incidence and mortality in patients with type 2 diabetes: A GRADE assessed systematic review and meta-analysis. *Cancer Epidemiol. Biomark. Prev.* **2018**, *27*, 627–635. [CrossRef] [PubMed]
58. Chen, M.J.; Wu, W.Y.; Yen, A.M.; Fann, J.C.; Chen, S.L.; Chiu, S.Y.; Chen, H.H.; Chiou, S.T. Body mass index and breast cancer: Analysis of a nation-wide population-based prospective cohort study on 1,393,985 Taiwanese women. *Int. J. Obes.* **2016**, *40*, 524–530. [CrossRef] [PubMed]
59. Lahmann, P.H.; Hoffmann, K.; Allen, N.; Van Gils, C.H.; Khaw, K.T.; Tehard, B.; Berrino, F.; Tjønneland, A.; Bigaard, J.; Olsen, A. Body size and breast cancer risk: Findings from the European prospective investigation into cancer and nutrition. *Int. J. Cancer* **2004**, *111*, 762–771. [CrossRef] [PubMed]
60. Guo, W.; Key, T.J.; Reeves, G.K. Adiposity and breast cancer risk in postmenopausal women: Results from the UK biobank prospective cohort. *Int. J. Cancer* **2018**, *143*, 1037–1046. [CrossRef]
61. Pimentel, I.; Lohmann, A.E.; Goodwin, P.J. Normal weight adiposity and postmenopausal breast cancer risk. *JAMA Oncol.* **2019**, *5*, 150–151. [CrossRef]
62. Taylor, E.F.; Burley, V.J.; Greenwood, D.C.; Cade, J.E. Meat consumption and risk of breast cancer in the UK women's cohort study. *Br. J. Cancer* **2007**, *96*, 1139–1146. [CrossRef]
63. Sieri, S.; Krogh, V.; Ferrari, P.; Berrino, F.; Pala, V.; Thiébaut, A.C.; Tjønneland, A.; Olsen, A.; Overvad, K.; Jakobsen, M.U.; et al. Dietary fat and breast cancer risk in the European Prospective Investigation into Cancer and Nutrition. *Am. J. Clin. Nutr.* **2008**, *88*, 1304–1312.
64. Berkey, C.S.; Rockett, H.R.; Willett, W.C.; Colditz, G.A. Milk, dairy fat, dietary calcium, and weight gain: A longitudinal study of adolescents. *Arch. Pediatrics Adolesc. Med.* **2005**, *159*, 543–550. [CrossRef]

65. Hatse, S.; Lambrechts, D.; Verstuyf, A.; Smeets, A.; Brouwers, B.; Vandorpe, T.; Brouckaert, O.; Peuteman, G.; Laenen, A.; Verlinden, L. Vitamin D status at breast cancer diagnosis: Correlation with tumor characteristics, disease outcome, and genetic determinants of vitamin D insufficiency. *Carcinogenesis* **2012**, *33*, 1319–1326. [CrossRef]
66. O'Brien, K.M.; Sandler, D.P.; Taylor, J.A.; Weinberg, C.R. Serum vitamin D and risk of breast cancer within five years. *Environ. Health Perspect.* **2017**, *125*, 077004. [CrossRef]
67. Hamajima, N.; Hirose, K.; Tajima, K.; Rohan, T.; Calle, E.E.; Heath, C.W., Jr.; Coates, R.J.; Liff, J.M.; Talamini, R.; Chantarakul, N.; et al. Alcohol, tobacco and breast cancer—Collaborative reanalysis of individual data from 53 epidemiological studies, including 58,515 women with breast cancer and 95,067 women without the disease. *Br. J. Cancer* **2002**, *87*, 1234–1245. [PubMed]
68. Romieu, I.; Scoccianti, C.; Chajès, V.; de Batlle, J.; Biessy, C.; Dossus, L.; Baglietto, L.; Clavel-Chapelon, F.; Overvad, K.; Olsen, A.; et al. Alcohol intake and breast cancer in the European prospective investigation into cancer and nutrition. *Int. J. Cancer.* **2015**, *137*, 1921–1930. [CrossRef] [PubMed]
69. Luo, J.; Margolis, K.L.; Wactawski-Wende, J.; Horn, K.; Messina, C.; Stefanick, M.L.; Tindle, H.A.; Tong, E.; Rohan, T.E. Association of active and passive smoking with risk of breast cancer among postmenopausal women: A prospective cohort study. *BMJ* **2011**, *342*, d1016. [CrossRef] [PubMed]
70. Tong, J.H.; Li, Z.; Shi, J.; Li, H.M.; Wang, Y.; Fu, L.Y.; Liu, Y.P. Passive smoking exposure from partners as a risk factor for ER+/PR+ double positive breast cancer in never-smoking Chinese urban women: A hospital-based matched case control study. *PLoS ONE* **2014**, *9*, e97498. [CrossRef] [PubMed]
71. The Health Consequences of Involuntary Exposure to Tobacco Smoke: A Report of the Surgeon General. 2006. Available online: www.surgeongeneral.gov/library/secondhandsmoke/report/index.html (accessed on 11 October 2020).
72. Mctiernan, A.; Kooperberg, C.; White, E.; Wilcox, S.; Coates, R.; Adams-Campbell, L.L.; Woods, N.; Ockene, J. Women's health initiative cohort study recreational physical activity and the risk of breast cancer in postmenopausal women: The women's health initiative cohort study. *JAMA* **2003**, *290*, 1331–1336. [CrossRef] [PubMed]
73. Lee, J.A. Meta-analysis of the association between physical activity and breast cancer mortality. *Cancer Nurs.* **2019**, *42*, 271–285. [CrossRef] [PubMed]
74. Benabu, J.C.; Stoll, F.; Gonzalez, M.; Mathelin, C. Night work, shift work: Breast cancer risk factor? *Gynecol. Obstet. Fertil.* **2015**, *43*, 791–799. [CrossRef]
75. Stevens, R.G.; Davis, S. The melatonin hypothesis: Electric power and breast cancer. *Environ. Health Perspect.* **1996**, *104* (Suppl. 1), 135–140.
76. Megdal, S.P.; Kroenke, C.H.; Laden, F.; Pukkala, E.; Schernhammer, E.S. Night work and breast cancer risk: A systematic review and meta-analysis. *Eur. J. Cancer* **2005**, *41*, 2023–2032. [CrossRef]
77. Orsini, M.; Trétarre, B.; Daurès, J.P.; Bessaoud, F. Individual socioeconomic status and breast cancer diagnostic stages: A French case-control study. *Eur. J. Public Health* **2016**, *26*, 445–450. [CrossRef]
78. Lundqvist, A.; Andersson, E.; Ahlberg, I.; Nilbert, M.; Gerdtham, U. Socioeconomic inequalities in breast cancer incidence and mortality in Europe-a systematic review and meta-analysis. *Eur. J. Public Health* **2016**, *26*, 804–813. [CrossRef] [PubMed]
79. Abdulrahman, G.O.; Rahman, G.A. Epidemiology of Breast Cancer in Europe and Africa. *J. Cancer Epidemiol.* **2012**, *2012*, 915610. [CrossRef] [PubMed]
80. Hartmann, L.C.; Sellers, T.A.; Frost, M.H.; Lingle, W.L.; Degnim, A.C.; Ghosh, K.; Vierkant, R.A.; Maloney, S.D.; Pankratz, V.S.; Hillman, D.W.; et al. Benign breast disease and the risk of breast cancer. *N. Engl. J. Med.* **2005**, *353*, 229–237. [CrossRef] [PubMed]
81. Brinton, L.A.; Lubin, J.H.; Murray, M.C.; Colton, T.; Hoover, R.N. Mortality rates among augmentation mammoplasty patients: An update. *Epidemiology* **2006**, *17*, 162–169. [CrossRef]
82. Boyd, N.F.; Guo, H.; Martin, L.J.; Sun, L.; Stone, J.; Fishell, E.; Jong, R.A.; Hislop, G.; Chiarelli, A.; Minkin, S.; et al. Mammographic density and the risk and detection of breast cancer. *N. Engl. J. Med.* **2007**, *356*, 227–236. [CrossRef]
83. Nazari, S.S.; Mukherjee, P. An overview of mammographic density and its association with breast cancer. *Breast Cancer* **2018**, *25*, 259–267. [CrossRef]
84. Tamimi, R.M.; Byrne, C.; Colditz, G.A.; Hankinson, S.E. Endogenous hormone levels, mammographic density, and subsequent risk of breast cancer in postmenopausal women. *J. Natl. Cancer Inst.* **2007**, *99*, 1178–1187. [CrossRef]
85. Land, C.E.; Tokunaga, M.; Koyama, K.; Soda, M.; Preston, D.L.; Nishimori, I.; Tokuoka, S. Incidence of female breast cancer among atomic bomb survivors, Hiroshima and Nagasaki, 1950–1990. *Radiat. Res.* **2003**, *160*, 707–717. [CrossRef]
86. Henderson, T.O.; Moskowitz, C.S.; Chou, J.F.; Bradbury, A.R.; Neglia, J.P.; Dang, C.T.; Onel, K.; Friedman, D.N.; Bhatia, S.; Strong, L.C.; et al. Breast cancer risk in childhood cancer survivors without a history of chest radiotherapy: A report from the Childhood Cancer Survivor Study. *J. Clin. Oncol.* **2016**, *34*, 910–918. [CrossRef]
87. Horwich, A.; Swerdlow, A.J. Second primary breast cancer after Hodgkin's disease. *Br. J. Cancer* **2004**, *90*, 294–298. [CrossRef]
88. Perou, C.M.; Sørlie, T.; Eisen, M.B.; Van de Rijn, M.; Jeffrey, S.S.; Rees, C.A.; Pollack, J.R.; Ross, D.T.; Johnsen, H.; Akslen, L.A.; et al. Molecular portraits of human breast tumors. *Nature* **2000**, *17*, 747–752. [CrossRef] [PubMed]
89. Howlader, N.; Altekruse, S.F.; Li, C.I.; Chen, V.W.; Clarke, C.A.; Ries, L.A.; Cronin, K.A. US incidence of breast cancer subtypes defined by joint hormone receptor and HER2 status. *J. Natl. Cancer Inst.* **2014**, *106*, dju055. [CrossRef] [PubMed]
90. Dwivedi, S.; Purohit, P.; Misra, R.; Lingeswaran, M.; Vishnoi, J.R.; Pareek, P.; Sharma, P.; Misra, S. Application of single-cell omics in breast cancer in single-cell omics. *Appl. Biomed. Agric.* **2019**, *2*, 69–103.

91. Aftimos, P.; Azim, H.A., Jr.; Sotiriou, C. Molecular biology of breast cancer. In *Molecular Pathology*, 2nd ed.; Coleman, W.B., Tsongalis, G.J., Eds.; Academic Press: Cambridge, MA, USA, 2018; pp. 569–588.
92. Lehmann, B.D.; Bauer, J.A.; Chen, X.; Sanders, M.E.; Chakravarthy, A.B.; Shyr, Y.; Pietenpol, J.A. Identification of human triple-negative breast cancer subtypes and preclinical models for selection of targeted therapies. *J. Clin. Invest.* **2011**, *121*, 2750–2767. [CrossRef]
93. Colozza, M.; Azambuja, E.; Cardoso, F.; Sotiriou, C.; Larsimont, D.; Piccart, M.J. Proliferative markers as prognostic and predictive tools in early breast cancer: Where are we now? *Ann. Oncol.* **2005**, *16*, 1723–1739. [CrossRef]
94. Lim, W.; Mayer, B.; Pawson, T. *Cell Signaling: Principles and Mechanisms*; Garland Science: New York, NY, USA, 2015.
95. Hancock, J.F. Ras proteins: Different signals from different locations. *Nat. Rev. Mol. Cell Biol.* **2003**, *4*, 373–384. [CrossRef]
96. Bourne, H.R.; Sanders, D.A.; McCormick, F. The GTPase superfamily: A conserved switch for diverse cell functions. *Nature* **1990**, *348*, 125–132. [CrossRef]
97. Paduch, M.; Jelen, F.; Otlewski, J. Structure of small G proteins and their regulators. *Acta Biochim. Pol.* **2001**, *48*, 829–850. [CrossRef]
98. Yudushkin, I. Getting the Akt together: Guiding intracellular Akt activity by PI3K. *Biomolecules* **2019**, *9*, 67. [CrossRef]
99. Yu, X.; Long, Y.C.; Shen, H.M. Differential regulatory functions of three classes of phosphatidylinositol and phosphoinositide 3-kinases in autophagy. *Autophagy* **2015**, *11*, 1711–1728. [CrossRef]
100. Lehninger, A.; Nelson, D.L.; Cox, M.C.; Freeman, W.H. *Lehninger Principles of Biochemistry*; W.H. Freeman: New York, NY, USA, 2012.
101. Balla, T. Phosphoinositides: Tiny lipids with giant impact on cell regulation. *Physiol. Rev.* **2013**, *93*, 1019–1137. [CrossRef] [PubMed]
102. Braccini, L.; Ciraolo, E.; Campa, C.C.; Perino, A.; Longo, D.L.; Tibolla, G.; Pregnolato, M.; Cao, Y.; Tassone, B.; Damilano, F.; et al. PI3K-C2γ is a Rab5 effector selectively controlling endosomal Akt2 activation downstream of insulin signalling. *Nat. Commun.* **2015**, *6*, 7400. [CrossRef] [PubMed]
103. Falasca, M.; Hughes, W.E.; Dominguez, V.; Sala, G.; Fostira, F.; Fang, M.Q.; Cazzolli, R.; Shepherd, P.R.; James, D.E.; Maffucci, T. The role of phosphoinositide 3-kinase C2α in insulin signaling. *J. Biol. Chem.* **2007**, *282*, 28226–28236. [CrossRef] [PubMed]
104. Backer, J. The intricate regulation and complex functions of the Class III phosphoinositide 3-kinase Vps34. *Biochem. J.* **2016**, *473*, 2251–2271. [CrossRef]
105. Manning, B.D.; Toker, A. AKT/PKB signaling: Navigating the network. *Cell* **2017**, *169*, 381–405. [CrossRef]
106. Dummler, B.; Hemmings, B.A. Physiological roles of PKB/Akt isoforms in development and disease. *Biochem. Soc. Trans.* **2007**, *35*, 231–235. [CrossRef]
107. Szymonowicz, K.; Oeck, S.; Malewicz, N.M.; Jendrossek, V. New insights into protein kinase B/Akt signaling: Role of localized Akt activation and compartment-specific target proteins for the cellular radiation response. *Cancers* **2018**, *10*, 78. [CrossRef]
108. Revathidevi, S.; Munirajan, A.K. Akt in cancer: Mediator and more. *Semin. Cancer Biol.* **2019**, *59*, 80–91. [CrossRef]
109. Risso, G.; Blaustein, M.; Pozzi, B.; Mammi, P.; Srebrow, A. Akt/PKB: One kinase, many modifications. *Biochem. J.* **2015**, *468*, 203–214. [CrossRef]
110. Luo, C.T.; Li, M. Foxo transcription factors in T cell biology and tumor immunity. *Semin. Cancer Biol.* **2018**, *50*, 13–20. [CrossRef]
111. Arcaro, A.; Guerreiro, A.S. The phosphoinositide 3-kinase pathway in human cancer: Genetic alterations and therapeutic implications. *Curr. Genom.* **2007**, *8*, 271–306. [CrossRef] [PubMed]
112. Patel, P.; Woodgett, J.R. Glycogen Synthase Kinase 3: A Kinase for All Pathways? *Curr. Top. Dev. Biol.* **2017**, *123*, 277–302. [PubMed]
113. Dokken, B.B.; Sloniger, J.A.; Henriksen, E.J. Acute selective glycogen synthase kinase-3 inhibition enhances insulin signaling in prediabetic insulin-resistant rat skeletal muscle. *Am. J. Physiol. Endocrinol. Metab.* **2005**, *288*, E1188–E1194. [CrossRef] [PubMed]
114. Lochhead, P.A.; Coghlan, M.; Rice, S.Q.; Sutherland, C. Inhibition of GSK-3 selectively reduces glucose-6-phosphatase and phosphatase and phosphoenolypyruvate carboxykinase gene expression. *Diabetes* **2001**, *50*, 937–946. [CrossRef]
115. Wei, X.; Luo, L.; Chen, J. Roles of mTOR signaling in tissue regeneration. *Cells* **2019**, *8*, 1075. [CrossRef]
116. Kakumoto, K.; Ikeda, J.; Okada, M.; Morii, E.; Oneyama, C. mLST8 promotes mTOR-mediated tumor progression. *PLoS ONE* **2015**, *10*, e0119015. [CrossRef]
117. Mahoney, R.E.; Azpurua, J.; Eaton, B.A. Insulin signaling controls neurotransmission via the 4eBP-dependent modification of the exocytotic machinery. *eLife* **2016**, *5*, e16807. [CrossRef]
118. Berchtold, D.; Walther, T.C. TORC2 Plasma membrane localization is essential for cell viability and restricted to a distinct domain. *Mol. Biol. Cell* **2009**, *20*, 1565–1575. [CrossRef]
119. Liu, P.; Gan, W.; Inuzuka, H.; Lazorchak, A.S.; Gao, D.; Arojo, O.; Liu, D.; Wan, L.; Zhai, B.; Yu, Y.; et al. Sin1 phosphorylation impairs mTORC2 complex integrity and inhibits downstream Akt signalling to suppress tumorigenesis. *Nat. Cell Biol.* **2013**, *15*, 1340–1350. [CrossRef]
120. Hollenhorst, P.C.; Bose, M.E.; Mielke, M.R.; Müller, U.; Fox, C.A. Forkhead genes in transcriptional silencing, cell morphology and the cell cycle. Overlapping and distinct functions for FKH1 and FKH2 in *Saccharomyces cerevisiae*. *Genetics* **2000**, *154*, 1533–1548.
121. Cabrera-Ortega, A.; Feinberg, D.; Liang, Y.; Rossa, J.C.; Graves, D.T. The role of Forkhead Box 1 (FOXO1) in the immune system: Dendritic cells, T cells, B cells, and hematopoietic stem cells. *Crit. Rev. Immunol.* **2017**, *37*, 1–13. [CrossRef] [PubMed]
122. Ma, Z.; Xin, Z.; Hu, W.; Jiang, S.; Yang, Z.; Yan, X.; Li, X.; Yang, Y.; Chen, F. Forkhead box O proteins: Crucial regulators of cancer EMT. *Semin. Cancer Biol.* **2018**, *50*, 21–31. [CrossRef] [PubMed]

123. Maiese, K. Forkhead transcription factors: Formulating a FOXO target for cognitive loss. *Curr. Neurovascular Res.* **2017**, *14*, 415–420. [CrossRef] [PubMed]
124. Cretella, D.; Digiacomo, G.; Giovannetti, E.; Cavazzoni, A. PTEN alterations as a potential mechanism for tumor cell escape from PD-1/PD-L1 inhibition. *Cancers* **2019**, *11*, 1318. [CrossRef]
125. Luongo, F.; Colonna, F.; Calapà, F.; Vitale, S.; Fiori, M.E.; De Maria, R. PTEN tumor-suppressor: The dam of stemness in cancer. *Cancers* **2019**, *11*, 1076. [CrossRef]
126. Naderali, E.; Khaki, A.A.; Rad, J.S.; Alihemmati, A.; Rahmati, M.; Nozad-Charoudeh, H. Regulation and modulation of PTEN activity. *Mol. Biol. Rep.* **2018**, *45*, 2869–2881. [CrossRef]
127. Maehama, T.; Taylor, G.S.; Dixon, J.E. PTEN and myotubularin: Novel phosphoinositide phosphatases. *Annu. Rev. Biochem.* **2001**, *70*, 247–279. [CrossRef]
128. Nguyen, K.T.; Tajmir, P.; Lin, C.H.; Liadis, N.; Zhu, X.D.; Eweida, M.; Tolasa-Karaman, G.; Cai, F.; Wang, R.; Kitamura, T.; et al. Essential role of PTEN in body size determination and pancreatic beta-cell homeostasis in vivo. *Mol. Cell. Biol.* **2006**, *26*, 4511–4518. [CrossRef]
129. Abraham, J. PI3K/AKT/mTOR pathway inhibitors: The ideal combination partners for breast cancer therapies? *Expert Rev. Anticancer Ther.* **2015**, *15*, 51–68. [CrossRef]
130. Miller, T.W.; Rexer, B.N.; Garrett, J.T.; Arteaga, C.L. Mutations in the phosphatidylinositol 3-kinase pathway: Role in tumor progression and therapeutic implications in breast cancer. *Breast Cancer Res.* **2011**, *13*, 224. [CrossRef]
131. Chalhoub, C.; Baker, S.J. PTEN and the PI3-kinase pathway in cancer. *Annu. Rev. Pathol.* **2009**, *4*, 127–150. [CrossRef] [PubMed]
132. Martínez-Sáez, O.; Chic, N.; Pascual, T.; Adamo, B.; Vidal, M.; González-Farré, B.; Sanfeliu, E.; Schettini, F.; Conte, B.; Brasó-Maristany, F.; et al. Frequency and spectrum of PIK3CA somatic mutations in breast cancer. *Breast Cancer Res.* **2020**, *22*, 45. [CrossRef] [PubMed]
133. Vasan, N.; Razavi, P.; Johnson, J.L.; Shao, H.; Shah, H.; Antoine, A.; Ladewig, E.; Gorelick, A.N.; Lin, T.-Y.; Toska, E.; et al. Double PIK3CA mutations in cis increase oncogenicity and sensitivity to PI3Kα inhibitors. *Science* **2019**, *366*, 714–723. [CrossRef] [PubMed]
134. Xing, Y.; Lin, N.U.; Maurer, M.A.; Chen, H.; Mahvash, A.; Sahin, A.; Akcakanat, A.A.; Yisheng, L.; Abramson, V.; Litton, J.; et al. Phase II trial of AKT inhibitor MK-2206 in patients with advanced breast cancer who have tumors with PIK3CA or AKT mutations, and/or PTEN loss/PTEN mutation. *Breast Cancer Res.* **2019**, *21*, 78. [CrossRef] [PubMed]
135. Samuels, Y.; Wang, Z.; Bardelli, A.; Ptak, N.J.; Szabo, S.; Yan, H.; Gazdar, A.; Powell, S.M.; Riggins, G.J.; Willson, J.K.V.; et al. High frequency of mutations of the PIK3CA gene in human cancers. *Science* **2004**, *304*, 554. [CrossRef] [PubMed]
136. Samuels, Y.; Diaz, L.A.; Schmidt-Kittler, O.; Cummins, J.M.; Delong, L.; Cheong, I.; Rago, C.; Huso, D.L.; Lengauer, C.; Kinzler, K.W.; et al. Mutant PIK3CA promotes cell growth and invasion of human cancer cells. *Cancer Cell* **2005**, *7*, 561–573. [CrossRef]
137. Karakas, B.; Bachman, K.E.; Park, B.H. Mutation of the PIK3CA oncogene in human cancers. *Br. J. Cancer* **2006**, *94*, 455–459. [CrossRef]
138. Rand, A.; Yardena, S. PIK3CA in cancer: The past 30 years. *Semin. Cancer. Biol.* **2019**, *59*, 36–49.
139. Saal, L.H.; Holm, K.; Maurer, M.; Memeo, L.; Su, T.; Wang, X.; Yu, J.S.; Malmström, P.O.; Mansukhani, M.; Enoksson, J.; et al. PIK3CA mutations correlate with hormone receptors, node metastasis, and ERBB2, and are mutually exclusive with PTEN loss in human breast carcinoma. *Cancer Res.* **2005**, *1*, 2554–2559. [CrossRef]
140. Bachman, K.E.; Argani, P.; Samuels, Y.; Silliman, N.; Ptak, J.; Szabo, S.; Konishi, H.; Karakas, B.; Blair, B.G.; Lin, C.; et al. The PIK3CA gene is mutated with high frequency in human breast cancers. *Cancer Biol. Ther.* **2004**, *3*, 772–775. [CrossRef]
141. Stemke-Hale, K.; Gonzalez-Angulo, A.M.; Lluch, A.; Neve, R.M.; Kuo, W.-L.; Davies, M.; Carey, M.; Yinghui, G.; Guan, Y.; Sahin, A.; et al. An Integrative genomic and proteomic analysis of PIK3CA, PTEN, and AKT mutations in breast cancer. *Cancer Res.* **2008**, *68*, 6084–6091. [CrossRef] [PubMed]
142. Li, S.Y.; Rong, M.; Grieu, F.; Iacopetta, B. PIK3CA mutations in breast cancer are associated with poor outcome. *Breast Cancer Res. Treat.* **2005**, *96*, 91–95. [CrossRef] [PubMed]
143. Tan, M.-H.; Mester, J.L.; Ngeow, J.; Rybicki, L.A.; Orloff, M.S.; Eng, C. Lifetime cancer risks in individuals with germline PTEN mutations. *Clin. Cancer Res.* **2012**, *18*, 400–407. [CrossRef] [PubMed]
144. Carbognin, L.; Miglietta, F.; Paris, I.; Dieci, M.V. Prognostic and predictive implications of PTEN in breast cancer: Unfulfilled promises but intriguing perspectives. *Cancers* **2019**, *11*, 1401. [CrossRef] [PubMed]
145. Sobral-Leite, M.; Salomon, I.; Opdam, M.; Kruger, D.T.; Beelen, K.J.; Van Der Noort, V.; Van Vlierberghe, R.L.P.; Blok, E.J.; Giardiello, D.; Sanders, J.; et al. Cancer-immune interactions in ER-positive breast cancers: PI3K pathway alterations and tumor-infiltrating lymphocytes. *Breast Cancer Res.* **2019**, *21*, 90. [CrossRef] [PubMed]
146. Zardavas, D.; Te Marvelde, L.; Milne, R.L.; Fumagalli, D.F.; Fountzilas, G.; Kotoula, V.; Razis, E.; Papaxoinis, G.; Joensuu, H.; Moynahan, M.E.; et al. Tumor PIK3CA genotype and prognosis in early-stage breast cancer: A pooled analysis of individual patient data. *J. Clin. Oncol.* **2018**, *1*, 981–990. [CrossRef] [PubMed]
147. Ling, D.; Xuehua, Z.; Yun, S.; Jiemin, W.; Xiaorong, Z.; Jiayuan, L.; Min, H.; Hong, Z. Prevalence and prognostic role of PIK3CA/AKT1 mutations in chinese breast cancer patients. *Cancer. Res. Treat.* **2019**, *51*, 128–140.

148. Anderson, A.J.; Mollon, L.E.; Dean, J.L.; Warholak, T.L.; Aizer, A.; Platt, E.A.; Tang, D.H.; Lisa, E.; Davis, L.E. A systematic review of the prevalence and diagnostic workup of PIK3CA mutations in HR+/HER2− metastatic breast cancer. *Int. J. Breast. Cancer* **2020**, *2020*, 3759179. [CrossRef]
149. Lehmann, B.D.; Bauer, J.A.; Schafer, J.M.; Pendleton, C.S.; Tang, L.; Johnson, K.C.; Chen, X.; Balko, J.M.; Gómez, H.L.; Arteaga, C.L.; et al. PIK3CA mutations in androgen receptor-positive triple negative breast cancer confer sensitivity to the combination of PI3K and androgen receptor inhibitors. *Breast Cancer Res.* **2014**, *8*, 406. [CrossRef]
150. Dieci, M.V.; Miglietta, F.; Griguolo, G.; Guarneri, V. Biomarkers for HER2-positive metastatic breast cancer: Beyond hormone receptors. *Cancer Treat. Rev.* **2020**, *88*, 102064. [CrossRef]
151. Toss, A.; Cristofanilli, M. Molecular characterization and targeted therapeutic approaches in breast cancer. *Breast Cancer Res.* **2015**, *17*, 60. [CrossRef] [PubMed]
152. Hsu, J.H.; Hung, M.C. The role of HER2, EGFR, and other receptor tyrosine kinases in breast cancer. *Cancer Metastasis Rev.* **2016**, *35*, 575–588. [CrossRef] [PubMed]
153. Mitri, Z.; Constantine, T.; O'Regan, R. The HER2 receptor in breast cancer: Pathophysiology, clinical use, and new advances in therapy. *Chemother. Res. Pract.* **2012**, *2012*, 743193. [CrossRef] [PubMed]
154. Gagliato, D.D.M.; Jardim, D.L.F.; Marchesi, M.S.P.; Hortobagyi, G.N. Mechanisms of resistance and sensitivity to anti-HER2 therapies in HER2+ breast cancer. *Oncotarget* **2016**, *7*, 64431–64446. [CrossRef] [PubMed]
155. Schettini, F.; Buono, G.; Cardalesi, C.; Desideri, I.; De Placido, S.; Del Mastro, L. Hormone receptor/human epidermal growth factor receptor 2-positive breast cancer: Where we are now and where we are going. *Cancer Treat. Rev.* **2016**, *46*, 20–26. [CrossRef] [PubMed]
156. Babak, N.; Hamid, M.; Zhixiang, W. Mechanisms underlying the action and synergism of trastuzumab and pertuzumab in targeting HER2-positive breast cancer. *Cancers* **2018**, *10*, 342.
157. Kechagioglou, P.; Papi, R.; Provatopoulou, X.; Kalogera, E.; Papadimitriou, E.; Grigoropoulos, P.; Nonni, A.; Zografos, G.; Kyriakidis, D.A.; Gounaris, A. Tumor suppressor PTEN in breast cancer: Heterozygosity, mutations and protein expression. *Anticancer Res.* **2014**, *34*, 1387–1400.
158. Ruiz-Saenz, A.; Dreyer, C.; Campbell, M.R.; Steri, V.; Gulizia, N.; Moasser, M. HER2 amplification in tumors activates PI3K/AKT signaling independent of HER3. *Cancer Res.* **2018**, *78*, 3655–3658. [CrossRef]
159. Yang, Z.; Li, N.; Li, X.; Lei, L.; Wang, X. The prognostic impact of hormonal receptor and HER-2 expression discordance in metastatic breast cancer patients. *OncoTargets Ther.* **2020**, *13*, 853–863. [CrossRef]
160. Clarke, R.; Tyson, J.J.; Dixon, J.M. Endocrine resistance in breast cancer-An overview and update. *Mol. Cell. Endocrinol.* **2015**, *418*, 220–234. [CrossRef]
161. García-Becerra, R.; Santos-Martínez, N.; Díaz, L.; Camacho, J. Mechanisms of resistance to endocrine therapy in breast cancer: Focus on signaling pathways, miRNAs and genetically based resistance. *Int. J. Mol. Sci.* **2013**, *14*, 108–145. [CrossRef] [PubMed]
162. Osborne, C.K.; Schiff, R. Mechanisms of endocrine resistance in breast cancer. *Annu. Rev. Med.* **2011**, *62*, 233–247. [CrossRef] [PubMed]
163. Ghayad, S.E.; Vendrell, J.A.; Larbi, S.B.; Dumontet, C.; Bieche, I.; Cohen, P.A. Endocrine resistance associated with activated ErbB system in breast cancer cells is reversed by inhibiting MAPK or PI3K/Akt signaling pathways. *Int. J. Cancer* **2010**, *126*, 545–562. [CrossRef] [PubMed]
164. Johnston, S.R. Enhancing endocrine therapy for hormone receptor-positive advanced breast cancer: Cotargeting signaling pathways. *J. Natl. Cancer Inst.* **2015**, *107*, djv212. [CrossRef]
165. Lange, C.A.; Yee, D. Killing the second messenger: Targeting loss of cell cycle control in endocrine-resistant breast cancer. *Endocr.-Relat. Cancer* **2011**, *18*, C19–C24. [CrossRef]
166. Jeselsohn, R.; Buchwalter, G.; De Angelis, C.; Brown, M.; Schiff, R. ESR1 mutations-a mechanism for acquired endocrine resistance in breast cancer. *Nat. Rev. Clin. Oncol.* **2015**, *12*, 573. [CrossRef]
167. Giuliano, M.; Schiff, R.; Osborne, C.K.; Trivedi, M.V. Biological mechanisms and clinical implications of endocrine resistance in breast cancer. *Breast* **2011**, *20*, S42–S49. [CrossRef]
168. Cook, K.L.; Shajahan, A.N.; Clarke, R. Autophagy and endocrine resistance in breast cancer. *Expert Rev. Anticancer Ther.* **2011**, *11*, 1283–1294. [CrossRef]
169. Mackey, J.; Kaufman, B.; Clemens, M.; Bapsy, P.P. Trastuzumab prolongs progression free survival in hormone-dependent and HER2-positive metastatic breast cancer. In *Breast Cancer Research and Treatment, Proceedings of the 29th Annual San Antonio Breast Cancer Symposium, San Antonio, TX, USA, 14–17 December 2006*; Springer: Berlin/Heidelberg, Germany, 2006; Volume 100.
170. Merenbakh-Lamin, K.; Ben-Baruch, N.; Yeheskel, A.; Dvir, A.; Soussan-Gutman, L.; Jeselsohn, R.; Rizel, S. D538G mutation in estrogen receptor-α: A novel mechanism for acquired endocrine resistance in breast cancer. *Cancer Res.* **2013**, *73*, 6856–6864. [CrossRef]
171. O'Regan, R.M.; Paplomata, E. New and emerging treatments for estrogen receptor-positive breast cancer: Focus on everolimus. *Ther. Clin. Risk Manag.* **2013**, *9*, 27–36. [CrossRef]
172. Hoppe, R.; Achinger-Kawecka, J.; Winter, S.; Fritz, P.; Lo, W.-Y.; Schroth, W.; Brauch, H. Increased expression of miR-126 and miR-10a predict prolonged relapse-free time of primary oestrogen receptor-positive breast cancer following tamoxifen treatment. *Eur. J. Cancer* **2013**, *49*, 3598–3608. [CrossRef] [PubMed]

173. Kovalchuk, O.; Filkowski, J.; Meservy, J.; Ilnytskyy, Y.; Tryndyak, V.P.; Chekhun, V.F.; Pogribny, I.P. Involvement of microRNA-451 in resistance of the MCF-7 breast cancer cells to chemotherapeutic drug doxorubicin. *Mol. Cancer Ther.* **2008**, *7*, 2152–2159. [CrossRef] [PubMed]
174. Luqmani, Y.A.; Al Saleh, S.; Sharaf, L.H. Signalling pathways involved in endocrine resistance in breast cancer and associations with epithelial to mesenchymal transition. *Int. J. Oncol.* **2011**, *38*, 1197–1217. [CrossRef] [PubMed]
175. Creighton, C.J.; Fu, X.; Hennessy, B.T.; Casa, A.J.; Zhang, Y.; Gonzalez-Angulo, A.M.; Lluch, A.; Gray, J.W.; Brown, P.H.; Hilsenbeck, S.G.; et al. Proteomic and transcriptomic profiling reveals a link between the PI3K pathway and lower estrogen-receptor (ER) levels and activity in ER+ breast cancer. *Breast Cancer Res.* **2010**, *12*, R40. [CrossRef] [PubMed]
176. Cavazzoni, A.; Bonelli, M.; Fumarola, C.; La Monica, S.; Airoud, K.; Bertoni, R.; Alfieri, R.; Galetti, M.; Tramonti, S.; Galvani, E.; et al. Overcoming acquired resistance to letrozole by targeting the PI3K/AKT/mTOR pathway in breast cancer cell clones. *Cancer Lett.* **2012**, *323*, 77–87. [CrossRef] [PubMed]
177. Johnston, S.R. New strategies in estrogen receptor-positive breast cancer. *Clin. Cancer Res.* **2010**, *16*, 1979–1987. [CrossRef]
178. Houghton, P.J. Everolimus. *Clin. Cancer Res.* **2010**, *16*, 1368–1372. [CrossRef]
179. Lee, J.J.X.; Loh, K.; Yap, Y.-S. PI3K/Akt/mTOR inhibitors in breast cancer. *Cancer Biol. Med.* **2015**, *12*, 342–354.
180. Baselga, J.; Campone, M.; Piccart, M.; Burris, H.A., III; Rugo, H.S.; Sahmoud, T.; Noguchi, S.; Gnan, M.; Pritchard, K.I.; Lebrun, F.; et al. Everolimus in postmenopausal hormone-receptor-positive advanced breast cancer. *N. Engl. J. Med.* **2012**, *366*, 520–529. [CrossRef]
181. Schmid, P.; Zaiss, M.; Harper-Wynne, C.; Ferreira, M.; Dubey, S.; Chan, S.; Ruiz, I. Fulvestrant plus vistusertib vs. fulvestrant plus everolimus vs. fulvestrant alone for women with hormone receptor–positive metastatic breast cancer: The MANTA phase 2 randomized clinical trial. *JAMA Oncol.* **2019**, *5*, 1556–1563. [CrossRef]
182. Bachelot, T.; Bourgier, C.; Cropet, C.; Ray-Coquard, I.; Ferrero, J.-M.; Freyer, G.; Abadie-Lacourtoisie, S.; Eymard, J.-C.; Debled, M.; Spaëth, D.; et al. Randomized phase II trial of everolimus in combination with tamoxifen in patients with hormone receptor-positive, human epidermal growth factor receptor 2–negative metastatic breast cancer with prior exposure to aromatase inhibitors: A GINECO Study. *J. Clin. Oncol.* **2012**, *30*, 2718–2724. [CrossRef] [PubMed]
183. André, F.; O'Regan, R.; Ozguroglu, M.; Toi, M.; Xu, B.; Jerusalem, G.; Masuda, N.; Wilks, S.; Arena, F.; Isaacs, C.; et al. Everolimus for women with trastuzumab-resistant, HER2-positive, advanced breast cancer (BOLERO-3): A randomised, double-blind, placebo-controlled phase 3 trial. *Lancet Oncol.* **2014**, *15*, 580–591. [CrossRef]
184. Hurvitz, S.A.; Dalenc, F.; Campone, M.; O'Regan, R.M.; Tjan-Heijnen, V.C.; Gligorov, J.; Llombart, A.; Jhangiani, H.; Mirshahidi, H.R.; Tan-Chiu, E.; et al. A phase 2 study of everolimus combined with trastuzumab and paclitaxel in patients with HER2-overexpressing advanced breast cancer that progressed during prior trastuzumab and taxane therapy. *Breast Cancer Res. Treat.* **2013**, *141*, 437–446. [CrossRef] [PubMed]
185. Boulay, A.; Rudloff, J.; Ye, J.; Zumstein-Mecker, S.; O'Reilly, T.; Evans, D.B.; Chen, S.; Lane, H.A. Dual inhibition of mTOR and estrogen receptor signaling in vitro induces cell death in models of breast cancer. *Clin. Cancer Res.* **2005**, *11*, 5319–5328. [CrossRef] [PubMed]
186. Yi, Z.; Ma, F.; Liu, B.; Guan, X.; Li, L.; Li, C.; Qian, H.; Xu, B. Everolimus in hormone receptor-positive metastatic breast cancer: PIK3CA mutation H1047R was a potential efficacy biomarker in a retrospective study. *BMC Cancer* **2019**, *19*, 442. [CrossRef] [PubMed]
187. Yu, K.; Toral-Barza, L.; Discafani, C.; Zhang, W.G.; Skotnicki, J.; Frost, P.; Gibbons, J.J. mTOR, a novel target in breast cancer: The effect of CCI-779, an mTOR inhibitor, in preclinical models of breast cancer. *Endocr.-Relat. Cancer* **2001**, *8*, 249–258. [CrossRef]
188. Wolff, A.C.; Lazar, A.A.; Bondarenko, I.; Garin, A.M.; Brincat, S.; Chow, L.; Sun, Y.; Neskovic-Konstantinovic, Z.; Guimaraes, R.C.; Fumoleau, P.; et al. Randomized phase III placebo-controlled trial of letrozole plus oral temsirolimus as first-line endocrine therapy in postmenopausal women with locally advanced or metastatic breast cancer. *J. Clin. Oncol.* **2013**, *31*, 195–202. [CrossRef]
189. Fleming, G.F.; Ma, C.X.; Huo, D.; Sattar, H.; Tretiakova, M.; Lin, L.; Hahn, O.M.; Olopade, F.O.; Nanda, R.; Hoffman, P.C.; et al. Phase II trial of temsirolimus in patients with metastatic breast cancer. *Breast Cancer Res. Treat.* **2012**, *136*, 355–363. [CrossRef]
190. Sadler, T.M.; Gavriil, M.; Annable, T.; Frost, P.; Greenberger, L.M.; Zhang, Y. Combination therapy for treating breast cancer using antiestrogen, ERA-923, and the mammalian target of rapamycin inhibitor, temsirolimus. *Endocr.-Relat. Cancer* **2006**, *13*, 863–873. [CrossRef]
191. Bhattacharvva, G.S.; Biswas, J.; Singh, J.K.; Singh, M.; Govindbabu, K.; Ranade, A.; Malhotra, H.; Parikh, P.; Shahid, T.; Basu, S. Reversal of tamoxifen resistance (hormone resistance) by addition of Sirolimus (mTOR Inhibitor) in metastatic breast cancer. *Eur. J. Cancer* **2011**, *47*, 9. [CrossRef]
192. Seiler, M.; Ray-Coquard, I.; Melichar, B.; Yardley, D.A.; Wang, R.X.; Dodion, P.F.; Lee, M.A. Oral Ridaforolimus plus Trastuzumab for patients with HER2+ trastuzumab-refractory metastatic breast cancer. *Clin. Breast Cancer* **2015**, *15*, 60–65. [CrossRef] [PubMed]
193. Shi, J.J.; Chen, S.M.; Guo, C.L.; Li, Y.X.; Ding, J.; Meng, L.H. The mTOR inhibitor AZD8055 overcomes tamoxifen resistance in breast cancer cells by down-regulating HSPB8. *Acta Pharmacol. Sin.* **2018**, *39*, 1338–1346. [CrossRef] [PubMed]
194. Jordan, N.J.; Dutkowski, C.M.; Barrow, D.; Mottram, H.J.; Hutcheson, I.R.; Nicholson, R.I.; Guichard, S.M.; Gee, J.M.W. Impact of dual mTORC1/2 mTOR kinase inhibitor AZD8055 on acquired endocrine resistance in breast cancer in vitro. *Breast Cancer Res.* **2014**, *23*, R12. [CrossRef] [PubMed]
195. Petrossian, K.; Nguyen, D.; Lo, C.; Kanaya, N.; Somlo, G.; Cui, Y.X.; Huang, C.-S.; Chen, S. Use of dual mTOR inhibitor MLN0128 against everolimus-resistant breast cancer. *Breast Cancer Res. Treat.* **2018**, *170*, 499–506. [CrossRef] [PubMed]

196. Bostner, J.; Alayev, A.; Berman, A.Y.; Fornander, T.; Nordenskjöld, B.; Holz, M.K.; Stål, O. Raptor localization predicts prognosis and tamoxifen response in estrogen receptor-positive breast cancer. *Breast Cancer Res. Treat.* **2018**, *168*, 17–27. [CrossRef]
197. Zhu, L.; Li, X.; Shi, L.; Wu, J.; Qian, J.; Xia, T.; Zhou, W.-B.; Sun, X.; Xu-Jie, Z.; Wei, J.-F.; et al. Rapamycin enhances the sensitivity of ER-positive breast cancer cells to tamoxifen by upregulating p73 expression. *Oncol. Rep.* **2019**, *41*, 455–464. [CrossRef]

International Journal of
Molecular Sciences

MDPI

Article

SATB1-Mediated Upregulation of the Oncogenic Receptor Tyrosine Kinase HER3 Antagonizes MET Inhibition in Gastric Cancer Cells

Robert Jenke [1,2], Miriam Holzhäuser-Rein [1,2], Stefanie Mueller-Wilke [1,2], Florian Lordick [1], Achim Aigner [2,*] and Thomas Büch [2]

1 University Cancer Center Leipzig (UCCL), University Hospital Leipzig, D-04103 Leipzig, Germany; robert.jenke@medizin.uni-leipzig.de (R.J.); mr41refa@studserv.uni-leipzig.de (M.H.-R.); Stefanie.Mueller3@medizin.uni-leipzig.de (S.M.-W.); florian.lordick@medizin.uni-leipzig.de (F.L.)
2 Clinical Pharmacology, Rudolf-Boehm-Institute for Pharmacology and Toxicology, Medical Faculty, Leipzig University, D-04107 Leipzig, Germany; thomas.buech@medizin.uni-leipzig.de
* Correspondence: achim.aigner@medizin.uni-leipzig.de; Tel.: +49-(0)341-97-24661

Abstract: MET-amplified gastric cancer cells are extremely sensitive to MET inhibition in vitro, whereas clinical efficacy of MET inhibitors is disappointing. The compensatory activation of other oncogenic growth factor receptors may serve as an underlying mechanism of resistance. In this study, we analyzed the role of HER receptors, in particular HER3 and its ligand heregulin, in this respect. This also included the chromatin-organizer protein SATB1, as an established regulator of HER expression in other tumor entities. In a panel of MET-amplified gastric carcinoma cell lines, cell growth under anchorage-dependent and independent conditions was studied upon inhibitor treatment or siRNA-mediated knockdown. Expression analyses were performed using RT-qPCR, FACS, and immunoblots. Signal transduction was monitored via antibody arrays and immunoblots. As expected, MET inhibition led to a growth arrest and inhibition of MAPK signaling. Strikingly, however, this was accompanied by a rapid and profound upregulation of the oncogenic receptor HER3. This finding was determined as functionally relevant, since HER3 activation by HRG led to partial MET inhibitor resistance, and MAPK/Akt signaling was even found enhanced upon HRG+MET inhibitor treatment compared to HRG alone. SATB1 was identified as mediator of HER3 upregulation. Concomitantly, SATB1 knockdown prevented upregulation of HER3, thus abrogating the HRG-promoted rescue from MET inhibition. Taken together, our results introduce the combined HER3/MET inhibition as strategy to overcome resistance towards MET inhibitors.

Keywords: gastric cancer; HER3; heregulin; MET; SATB1

Citation: Jenke, R.; Holzhäuser-Rein, M.; Mueller-Wilke, S.; Lordick, F.; Aigner, A.; Büch, T. SATB1-Mediated Upregulation of the Oncogenic Receptor Tyrosine Kinase HER3 Antagonizes MET Inhibition in Gastric Cancer Cells. *Int. J. Mol. Sci.* **2021**, *22*, 82. https://dx.doi.org/10.3390/ijms22010082

Received: 26 November 2020
Accepted: 17 December 2020
Published: 23 December 2020

Publisher's Note: MDPI stays neutral with regard to jurisdictional claims in published maps and institutional affiliations.

1. Introduction

Gastric cancer is one of the most common cancer types [1,2]. Worldwide, it represents the second or third most common cause of cancer-related deaths [3,4], and the lifetime risk of developing gastric cancer is about one case per 100 persons. Until now, complete surgical resection of the tumor is a prerequisite for curative treatment [5,6]. However, gastric cancer is often advanced and inoperable at the time point of diagnosis, and conventional cytoreductive chemotherapy is of rather limited efficacy. Thus, there is a desperate need for novel systemic treatment approaches to improve prognosis, especially in metastatic gastric cancer [7]. Recently, targeted therapies against oncogenic receptor tyrosine kinases (RTKs), e.g., FGFR, HER1, HER2, HER3, or MET, have been tested in patients with gastric cancer [8]. Despite promising findings in cell culture, the clinical efficacy of these novel therapeutics has been rather limited in most cases, with the partial exception of HER2 inhibition, showing a statistically significant albeit small survival advantage in a subset of patients with HER2 overexpression [9].

Discrepancies between preclinical data and the clinical reality are particularly striking in the case of the inhibition of the HGF/MET axis in gastric cancer cells harboring a MET amplification, which occurs in 3–7% of gastric tumors [10,11]. In these cells, marked anti-proliferative effects are observed in vitro after inhibition of MET; however, until now clinical trials with HGF or MET inhibitors have not produced any breakthrough [12]. The difficulties of translating positive preclinical data into favorable clinical outcomes can be attributed in part to the problem of identifying the correct subgroup of patients suitable for a specific molecularly defined targeted therapy. While it appears reasonable to pre-select patients with tumors showing high expression levels of the respective target molecule, it should be noted that the overexpression of a given oncogene does not necessarily translate into high sensitivity of tumor cells towards its inhibition. This indicates that expression levels may be a poor predictor of therapy response, with the potential redundancy of oncogenic signaling pathways in tumors being one explanation for this discrepancy. Indeed, upon inhibition of a distinct critical pathway, for example a specific RTK, the activation of other signaling molecules can compensate for its reduced function [13,14].

It has been proposed that members of the HER family of RTKs, especially HER2 or HER3, could compensate for a reduced MET function, thus contributing to tumor resistance against MET inhibitors. In fact, stimulation of MET-amplified gastric cancer cells with the HER agonist heregulin (HRG) could ameliorate the cytotoxic effects of MET inhibitors [15–17]. From these findings, the questions arise (1) whether the HRG rescue effect is relevant for all MET-amplified gastric cancer cells, (2) if alterations in HER receptor expression and/or signaling are observed upon MET inhibition, (3) whether HRG elicits its positive effects via HER1/HER3 or HER2/HER3-promoted survival signaling in this context, and (4) which other molecules are involved in MET resistance. The chromatin organizer protein SATB1 has been shown to be upregulated in many solid tumors and, as proto-oncogene, to affect the expression of many tumor-relevant gene products including HER receptors [18–20]. Thus, we hypothesized that SATB1 could also be involved in HER-dependent resistance mechanisms upon MET inhibition in gastric cancer cells.

In this study, we address the functional relevance of HER receptors, and in particular of HER3, in MET-amplified gastric cancer cell lines. This also includes the role of SATB1 in this process. We show a rapid and substantial increase in HER3 expression upon inhibiting MET, which is mediated by SATB1 and leads to an even enhanced heregulin (HRG)/HER3 signaling. This establishes the role of HRG/HER3 signaling in mediating resistance of gastric cancer cells towards MET inhibition. Thus, our findings provide an avenue towards increasing the efficacy of MET-directed therapeutic interventions.

2. Results

2.1. MET-Amplified Gastric Cancer Cells Are Highly Sensitive to MET Inhibition or siRNA-Mediated MET Knockdown

To investigate the role of HER receptors in resistance of gastric cancer cells against MET inhibitors, we used a panel of five gastric cancer cell lines, three of which (MKN45, Hs746T, and SNU5) were described as MET-amplified and sensitive to MET inhibition, and two (MKN7 and MKN74) are not MET-amplified. Expression analyses on the mRNA level confirmed the exceptionally high MET expression in the three MET-amplified cell lines (Figure 1A). MET-amplified cells showed high sensitivity towards MET inhibition via the specific inhibitor PF04217903 (0.2 μM) (Figure 1B, Supplementary Materials Figure S1A,B) or siRNA-mediated downregulation of MET (Figure 1C). In contrast, no anti-proliferative effects were observed in cell lines MKN74 or MKN7 without MET amplification, even for MET inhibitor concentrations of up to 2 μM or following siRNA-mediated MET knockdown (Supplementary Materials Figure S2A–D).

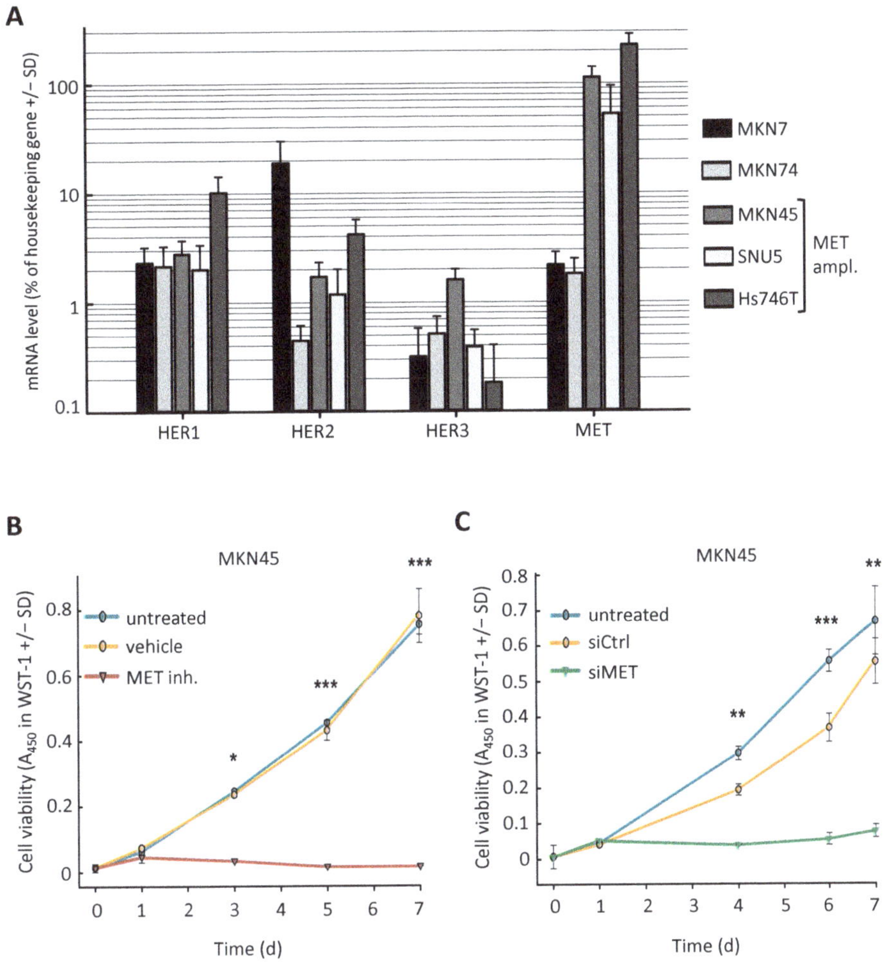

Figure 1. (**A**) Expression levels of MET and HER receptor mRNA in different gastric carcinoma cell lines. mRNA expression was determined using quantitative RT-PCR using RPLP0 as housekeeping gene. (**B**,**C**) Effects of MET inhibition via small molecule inhibitor or siRNA on tumor cell proliferation in MET-amplified MKN45 cells. Using WST-1 reagent cell proliferation was monitored on day 0 (prior to treatment) and day 3, 5, and 7. Upon treatment with 0.2 µM MET inhibitor PF04217903 (**B**) or siRNA-mediated MET knockdown (**C**), profound inhibition was observed. Level of significance: *, $p < 0.05$; **, $p < 0.01$, and ***, $p < 0.001$.

2.2. Downregulation or Inhibition of MET Leads to Upregulation of HER3

It has been described previously in non-gastric cancer cell lines that resistance of HER receptor overexpressing cells towards inhibition or knockdown can be attributed to the adaptive activation of other HER family members ([13,14] for review). Thus, we next asked the question whether the targeting of MET, despite its profound cell-inhibitory effects, may lead to similar alterations. Of note, a very strong > 6-fold upregulation of HER3 was detected in MKN45 cells on the mRNA (Figure 2A,B) and protein level (Figure 2C). Western blot data were also confirmed by flow cytometry (Supplementary Materials Figure S3). Since this method is very quantitative and also allows for specifically monitoring cell surface levels, we sticked to flow cytometry for measuring HER3 protein in subsequent experiments. This HER3 upregulation was independent of whether MET inhibition was achieved by siRNA-mediated knockdown or using the inhibitor PF04217903. The same increase in HER3 levels was observed in SNU5 cells on mRNA (Figure 2D) and protein level (Figure 2E). In contrast, in Hs746T cells a less pronounced ~1.5 increase in HER3 was observed, but in this cell line, it was accompanied by a concomitant induction of HER1 and HER2 in the same range (Figure 2F).

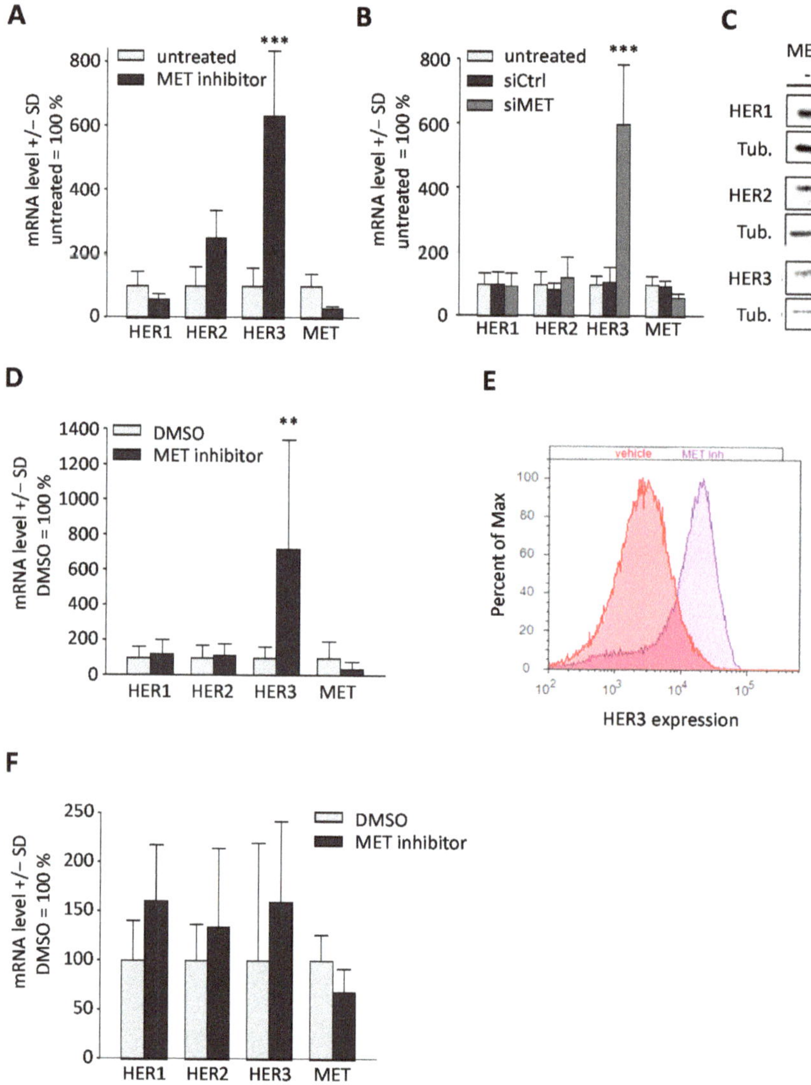

Figure 2. Inhibition of MET leads to HER3 receptor upregulation in MET-amplified MKN45 and SNU5 cells, but not in Hs746T cells. (**A**) After treatment of MKN45 cells, with MET inhibitor PF04217903 (0.2 μM for 48 h) a pronounced upregulation of HER3 was traceable on mRNA level. (**B**) Transfection of MKN45 cells with specific siRNA against MET for 48 h yielded similar HER3 upregulation results. (**C**) Accordingly, 48 h treatment of MKN45 cells with 0.2 μM of PF04217903 also led to upregulation of HER3 on protein level, whereas differential effects occurred for HER1 and HER2. (**D**) In SNU5 cells, treatment with 0.2 μM PF04217903 also showed marked HER3 upregulation on mRNA level. (**E**) Moreover, a shift in expression of HER3 protein level was observed after 0.2 μM PF04217903 treatment (48 h). (**F**) Contrastingly, no HER3 upregulation was traceable in Hs746T cells under these conditions. Level of significance: **, $p < 0.01$, and ***, $p < 0.001$.

Additionally, various responses were noted with regard to HER1 and HER2 levels: in MKN45 cells, HER1 mRNA was slightly reduced upon MET inhibitor treatment (Figure 2A), but not after RNAi-mediated knockdown of MET (Figure 2B). Of note, these HER1 effects upon MET inhibitor treatment were also discernible on protein level (Figure 2C). In contrast, in SNU5 cells no major effects were found (Figure 2D), and in Hs746T cells, even a minor HER1 induction occurred (Figure 2F). Regarding HER2 expression, a strong mRNA induction was discernible in MKN45 cells (Figure 2A), which was, however, not seen on protein levels (Figure 2C) and may be, therefore, of minor relevance. For the other cell lines, only weak effects on HER2 were found (Figure 2D,F). Additionally, the determination of mRNA levels also revealed that treatment of cells with the MET inhibitor led to a marked reduction in MET after 48 h, indicating an inhibitory effect of PF04217903 on the transcription of its target (Figure 2A,D,F). Taken together, this identifies HER3 as a candidate oncogene for mediating resistance towards MET inhibition.

2.3. Anti-Proliferative Effects of MET Inhibition Are Partially Abolished by Treatment with HER3 Activator Heregulin

The interplay between MET inhibition and alterations in HER receptor expression levels suggested the possibility that the very profound anti-proliferative effects of the MET inhibitor may be counteracted by HER3 activation in the presence HER receptor ligands. Indeed, addition of heregulin (HRG) in the physiological concentration of 20 ng/mL to the culture media led to a partial rescue of MET inhibitor-mediated (0.2 μM of PF04217903) arrest in proliferation in MKN45 cells. This was even true in the constant presence of the inhibitor and thus under conditions of sustained MET inhibition (Figure 3A). In the absence of HRG, earlier removal of the inhibitor after 48 h did not lead to reduced inhibition of cell proliferation over time but resulted in a further enhanced HRG-mediated rescue effect (Figure 3A). Thus, albeit HRG could not fully compensate for the MET inhibitor effects, a major recovery of cell proliferation was observed. Likewise, in colony formation assays MET inhibitor (0.2 μM of PF04217903 for 48 h) severely impaired MKN45 colony formation, an effect that was reversed by HRG treatment (Supplementary Materials Figure S4). This protective effect was also seen in a spheroid growth assay, where the MET inhibitor alone almost completely abrogated spheroid growth, whereas in the presence of HRG, the three-dimensional growth was partially retained with spheroid sizes reaching ~30% of the control cells (Figure 3B). In SNU5 cells, comparable results were obtained (data not shown). In contrast, in Hs476T cells lacking the very profound HER3 induction upon MET inhibition (see Figure 2D), HRG treatment could not antagonize the growth inhibition of the MET inhibitor in WST-1 assays (Figure 3C) or spheroid outgrowth assays (Figure 3D). This identifies HER3 upregulation, in combination with the presence of its ligand HRG, as a mediator of resistance towards MET inhibition.

When analyzing the percentages of viable and dead cells in a live/dead cell assay, a substantial > 4-fold increase in apoptotic cells upon exposure to the MET inhibitor was observed. Again, however, this was markedly reduced in the combined treatment scenario with MET inhibitor plus HRG, indicating the pro-survival effects of HER3/HRG signaling under MET inhibition (Figure 3E).

The functional relevance of HER3 was further explored by siRNA-mediated parallel HER3 knockdown. In negative control transfected cells and in the absence of MET inhibitor, no further stimulation of cell proliferation was obtained upon addition of HRG or the HER1 ligand EGF (Figure 3F, left). In contrast, the proliferation arrest exerted by the MET inhibitor could again be rescued by >50% upon addition of HRG, while treatment with the HER1 ligand EGF was without effect (Figure 3F, left). This identifies HER3 rather than HER1 as relevant in this context. Upon siRNA-mediated transient HER3 knockdown, a marked reduction in cell proliferation was seen (Figure 3G; note the y-axis scale different to Figure 3F). This was further augmented by parallel treatment with the MET inhibitor. Notably, the HER3 knockdown abolished the recovery of cell proliferation upon addition of HRG (Figure 3F, right), indicating that the HRG-mediated rescue described above is indeed dependent on HER3.

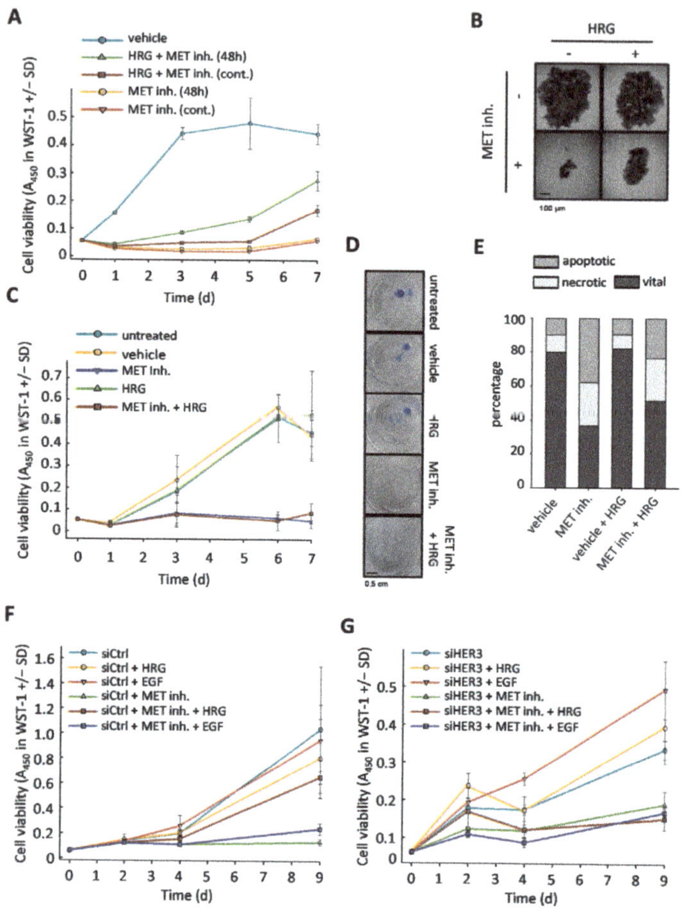

Figure 3. (**A**) In WST-1 assay, impaired MKN45 cell proliferation (vs. vehicle DMSO) upon MET inhibition is partially rescued by treatment with 20 ng/mL of heregulin (HRG); (cont.): continuous inhibitor (0.2 µM PF04217903) exposure, (48 h): removal of MET inhibitor after 48 h. (**B**) HRG (20 ng/mL) also partially reversed antiproliferative effects of MET inhibition in MKN45 cells on spheroid formation. In contrast, the severe effects upon MET inhibition could not be reversed by HRG in Hs746T cells as shown (**C**) in WST-1 assay or (**D**) spheroid outgrow formation; of note, these cells did not show compensatory HER3 upregulation upon MET inhibition, as has been shown in the previous Figure 2F. (**E**) Addition of HRG (20 ng/mL) also reduced the number of apoptotic cells in SNU5 cells treated with 0.2 µM PF04217903. HER3 displays the crucial factor mediating resistance against MET inhibition, as (**F**) the HER1 ligand EGF (50 ng/mL) could not reverse the antiproliferative effects of PF04217903 in contrast to HRG treatment, and (**G**) HER3 knockdown abrogated the HRG-induced rescue effects.

2.4. Cellular Effects Are Mediated by Alterations in MAPK Signaling

To further characterize the pathways involved in the cellular effects of MET inhibition and HRG stimulation/rescue, phospho-antibody arrays were performed for analyzing changes in MAPK activities (Figure 4A,B, Figure S5). Upon addition of the MET inhibitor to MKN45 cells, reduced phosphorylation of Akt (especially Akt2) and of ERK1/2 was observed (Figure 4A). In contrast, HRG stimulation of the cells led to further enhancement of Akt signaling, with little effects on ERK1/2 phosphorylation. Notably, the inhibitory effects of the MET inhibitor on phosphorylation/activation were, except for Akt3, reversed upon HRG addition (Figure 4A, lower panel). The heat map analysis confirmed Akt and

ERK to be most profoundly affected. In fact, the quantitation of the signals revealed an even increase in Akt signaling upon combined MET inhibitor + HRG treatment and very profound ~3–4-fold higher ERK1/2 phosphorylation (Figure 4C). This further increase in Akt and ERK1/2 signaling to values above those obtained by HRG stimulation alone without MET inhibitor was also confirmed in independent Western blot experiments (Figure 4D, right panel). The profoundly increased p-ERK1/2 and p-Akt levels in MET + HRG treated cells over HRG single treatment can be explained by the upregulation of HER3 (and perhaps HER2) described above.

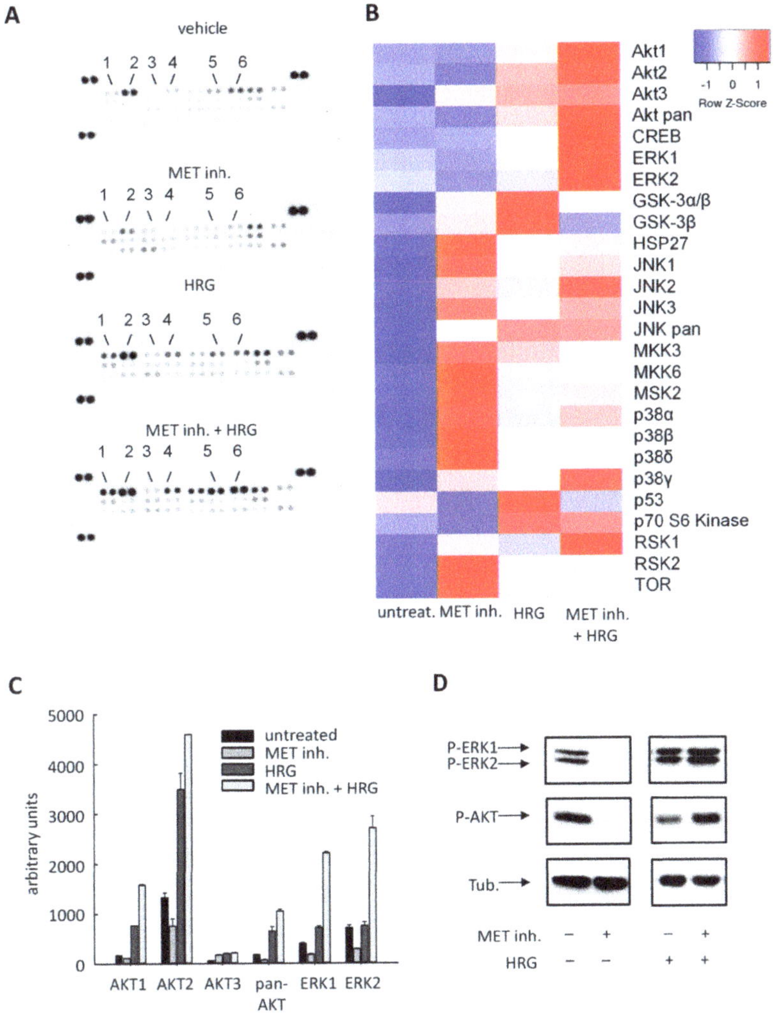

Figure 4. (**A**) Phospho-antibody arrays to elucidate downstream signal transduction effects of MET inhibition, heregulin stimulation, and the combination of both as compared to untreated (upper panel). 1, Akt1; 2, Akt2; 3, Akt3; 4, pan Akt; 5, ERK1; 6, ERK2. For the definition of all spots, see Supplementary Materials Figure S5. Cells were serum-starved for 18 h and subsequently treated with vehicle DMSO or inhibitor/HRG for 24 h. Interestingly, MET inhibition followed by HER3 stimulation via HRG showed the strongest phosphorylation levels. (**B**) Quantification of signal intensities from antibody arrays. The heat map depicts alterations upon treatment as indicated towards lower (blue) or higher signals (red). (**C**) Bar diagram showing the intensities of signaling molecules with most profound alterations. (**D**) Confirmation of alterations in ERK1/2 and Akt phosphorylation by Western blotting. Again, cells were serum-starved for 18 h and treated as indicated for 24 h.

2.5. Upregulation of HER3 upon MET Inhibition Is Dependent on PKC and SATB1

We further analyzed the underlying molecular mechanism of this counter-upregulation. We did not find any evidence that the blockade of MAPK or AKT signaling induced by MET inhibition was involved in HER3 regulation, since the inhibition of MAPK by the MAPKK inhibitor PD98059 (10 μM for 48 h) or the blocking of AKT signaling via PI3K inhibition with LY294002 (10 μM for 48 h) did not reproduce the effects of MET inhibition on HER3 levels (Supplementary Materials Figure S6).

Since PKC is a known MET target, we next tested its role by pretreating MKN45 or SNU5 cells with the PKC inhibitor BIM II (10 μM) for 24 h, prior to the addition of the MET inhibitor PF 04217903 (0.2 μM for 48 h). Under these conditions the upregulation of HER3 was abrogated in MKN45 (Figure 5A, left) and markedly inhibited in SNU5 cells (Figure 5A, right). For HER1 and HER2, only minor effects were discernible. Vice versa, the PKC activator PMA (1 μM, for 48 h) led to a marked upregulation of HER3 in both MKN45 and SNU5 cells (Figure 5B). Of note, we also observed a strong upregulation of the transcriptional HER regulator SATB1 in SNU5, but not in MKN45 cells (Figure 5A). SATB1 has been shown in breast carcinoma to affect the expression of HER receptors [18–20] and was found upregulated in gastric cancer [21]. Thus, we next assessed its role in HER3 upregulation. To this end, we employed a specific siRNA, which was described previously to efficiently reduce SATB1 expression [19].

Treatment of MKN45 cells with MET inhibitor again led to ~ 50% decreased HER1 mRNA levels, independent of prior transient transfection with SATB1 siRNA or a non-specific negative control siRNA (Figure 5C, left). Slight effects of SATB1 knockdown on basal (i.e., in the absence of MET inhibitor) expression of HER1 and HER3 were observed. In contrast, the marked upregulation of HER3 under MET inhibitor treatment was almost fully abolished upon SATB1 knockdown (Figure 5C, left). The strong dependence of the HER3 counter-upregulation on SATB1 expression thus indicates that it is mediated through SATB1. Similarly, in SNU5 cells, knockdown of SATB1 abrogated the strong HER3 induction upon MET inhibition (Figure 5C, right). Of note, in this cell line, MET inhibition per se again led to a marked upregulation of SATB1 reproducing the data shown in Figure 5A, right, and underlining the potential interplay between MET signaling, SATB1 function, and HER3 expression. Note that in both cell lines MET inhibition again reduced the expression of MET receptor itself (Figure 5C), indicating a yet unknown putative transcriptional inhibitory mechanism on MET activity. This MET downregulation after MET inhibition was not affected by SATB1 knockdown.

The SATB1-dependent regulation of HER3 expression upon MET inhibition was also found on the protein level. While the MET inhibitor led to a pronounced increase in HER3 expression in control transfected cells (Figure 5D, upper panel), this effect was markedly reduced upon SATB1 knockdown (Figure 5D, lower panel). Addressing the possible consequences of SATB1 affecting HER3 expression, we analyzed cell viabilities. The RNAi-mediated reduction in SATB1 expression did not lead to major alterations of viable cell numbers in untreated or HRG-stimulated cells, or in cells treated with MET inhibitor (Figure 5E). Notably, however, the HRG-mediated partial restoration of cell proliferation under MET inhibition was almost completely abolished upon SATB1 knockdown (Figure 5F, right bars), demonstrating the dependence of this effect on SATB1-mediated HER3 upregulation.

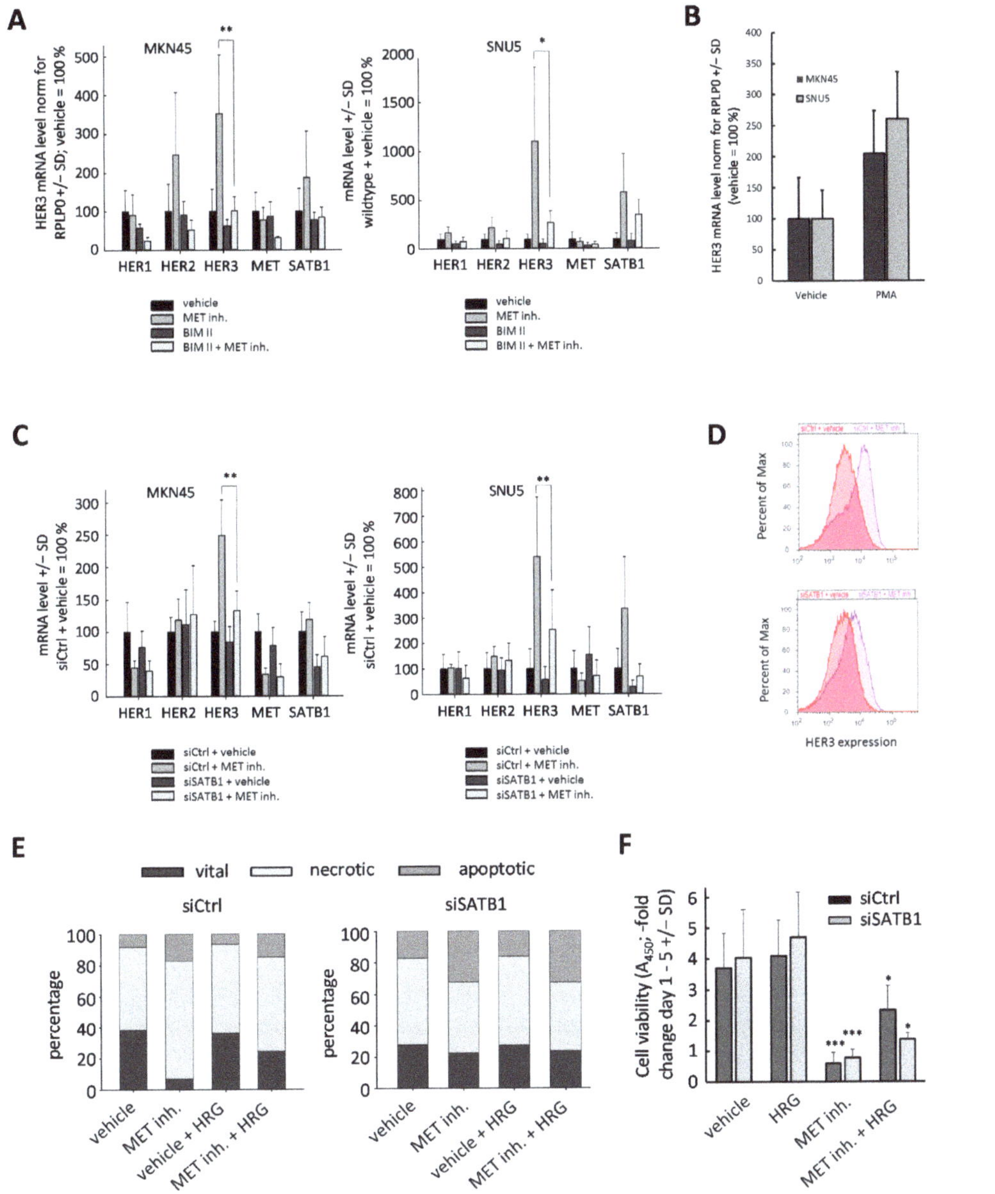

Figure 5. (**A**) MKN45 cells (left) and SNU5 cells (right) were pretreated with PKC inhibitor BIM II (10 μM) for 24 h; afterwards, cells were treated with MET inhibitor PF04217903 (0.2 μM for 48 h). (**B**) MKN45 and SNU5 cells were treated for 48 h with the PKC activator PMA (1 μM) before analyzing HER3 mRNA expression. (**C**) MKN45 cells (left) and SNU5 cells (right) were pretreated with SATB1 siRNA for 48 h prior to MET inhibition with 0.2 μM PF04217903 to induce compensatory upregulation effects, as shown previously. siSATB1 significantly reduced the magnitude of observed HER3 mRNA upregulation. (**D**) SNU5 cells also showed diminished HER3 upregulation on the protein level after SATB1 knockdown (48 h after transfection) in comparison with control siRNA. (**E**,**F**) While siSATB1 showed no antiproliferative effect itself on MKN45 cells, the HRG-associated rescue effect was reduced upon SATB1 knockdown as compared to siCtrl. Level of significance: *, $p < 0.05$; **, $p < 0.01$, and ***, $p < 0.001$.

3. Discussion

In the present study, we demonstrate that treatment of MET-amplified gastric cancer cells with a MET inhibitor leads to a SATB1-mediated upregulation of HER3. In the absence of HER3 ligands in cell culture, which are not endogenously produced by the tumor cells, this adaptive and rapid induction of HER3 did not confer resistance towards MET inhibition. In contrast, in the presence of the HER3 ligand heregulin, a scenario, which resembles more closely the in vivo situation in tumors (see below), a partial rescue of the cancer cells from the detrimental effects of MET inhibition was observed.

It has been shown previously that MET inhibition in monocultured gastric cancer cells with MET amplification exerts dramatic anti-proliferative effects, in parallel with abrogation of ERK and Akt phosphorylation. This can be overcome in part by HER activation [15,16,22]. Heregulin can activate HER3-promoted signaling pathways via HER1/HER3 or HER2/HER3 heterodimers [23]; however, it has not been elucidated so far which of these heterodimers mediate these pro-survival effects in MET addicted gastric cancer cells. We demonstrate that inhibition of HER2 or HER3 via siRNA-mediated knockdown or a small molecule inhibitor abrogated the rescue effect of heregulin, giving proof of the relevance of intact HER2/HER3 signaling. In contrast, we found the treatment of the cells with the HER1 ligand EGF ineffective in mediating any resistance against MET inhibition. This effect cannot be explained by an insufficient dosing of EGF, since EGF treatment led to a comparable reactivation of ERK phosphorylation as did HRG (Supplementary Materials Figure S7).

The fact that Akt phosphorylation after MET inhibition is more efficiently restored by HRG treatment of gastric cancer cells than by EGF treatment is in line with previous findings [15] and indicates that PI3K-Akt signaling is of particular importance for survival signaling in gastric cancer cells. Somewhat contrasting previous findings that demonstrated the ability of EGF (in a concentration comparable to our study) to confer resistance against MET inhibition as well [15,16] may be attributable to the fact that different MET inhibitors were used. Of note, the inhibitor PHA 665752 used previously at a concentration of 250 nM would also inhibit Ron and at least partially Flk-1 (IC50: 200 nM), whereas PF 04217903 employed here offers greater selectivity towards MET [24].

Remarkably, MET inhibition led to a substantial upregulation of HER3, the critical signaling molecule responsible for heregulin-promoted survival. Of note, HER3 has been characterized as a significant factor for tumor progression in gastric cancer and is often upregulated in this tumor entity (see [25] for review). On the transcriptional level, HER3 expression in gastric cancer is critically regulated by the transcription factor EHF and overexpression of EHF leads to increased HER3 levels [26]. With respect to the adaptive response upon MET inhibition observed here, it is noteworthy that PI3K-AKT inhibition, which is a consequence of MET inhibition in MET-amplified gastric cancer cells, can induce HER3 upregulation in other tumor entities via a FOXO-dependent mechanism [27]. However, PI3K-AKT signaling does not seem to play a crucial role in the present context, since PI3K inhibition in gastric cancer cells had no impact on basal HER3 expression or on HER3 induction after MET inhibitor treatment (Supplementary Materials Figure S6).

On the mechanistic level, we identify SATB1 as a mediator of this HER3 upregulation. Concomitantly, SATB1 knockdown abrogated the HRG-promoted rescue of gastric cancer cells after MET inhibition. SATB1 acts as a chromatin organizer, and dependent on the cellular context and on post-translational modifications, SATB1 has been shown to act as a repressor or activator of gene expression [28,29]. SATB1 affects the expression of a large number of oncogenic signaling molecules, and consequently, its function has been studied in several tumor entities [30,31]. While in gastric cancer the role of SATB1 for the regulation of oncogene expression is still elusive, a meta-analysis has revealed that SATB1 expression itself represents a potential marker for unfavorable prognosis, emphasizing its putative relevance in this tumor type [21]. In line with this, SATB1 has been found to increase viability, invasiveness, and chemoresistance of gastric cancer cells and to promote tumor growth in vivo [32,33].

In other tumor entities, a critical role of SATB1 in regulating the expression of receptors of the HER family has been described. More specifically, SATB1 has been shown to be involved in the upregulation of EGFR (HER1), HER2, HER3, and HER4 in breast cancer cells [18]. In contrast, in colorectal cancer cells, SATB1 induced HER3 expression but exerted only mild effects on HER2 and no effect on HER1 expression [19]. In glioma cells, SATB1 was found to even act as a repressor of HER2, since SATB1 knockdown led to an induction of HER2 expression [20]. Collectively, these results support the notion that the role of SATB1 in regulating the expression of different HER receptors strongly depends on the cellular and/or tumor context. Notably, in the gastric cancer cells investigated here, SATB1 knockdown had no impact on the basal expression of any HER receptor; however, the upregulation of HER3 upon MET inhibition was prevented. This further emphasizes the dependence of SATB1 effects on the cellular context.

As mentioned above, in the absence of HER3 ligand MET inhibition was found to decrease ERK and Akt phosphorylation and to completely abrogate cellular proliferation despite of the elevated HER3 expression. This indicates that HER3 overexpression, even after the pronounced increase in HER3 levels upon treatment with MET inhibitors or siRNA, is insufficient to compensate for the blocking of MET-dependent pathways in tumor cells that do not endogenously express heregulin. Thus, 2D cell culture insufficiently reflects the in vivo situation where heregulin expressing and secreting stroma cells are present within the gastric tumor. Notably, heregulin stimulation of MKN45 cells pretreated with MET inhibitor yielded even a higher activation of ERK and Akt signaling than heregulin stimulation of cells with intact MET signaling. Taken together, the induction of HER3 after MET inhibition represents a critical factor in HRG-promoted resistance against MET inhibitors. This can severely impair the effect of MET inhibition, even in tumors with an amplification of MET, and may well explain—at least in part—the poor clinical outcomes of MET inhibitor treatment. The further elucidation of the mechanisms involved in regulation of HER3 expression in gastric cancer could provide the basis for novel strategies improving the efficacy of RTK-targeted therapies. It is noteworthy in this context that heregulin secretion by fibroblasts is a critical homeostatic signal to maintain the integrity of the gastric epithelial lining [34] and that inflammatory processes in the stomach lead to an upregulation of heregulin production of gastric fibroblasts [35]. Concomitantly, we could detect significant heregulin mRNA expression in cancer-associated gastric fibroblasts (Figure S4). This highlights the potential involvement of stromal cells in tumor resistance, as shown here through the expression of heregulin. Many studies on the tumor biology of oncogenic growth factor receptors focus on their expression, basal activity, and downstream signaling in tumor cells. Our findings underline the importance of extending the analyses towards the possible impact of the respective receptor ligands, to better understand and predict the effects of targeted therapeutics in the actual in vivo context.

4. Material and Methods

4.1. Materials

Cell culture media, phosphate buffered saline, and fetal bovine serum were obtained from Invitrogen (Gibco, Karlsruhe, Germany). Antibodies against Akt, phospho-Akt, p44/42 MAPK, phospho-p44/42 MAPK, actin, and anti-mouse IgG Alexa Fluor® 647 (#4410S) were purchased from Cell Signaling (Danvers, MA, USA). Anti-HER3 purified antibody (clone 1B4C3) was purchased from BioLegend®. Secondary antibodies were from Sigma–Aldrich (Steinheim, Germany). Protran Nitrocellulose Transfer membranes were purchased from Whatman (Dassel, Germany). The enhanced chemiluminescence systems (Super Signal West Femto Maximum Sensitivity Substrate and SuperSignal West Pico Chemiluminescent Substrate) were from Thermo-Scientific (Bonn, Germany). The WST-1 kit was from Roche Applied Science (Mannheim, Germany). The PCK inhibitor bisindolylmaleimide II (BIM II), the MET inhibitor PF04217903, and the HER1 inhibitor AG1478 were from Tocris (Wiesbaden, Germany). The HER2 inhibitor CP724714 was purchased from Selleckchem (Munich, Germany). Heregulin and 12-O-Tetradecanoylphorbol

13-acetate (PKC activator, PMA) were obtained from Sigma–Aldrich (Steinheim, Germany). The human phospho-MAPK Array Kit was from R&D (Minneapolis, MN, USA). All other chemicals used were purchased from Carl Roth (Karlsruhe, Germany) unless indicated otherwise.

4.2. Methods

4.2.1. Cell Culture

Human gastric cancer cell lines MKN7, MKN74, MKN45, SNU5, and Hs746T were obtained from the American Type Culture Collection (ATCC, Manassas, VA, USA). Cell line authentication was monitored regularly by genotyping (Genolytic, Leipzig, Germany). MKN cells were cultured in RPMI 1640 medium (Thermo-Fisher, Waltham, MA, USA) supplemented with 10% (*v*/*v*) heat-inactivated fetal bovine serum (FBS). SNU5 cells were cultured in Iscove's Modified Dulbecco's Medium plus 20% (*v*/*v*) FBS. The cell line Hs746T was cultivated in Dulbecco's Modified Eagle's Medium (4 mM L-glutamine, 4500 mg/L glucose, 1 mM sodium pyruvate, and 1500 mg/L sodium bicarbonate) supplemented with 10% (*v*/*v*) FBS. All media were used without antibiotics and cells were incubated at 37 °C in a humidified atmosphere containing 5% CO_2 and passaged every 2–3 days.

4.2.2. Cell Transfection and Treatment

siRNAs (see Table S1 for sequences) were purchased from Eurofins MWG Operon (Ebersberg, Germany). In all knockdown experiments, irrelevant siRNAs targeting luciferase (pGL3) were used as negative control. Prior to transfection, cells were seeded in appropriate cell culture plates and maintained overnight under standard conditions. An amount of 10 nM siRNA (50 nm for SNU5, respectively) were transfected using INTERFERin (Polyplus, Illkirch, France), at 1 µL INTERFERin™/pmol siRNA according to the manufacturer's protocol. For inhibitor and heregulin treatment, the following concentrations were used: 0.2 µM MET inhibitor unless otherwise state and 20 ng/mL heregulin.

4.2.3. WST-1 Assay

Cell viability was quantified by measuring the metabolically activated formazan dye from the water-soluble tetrazolium salt WST-1 according to the manufacturer's protocol. Briefly, cells were seeded into 96-well microplates (Sarstedt, Nümbrecht, Germany) at 1000 cells/well and incubated overnight, prior to siRNA transfection or incubation with inhibitors. At the time points indicated, 10 µl WST solution was added to each well, and after incubation at 37 °C for 60 min, absorbance at 450 nm was measured in a PolarSTAR plate reader from BMG (Offenburg, Germany).

4.2.4. Colony Forming Assay

Five × 10^5 cells growing in normal growth medium in 25 cm^2 cell culture flasks were treated with the respective agent for 48 h. Afterwards, cells were trypsinized and counted using a hemocytometer. One thousand cells per condition were re-seeded into a 6-well plate and incubated in normal growth medium (without any further treatment) for 7 days. Thereafter, the medium was aspirated. The colonies were gently washed with PBS, and then stained by use of 0.5% (*w*/*v*) methylene blue in a 1:1 mixture (*v*/*v*) of ethanol and water. The colonies were incubated for 15 min with the staining solution, then gently washed with deionized H_2O and dried at room temperature. Colonies of more than 50 cells were included in the evaluation.

4.2.5. RNA Isolation and RT-qPCR

Total RNA from cells was isolated using the guanidinium thiocyanate–phenol– chloroform extraction procedure (TRI Reagent, Sigma-Aldrich, Taufkirchen, Germany). The first-strand synthesis was carried out using the RevertAid H Minus First Strand cDNA Synthesis Kit from Fermentas (St Leon-Roth, Germany). Products were amplified using specific, intron-spanning primer pairs, with β-actin or RPLP0 serving as loading controls (for primer sequences, see Table S2). Real-time PCR was performed using the Absolute QPCR SYBR Green Mix from Thermo Fisher Scientific (Schwerte, Germany). To this end, 10 pmol of each primer pair and 4 µl from the 1:100 prediluted first-strand synthesis were added to the reaction mixture, and the PCR was carried out in a light cycler apparatus (LightCycler 2.0 System, Roche Applied Science) using the following conditions: 15 min of initial activation at 95 °C, followed by 55 cycles of 10 s at 95 °C, 10 s at 55 °C, and 10 s at 72 °C each. Fluorescence intensities were recorded after the extension step at 72 °C in each cycle. Crossing points were determined by the software, and the relative gene expression was quantified using the formula: %clearpage

$$2^{(\text{crossing point of } \beta\text{-actin} - \text{crossing point of gene of interest})} \times 100 = \text{relative expression of X vs. housekeeper (percentage of reference gene expression).}$$

4.2.6. Immunoblot

For Western blot analysis, cells were seeded in their respective medium at 2×10^5 cells/well into six-well plates. Cells were serum-starved for 18 h, then stimulated as described in the respective figure legends, washed with ice-cold phosphate-buffered saline, and lysed in 250 µl of lysis buffer containing 5 mM EDTA und 1% (*v*/*v*) NP-40 in PBS. Upon determination of protein concentration using DC™ Protein Assay (Bio-Rad Laboratories, Munich, Germany) lysates containing 25 µg of protein were dissolved in loading buffer (125 mM Tris (pH 6.8), 20% glycerol, 4% SDS, 2% β-mercaptoethanol, and 10 µg/mL bromophenol blue). Twenty microliters of lysate/lane was resolved on 9% SDS-polyacrylamid gels and electroblotted onto nitrocellulose membranes using transfer buffer (191 mM glycine, 25 mM Tris, 10% SDS, and 20% methanol). Blots were incubated for 1 h in Rotiblock to saturate non-specific binding sites, washed in Tris-buffered saline with Tween® 20 (TBST), and incubated in a 1:500 dilution of the phospho-specific anti-p44/42 MAPK or anti-Akt mouse monoclonal antibody in 5% milk powder (*w*/*v*) in TBST. An anti-p44/42 MAPK rabbit monoclonal antibody specific for total (phosphorylated and unphosphorylated) MAPK or Akt served for loading controls. Subsequently, blots were incubated with peroxidase-conjugated secondary antibodies (1:5000). Signals were revealed using enhanced chemiluminescence (Super Signal West Femto Maximum Sensitivity Substrate Kit and SuperSignal West Pico Chemiluminescent Substrate), and visualization was carried out using the Chemismart detection system from Peqlab Biotechnologie (Erlangen, Germany).

For monitoring the expression of a larger set of MAP kinases, the commercially available Human Phospho-MAPK Array Kit (Proteome Profiler™ Array, R&D, Minneapolis, MN, USA) was employed. Lysates were analyzed in the antibody array according to the manufacturer's protocol and visualized by chemiluminescence as described above. Signal intensities were quantitated using ImageJ and are shown as heat map (heat mapper software; http://www.heatmapper.ca/) and as a bar diagram.

4.2.7. Flow Cytometry

Cells were harvested and washed 2 times with phosphate buffered saline (PBS). Consequently, cells were resuspended in 100 µL staining buffer (0.5% BSA, 0.1% NaN_3 in PBS) with 0,125 µg HER3-antibody per sample and incubated at 4 °C overnight. Cells were washed 2 times in staining buffer and incubated with anti-mouse Alexa Fluor® 647 antibody for 1 h at room temperature in the dark before FACS analysis was carried out on a Attune® Acoustic Focusing Cytometer using Attune® Cytometric Software.

4.2.8. Spheroid Outgrow and Spheroid Formation

Tumor spheroids were generated by seeding 1000 cells into agarose-coated 96-well plates. Cells were incubated under normal conditions for 96 h and subsequently analyzed (3D growth) or were transferred into normal 12 well microtiter plates for determination of spheroid outgrowth. For this purpose, transferred spheroids were incubated for further

6 days in normal growth medium. Thereafter, the cells were fixated and stained using 0.5 mg/mL methylene blue in 50% (*v*/*v*) water/ethanol to visualize colony spread and formation of distant colonies.

4.2.9. Statistics

All assays were performed independently at least three times unless indicated otherwise, and either one representative experiment or means +/− S.E.M. of multiple experiments are shown. Densitometric analysis of MAPK array was performed using ImageJ software (NIH, Bethesda, MD, USA). Statistical significance of differences in all assays was assessed by ANOVA with Shapiro–Wilk test using SigmaPlot 13, with *, <0.05; **, <0.01, and ***, <0.001.

Supplementary Materials: The following are available online at https://www.mdpi.com/1422-0067/22/1/82/s1; Figure S1: MET inhibition by treatment of cells with 0.2 μM PF04217903 produced marked antiproliferative effects in MET-amplified (A) SNU5 and (B) Hs746T cells, Figure S2: Absence of inhibitory effects of 2 μM MET inhibitor PF04217903 in (A) MKN74 and (C) MKN7 cells. Similarly, no visible effects occurred upon siRNA-mediated MET knockdown in (B) MKN74 and (D) MKN7 cells, Figure S3: Upregulation of HER3 protein levels 48 h after treatment of MKN45 cells with 0.2 μM PF04217903, as determined by flow cytometry, Figure S4: Colony formation assay with MKN45 cells pretreated for 48 h with vehicle (DMSO), 0.2 μM PF04217903, 20 ng/mL HRG, or PF04217903 plus HRG. After the treatment phase cells 1000 cells of each preparation were seeded (without any further treatment) and colony growth was monitored. Number of colonies is presented (vehicle treated cells = 100%)., Figure S5: Definition of all spots on the phospho-antibody array shown in Figure 4. On the array, dots are always provided in duplicates, Figure S6: SNU5 cells and MKN45 cells were pretreated with PI3K inhibitor LY294002 (10 μM) or MEK inhibitor PD98059 (10 μM) for 24 h, prior to treatment with MET inhibitor PF04217903 (0.2 μM for 48 h) and analysis of HER3 mRNA expression via RT-qPCR. HER3 expression was normalized to reference gene expression (RPLP0) and is given as percentage of HER3 expression in vehicle treated cells, Figure S7: Western blot analyses of AKT and ERK1/2 activation (phosphorylation). MKN45 (MET amplified) or MKN74 (no MET amplification) cells were serum-starved for 18 h and then treated with MET inhibitor PF04217903 (2 μM for 24 h) plus 50 ng/mL EGF, where indicated. Phospho-AKT and Phospho-ERK1/2 are shown as compared to tubulin as loading control. Note that in MKN45 cells EGF stimulation led to reactivation of ERK1/2 phosphorylation in the presence of MET inhibitor as compared to MET inhibitor alone, whereas AKT phosphorylation is only mildly restored. In MKN74 cells, treatment with MET inhibitor did not affect AKT or ERK1/2 phosphorylation upon EGF stimulation, Figure S8: Expression of heregulin-B in cancer-associated fibroblasts as determined by RT-qPCR. In contrast, all investigated gastric cancer cell lines were found negative, Table S1: siRNA sequences used in the present study for RNAi experiments, Table S2: Primer sequences used in the present study for quantitative PCR analyses.

Author Contributions: Conceptualization, A.A., T.B., R.J., F.L.; methodology, A.A., T.B., R.J., S.M.-W.; formal analysis, A.A., T.B., R.J., M.H.-R., F.L.; investigation, T.B., R.J., M.H.-R., S.M.-W.; resources, A.A., F.L.; writing—original draft preparation, A.A., T.B., R.J., F.L.; writing—review and editing, A.A., R.J., T.B.; visualization, A.A., T.B., R.J., M.H.-R., S.M.-W.; supervision, A.A. and F.L. All authors have read and agreed to the published version of the manuscript.

Funding: This research received no external funding.

Institutional Review Board Statement: Not applicable.

Informed Consent Statement: Not applicable.

Acknowledgments: This work was supported by an intramural grant of the Medical Faculty, Leipzig University, to R.J. This research received no external funding.

Conflicts of Interest: The authors declare no competing financial interests.

References

1. Den Hoed, C.M.; Kuipers, E.J. Gastric Cancer: How Can We Reduce the Incidence of This Disease? *Curr. Gastroenterol. Rep.* **2016**, *18*, 1–8. [CrossRef] [PubMed]
2. Van Cutsem, E.; Sagaert, X.; Topal, B.; Haustermans, K.; Prenen, H. Gastric Cancer. *Lancet* **2016**, *388*, 2654–2664. [CrossRef]
3. Jemal, A.; Center, M.M.; DeSantis, C.; Ward, E.M. Global Patterns of Cancer Incidence and Mortality Rates and Trends. *Cancer Epidemiol. Biomarkers Prev.* **2010**, *19*, 1893–1907. [CrossRef] [PubMed]
4. Torre, L.A.; Bray, F.; Siegel, R.L.; Ferlay, J.; Lortet-Tieulent, J.; Jemal, A. Global Cancer Statistics, 2012. *CA Cancer J. Clin.* **2015**, *65*, 87–108. [CrossRef] [PubMed]
5. De Mestier, L.; Lardière-Deguelte, S.; Volet, J.; Kianmanesh, R.; Bouché, O. Recent Insights in the Therapeutic Management of Patients with Gastric Cancer. *Dig. Liver Dis.* **2016**, *48*, 984–994. [CrossRef]
6. Shum, H.; Rajdev, L. Multimodality Management of Resectable Gastric Cancer: A Review. *World J. Gastrointest. Oncol.* **2014**, *6*, 393. [CrossRef]
7. Obermannová, R.; Lordick, F. Management of Metastatic Gastric Cancer. *Hematol. Oncol. Clin. N. Am.* **2017**, *31*, 469–483. [CrossRef]
8. Lordick, F.; Janjigian, Y.Y. Clinical Impact of Tumour Biology in the Management of Gastroesophageal Cancer. *Nat. Rev. Clin. Oncol.* **2016**, *13*, 348–360. [CrossRef]
9. Bang, Y.J.; Van Cutsem, E.; Feyereislova, A.; Chung, H.C.; Shen, L.; Sawaki, A.; Lordick, F.; Ohtsu, A.; Omuro, Y.; Satoh, T.; et al. Trastuzumab in Combination with Chemotherapy versus Chemotherapy Alone for Treatment of HER2-Positive Advanced Gastric or Gastro-Oesophageal Junction Cancer (ToGA): A Phase 3, Open-Label, Randomised Controlled Trial. *Lancet* **2010**, *376*, 687–697. [CrossRef]
10. Lee, H.E.; Kim, M.A.; Lee, H.S.; Jung, E.-J.; Yang, H.-K.; Lee, B.L.; Bang, Y.-J.; Kim, W.H. MET in Gastric Carcinomas: Comparison between Protein Expression and Gene Copy Number and Impact on Clinical Outcome. *Br. J. Cancer* **2012**, *107*, 325–333. [CrossRef] [PubMed]
11. Liu, Y.J.; Shen, D.; Yin, X.; Gavine, P.; Zhang, T.; Su, X.; Zhan, P.; Xu, Y.; Lv, J.; Qian, J.; et al. HER2, MET and FGFR2 Oncogenic Driver Alterations Define Distinct Molecular Segments for Targeted Therapies in Gastric Carcinoma. *Br. J. Cancer* **2014**, *110*, 1169–1178. [CrossRef] [PubMed]
12. Ma, P.C. (Not Giving up) the Marathon Race of Met Targeting Therapy: Are We There Yet? *Clin. Cancer Res.* **2019**. [CrossRef] [PubMed]
13. Erjala, K.; Sundvall, M.; Junttila, T.T.; Zhang, N.; Savisalo, M.; Mali, P.; Kulmala, J.; Pulkkinen, J.; Grenman, R.; Elenius, K. Signaling via ErbB2 and ErbB3 Associates with Resistance and Epidermal Growth Factor Receptor (EGFR) Amplification with Sensitivity to EGFR Inhibitor Gefitinib in Head and Neck Squamous Cell Carcinoma Cells. *Clin. Cancer Res.* **2006**, *12*, 4103–4111. [CrossRef]
14. Yamaguchi, H.; Chang, S.S.; Hsu, J.L.; Hung, M.C. Signaling Cross-Talk in the Resistance to HER Family Receptor Targeted Therapy. *Oncogene* **2014**, *33*, 1073–1081. [CrossRef] [PubMed]
15. Bachleitner-Hofmann, T.; Sun, M.Y.; Chen, C.-T.; Tang, L.; Song, L.; Zeng, Z.; Shah, M.; Christensen, J.G.; Rosen, N.; Solit, D.B.; et al. HER Kinase Activation Confers Resistance to MET Tyrosine Kinase Inhibition in MET Oncogene-Addicted Gastric Cancer Cells. *Mol. Cancer Ther.* **2008**, *7*, 3499–3508. [CrossRef]
16. Corso, S.; Ghiso, E.; Cepero, V.; Sierra, J.R.; Migliore, C.; Bertotti, A.; Trusolino, L.; Comoglio, P.M.; Giordano, S. Activation of HER Family Members in Gastric Carcinoma Cells Mediates Resistance to MET Inhibition. *Mol. Cancer* **2010**, *9*, 121. [CrossRef]
17. Yun, C.; Gang, L.; Rongmin, G.; Xu, W.; Xuezhi, M.; Huanqiu, C. Essential Role of Her3 in Two Signaling Transduction Patterns: Her2/Her3 and MET/Her3 in Proliferation of Human Gastric Cancer. *Mol. Carcinog.* **2015**, *54*, 1700–1709. [CrossRef]
18. Han, H.J.; Russo, J.; Kohwi, Y.; Kohwi-Shigematsu, T. SATB1 Reprogrammes Gene Expression to Promote Breast Tumour Growth and Metastasis. *Nature* **2008**, *452*, 187–193. [CrossRef]
19. Frömberg, A.; Rabe, M.; Aigner, A. Multiple Effects of the Special AT-Rich Binding Protein 1 (SATB1) in Colon Carcinoma. *Int. J. Cancer* **2014**, *135*, 2537–2546. [CrossRef]
20. Frömberg, A.; Rabe, M.; Oppermann, H.; Gaunitz, F.; Aigner, A. Analysis of Cellular and Molecular Antitumor Effects upon Inhibition of SATB1 in Glioblastoma Cells. *BMC Cancer* **2017**, *17*. [CrossRef]
21. Zhang, S.; Tong, Y.X.; Xu, X.S.; Lin, H.; Chao, T.F. Prognostic Significance of SATB1 in Gastrointestinal Cancer: A Meta-Analysis and Literature Review. *Oncotarget* **2017**, *8*, 48410–48423. [CrossRef] [PubMed]
22. Shinomiya, N.; Chong, F.G.; Xie, Q.; Gustafson, M.; Waters, D.J.; Zhang, Y.W.; Vande Woude, G.F. RNA Interference Reveals That Ligand-Independent Met Activity Is Required for Tumor Cell Signaling and Survival. *Cancer Res.* **2004**, *64*, 7962–7970. [CrossRef] [PubMed]
23. Carraway, K.L.; Cantley, L.C. A Neu Acquaintance for ErbB3 and ErbB4: A Role for Receptor Heterodimerization in Growth Signaling. *Cell* **1994**, 5–8. [CrossRef]
24. Cui, J.J.; McTigue, M.; Nambu, M.; Tran-Dubé, M.; Pairish, M.; Shen, H.; Jia, L.; Cheng, H.; Hoffman, J.; Le, P.; et al. Discovery of a Novel Class of Exquisitely Selective Mesenchymal-Epithelial Transition Factor (c-MET) Protein Kinase Inhibitors and Identification of the Clinical Candidate 2-(4-(1-(Quinolin-6-Ylmethyl)-1H-[1–3]Triazolo[4,5-b]Pyrazin-6-Yl)-1H-Pyraz. *J. Med. Chem.* **2012**, *55*, 8091–8109. [CrossRef] [PubMed]

25. Ocana, A.; Vera-Badillo, F.; Seruga, B.; Templeton, A.; Pandiella, A.; Amir, E. HER3 Overexpression and Survival in Solid Tumors: A Meta-Analysis. *J. Natl. Cancer Inst.* **2013**, 266–273. [CrossRef]
26. Shi, J.; Qu, Y.; Li, X.; Sui, F.; Yao, D.; Yang, Q.; Shi, B.; Ji, M.; Hou, P. Increased Expression of EHF via Gene Amplification Contributes to the Activation of HER Family Signaling and Associates with Poor Survival in Gastric Cancer. *Cell Death Dis.* **2016**. [CrossRef]
27. Chandarlapaty, S.; Sawai, A.; Scaltriti, M.; Rodrik-Outmezguine, V.; Grbovic-Huezo, O.; Serra, V.; Majumder, P.K.; Baselga, J.; Rosen, N. AKT Inhibition Relieves Feedback Suppression of Receptor Tyrosine Kinase Expression and Activity. *Cancer Cell* **2011**. [CrossRef]
28. Pavan Kumar, P.; Purbey, P.K.; Sinha, C.K.; Notani, D.; Limaye, A.; Jayani, R.S.; Galande, S. Phosphorylation of SATB1, a Global Gene Regulator, Acts as a Molecular Switch Regulating Its Transcriptional Activity In Vivo. *Mol. Cell* **2006**, *22*, 231–243. [CrossRef]
29. Purbey, P.K.; Singh, S.; Notani, D.; Kumar, P.P.; Limaye, A.S.; Galande, S. Acetylation-Dependent Interaction of SATB1 and CtBP1 Mediates Transcriptional Repression by SATB1. *Mol. Cell. Biol.* **2009**, *29*, 1321–1337. [CrossRef]
30. Mir, R.; J. Pradhan, S.; Galande, S. Chromatin Organizer SATB1 As a Novel Molecular Target for Cancer Therapy. *Curr. Drug Targets* **2012**, *13*, 1603–1615. [CrossRef]
31. Frömberg, A.; Engeland, K.; Aigner, A. The Special AT-Rich Sequence Binding Protein 1 (SATB1) and Its Role in Solid Tumors. *Cancer Lett.* **2018**, *417*, 96–111. [CrossRef] [PubMed]
32. Sun, F.; Lu, X.; Li, H.; Peng, Z.; Wu, K.; Wang, G.; Tong, Q. Special AT Rich Sequence Binding Protein 1 Regulates the Multidrug Resistance and Invasion of Human Gastric Cancer Cells. *Oncol. Lett.* **2012**, *4*, 156–162. [CrossRef] [PubMed]
33. Peng, Z.; Wang, C.; Fang, E.; Lu, X.; Wang, G.; Tong, Q. Co-Delivery of Doxorubicin and SATB1 ShRNA by Thermosensitive Magnetic Cationic Liposomes for Gastric Cancer Therapy. *PLoS ONE* **2014**, *9*. [CrossRef] [PubMed]
34. Noguchi, H.; Sakamoto, C.; Wada, K.; Akamatsu, T.; Uchida, T.; Tatsuguchi, A.; Matsui, H.; Fukui, H.; Fujimori, T.; Kasuga, M. Expression of Heregulin α, ErbB2, and ErbB3 and Their Influences on Proliferation of Gastric Epithelial Cells. *Gastroenterology* **1999**, *117*, 1119–1127. [CrossRef]
35. Nagata, K.; Wada, K.; Tatsuguchi, A.; Futagami, S.; Gudis, K.; Miyake, K.; Tsukui, T.; Sakamoto, C. Heregulin-α and Heregulin-β Expression Is Linked to a COX-2-PGE2 Pathway in Human Gastric Fibroblasts. *Am. J. Physiol. Gastrointest. Liver Physiol.* **2006**, *290*, 1243–1251. [CrossRef]

International Journal of
Molecular Sciences

MDPI

Review

Autophagy—A Hidden but Important Actor on Oral Cancer Scene

Totan Alexandra [1], Imre Melescanu Marina [2,*], Miricescu Daniela [1], Stanescu Iulia Ioana [1], BencZe Maria [3,*], Radulescu Radu [1], Tancu Ana Maria [2], Spinu Tudor [4] and Greabu Maria [1]

1 Department of Biochemistry, Faculty of Dental Medicine, University of Medicine and Pharmacy Carol Davila, 8 Eroilor Sanitari Blvd, 050474 Bucharest, Romania; alexandra.totan@umfcd.ro (T.A.); daniela.miricescu@umfcd.ro (M.D.); iulia.stanescu@umfcd.ro (S.I.I.); radu.radulescu@umfcd.ro (R.R.); maria.greabu@umfcd.ro (G.M.)

2 Department of Complete Denture, Faculty of Dental Medicine, University of Medicine and Pharmacy Carol Davila, 17-23 Calea Plevnei, 010221 Bucharest, Romania; anamaria.tancu@umfcd.ro

3 Department of Orthodontics and Dental-Facial Orthopaedics, Faculty of Dental Medicine, University of Medicine and Pharmacy Carol Davila, 17-23 Calea Plevnei, 010221 Bucharest, Romania

4 Department of Fixed Prosthodontics and Occlusion, Faculty of Dental Medicine, University of Medicine and Pharmacy Carol Davila, 17-23 Calea Plevnei, 010221 Bucharest, Romania; tudor.spinu@umfcd.ro

* Correspondence: melim.marina@gmail.com or marina.imre@umfcd.ro (I.M.M.); maria.bencze@umfcd.ro (B.M.)

Received: 3 November 2020; Accepted: 3 December 2020; Published: 7 December 2020

Abstract: The duration of denture use, oral hygiene, smoking and male sex were identified as risk factors for oral mucosal lesions. As it is well known, all the oral mucosal lesions associated with risk factors have an important degree of malignity. Chronic mechanical irritation can be another cause of oral cancer and it is produced by the constant action of a deleterious agent from the oral cavity. Autophagy represents a complex evolutionary conserved catabolic process in which cells self-digest intracellular organelles in order to regulate their normal turnover and remove the damaged ones with compromised function to further maintain homeostasis. Autophagy is modulated by mTOR kinase and indirectly by PI3K/AKT survival pathway. Due to its dual capacity to either induce cell death or promote cell survival, important evidence pointed that autophagy has a two-faced role in response to chemotherapy in cancer. In conclusion, understanding how to overcome cytoprotective autophagy and how to take advantage of autophagic cell death is critical in order to enhance the cancer cells sensitivity to particular therapeutic agents.

Keywords: oral cancer; autophagy; PI3K/AKT/mTOR signaling pathway

1. Introduction

Oral mucosal lesions usually occur due to systemic diseases, nutritional disorders, medication side effects or wearing ill-fitting dentures in the elderly [1]. The most common oral mucosal lesions in the aging population can be caused by both poor oral hygiene and continuous use of dentures throughout the day and night [2]. Apart from the duration of denture use, smoking and male sex were also identified as risk factors for oral mucosal lesions, with fissured tongue and lingual varicosity being the most common forms [3,4].

As it is well known, all the oral mucosal lesions associated with risk factors have an important degree of malignity. The oral cavity is one of the most appropriate locations for the development of oncological diseases, especially in patients who are >40 years old. Malignant lesions were determined in only men in the study by Dundar and Ilhan Kal and in only women in a study by Cebeci et al. [5].

Autophagy represents a complex evolutionary conserved catabolic process in which cells self-digest intracellular organelles in order to regulate their normal turnover and remove the damaged ones with compromised function, to further maintain homeostasis [6–9].

Currently, the role of autophagy in cancer is still controversial. On the one hand, constitutive autophagy can be regarded as a cellular housekeeper that eliminates damaged organelles and protect cells against carcinogenesis, and moreover it has been shown that excess or persistent autophagy promotes cell death by inducing apoptosis or mediating "autophagic cell death". However, on the other hand, it can also act as a pro-survival signal in response to stress (like nutrient deprivation, hypoxia and the presence of chemotherapy or some other targeted therapies) that could induce resistance to anticancer therapies in advanced cancer [7–9].

Autophagy can control many cellular molecular pathways involved in tumor promotion and suppression, immune response intensity. A lot of studies have focused on its involvement in these processes as a modulator of pathogenesis and, consequently, as a potential therapeutic target.

In this paper, we review recent progress and provocation in our understanding of how to overcome cytoprotective autophagy and how to take advantage of autophagic cell death in order to enhance cancer cells' sensitivity to particular therapeutic agents.

2. Oral Cancer

Head and neck region cancers are one of the most common types of cancers, oral cancer being the sixth most common malignancy in the world, and is characterized by a very low five-year survival rate, about 50% due to late stage diagnosis, high degree of invasiveness and development of therapeutic resistance [10–13]. Almost all of the oral cancers (90%) are squamous cell carcinomas with various levels of cell differentiation and lymph nodes metastasis [10,13–16]. The other 10% of oral cancers originate from connective tissue, minor salivary glands, lymphoid tissue or melanocytes malignant processes [11,17].

According to the latest reports of the International Agency for Research on Cancer (IARC) for oral cancer (ICD-10 code C00-08: Lip, Oral Cavity), the annual incidence is higher over 300.000 diagnosed cases, and the annual mortality is about 145,000 death [18]. The regions characterized by a high incidence of oral cancer are found in South and Southeast Asia (Sri Lanka, India, Pakistan and Taiwan), areas of the West (France) and Eastern Europe (Hungary, Slovakia and Slovenia), Latin America and the Caribbean (Brazil, Uruguay and Puerto Rico) and Pacific regions (Papua New Guinea and Melanesia) [10,18].

Oral cancers can be located anywhere in the oral region that extends anatomically from the lip vermilion to the soft and hard palate junction and the circumvallate papillae of the tongue [18].

Oral cancer can be caused by genetic factors, epigenetic modifications (such as histones modifications; nucleosome integrity, DNA methylation and expression of non-coding RNAs (ncRNAs), tobacco and alcohol consumption, chronic infections such as human papilloma virus (HPV) or syphilis infections, dental factors, occupational risks [13,19].

Tobacco consumption is the main risk factor in oral cancer development and is responsible for other types of cancer also, such as lung, pharynx, larynx, esophagus, urinary bladder, renal, pelvis, and pancreas cancers [20,21]. The use of tobacco increases the risk of developing oral cancer by three times compared to non-smokers [22]. The main carcinogenetic factors found in tobacco smoke are nitrosamines, benzopyrenes and aromatic amines that undergo various enzymatic and non-enzymatic transformations resulting in molecules that are covalent bound to various regions of DNA resulting in DNA adducts and various mutations [22]. Tobacco consumption also generates a high oxidative stress via the high concentration of free radicals contained, both oxygen and nitrogen species, that deplete enzymatic and non-enzymatic cellular antioxidants resulting in cell damage leading to cancer [23].

Alcohol consumption can act as a local or systemic risk factor in oral cancer development. Systemic effects of alcohol consumption are related to the accumulation of acetaldehyde, the main metabolite of ethanol, that causes genetic alterations through disruption of DNA synthesis and repair mechanisms [19,24]. Acetaldehyde can also be produced by oral bacteria [1]. Locally, alcohol can increase the permeability of the oral mucosa for other carcinogenetic factors such as those found in tobacco and thus, working synergistically with tobacco carcinogens [16]. Alcohol can also induce epithelial atrophy, further increasing oral mucosa permeability, it's effects being directly linked with the intensity and duration of the chronic consumption [25].

In the oral cavity the human papilloma virus can be found near undifferentiated basal keratinocytes and mainly in the tonsillary crypts and the base of the tongue [16,26]. The two types of HPV involved in oral cancer development are HPV 16 and HPV 18 and they act by blocking or altering the expression of essential nuclear proteins such as 53, P21 and P16, thus inducing the transformation of normal cells into malignant cells [27].

Chronic mechanical irritation can be another cause of oral cancer and it is produced by the constant action of a deleterious agent from the oral cavity. The deleterious agents can be sharp and broken tooth, defective restorations, ill-fitting dentures with sharp or retentive edges or just constant biting of the oral mucosa that can sustain a chronic state of inflammation that induces epigenetic transformation of oral cells [28,29].

Oral cancer can evolve from a series of premalignant lesions, the most frequent being leukoplakia, submucosal fibrosis and erythroplakia [17]. Other less frequent premalignant lesions are oral lichen planus, actinic cheilitis, xeroderma pigmentosum and Fanconi's anemia [30]. Leukoplakia is the most frequent of the premalignant lesions, affecting any part of oral cavity. It has several clinical forms such as homogeneous and non-homogeneous lesions and verrucous leukoplakia, the rate of malignant transformation depending on the localization of the lesion, the size and the duration and is around 1% [17,30]. Erythroplakia has a lower incidence compared to leukoplakia, but a higher rate of malignancy, between 14–50% due to high levels of dysplasia that accompany these lesions [30].

All regions of the oral cavity can develop malignant processes, but the most frequently affected sites are the tongue and the floor of the mouth, followed by the lip or the alveolar process, and are closely related to risk factors prevalence and lifestyle conditions usually following the "field cancerization concept" [31,32]. Tongue cancers usually develop in elderly patients, chronic exposure to alcohol and tobacco being the most frequent causes, this type of cancer being more aggressive with high relapsing and high invasiveness [33]. Cancers are caused mainly by environmental factors such as solar radiation, followed by smoking and viral infectious factors, and are overwhelmingly located at the lower lip (90%). Early detection and treatment ensure a very high five years survival rate (almost 80%) with few functional and aesthetical complications [34].

3. Autophagy

Autophagy is a survival-promoting pathway that captures, degrades, and recycles intracellular proteins and organelles in lysosomes. Autophagy preserves organelle function, prevents the toxic buildup of cellular waste products, and provides substrates to sustain metabolism in starvation. Although in some context autophagy suppresses tumorigenesis, in most contexts autophagy facilitates tumorigenesis. Cancers can upregulate autophagy to survive microenvironmental stress and to increase growth and aggressiveness. Mechanisms by which autophagy promotes cancer include suppressing induction of the P53 tumor suppressor protein and maintaining metabolic function of mitochondria. Efforts to inhibit autophagy to improve cancer therapy have thereby attracted great interest.

There are 3 primary forms of autophagy: macroautophagy, microautophagy and chaperone-mediated autophagy (CMA). The main differences between them concern their patterns of delivery and physiological functions [35]. Macroautophagy (referred to here after as autophagy) involves the formation of multiple membrane structures starting from the phagophore to autophagosome and, finally, to the autolysosome [35]. Autophagosome's formation and consumption go through 4 steps: (1) induction and cargo packaging, (2) elongation of the phagophore, (3) autophagosome formation and completion, and (4) lysosomal fusion and breakdown [35].

The complex molecular process of autophagy is primarily dependent on the ATG (autophagy-related) family proteins [36].

Briefly, the molecular events sequence in autophagy is as follows:

(1) signals such as starvation activate the ULK complex, which will bind to the PtdIns3K complex following AMPK activation or mTOR suppression [35];
(2) following induction, the ULK complex, PtdIns3K complex and the ATG9 complex orchestrated action will trigger the phagophore assembly at the phagophore assembly site [35];
(3) ATG12 and LC3 conjugation systems are key players in regulating the phagophore elongation to the autophagosome. mTOR, the major autophagy inhibitory factor, suppresses autophagy as response to abundant nutrients conditions. This suppressive action is mediated by class I PI3K and AKT signaling [35];
(4) SQSTM1/p62 (sequestosome 1) receptor protein will consequently interact with both LC3 and ubiquitin chains [35];
(5) Further, the autophagosome will fuse with a lysosome, resulting the autolysosome formation. Inside autolysosome, the autophagosome constituents will be hydrolytically degraded. The trapped SQSTM1 will be degraded in the autolysosome, which highlight SQSTM1's role as an autophagy flux marker [35].

3.1. *Autophagy—An Important AKT/mTOR Pathway Target*

AKT or protein kinase B was discovered in 1987 by Stephan Staal as the v-AKT- transforming gene component of the AKT-8 provirus. Eight years later, Richard Roth and his co-workers discovered that this kinase is activated by insulin [37,38]. AKT/PKB are serine/threonine kinases belonging to the kinase superfamily together with cAMP-dependent protein kinases (c-AMP), protein kinase A (PKA), protein kinase G (PKG) and protein kinase C (PKC), presenting structural homology within the catalytic domain and similar mechanisms of action [39].

AKT/mTOR signaling pathway is activated by growth factors and cytokines binding to the insulin receptor, which will lead to the activation of phosphatidylinositol 3-kinase (PI3K) and phosphorylation of phosphatidylinositol 3,4 bisphosphate (PIP2) to phosphatidylinositol 3,4,5 trisphosphate (PIP3) (Figure 1) PIP3 is an important secondary messenger that will determine the localization of AKT in the plasma membrane and is further phosphorylated by phosphoinositide dependent protein kinase-1 at the threonine 308. AKT maximum activation is achieved by the second phosphorylation that takes place at the Serine 473 by mTORC2 (mammalian target of rapamycin complex-2) (Figure 1) [39,40].

mTOR is a serine/threonine multicomponent kinase complex consisting of mTOR complex1 (mTOR1) and mTOR complex 2 (mTOR2). After activation, AKT phosphorylates TSC 1 (tuberous sclerosis complex 1) and TSC 2 (tuberous sclerosis complex 2) and inhibits them leading to mTOR1 activation. mTOR1 further phosphorylates 40S ribosomal protein kinase S6 (S6K) and eukaryotic initiating factor 4E binding protein (4EP1) and stimulates protein synthesis, metabolism and cell growth. Subsequently, activated AKT phosphorylates a series of proteins that are involved in glucose metabolism, cell proliferation and survival, but in the apoptosis process as well (Figure 1) [40].

PTEN is a negative regulator of AKT signaling pathway, being involved in the dephosphorylation of PIP3 to PIP2. Overactivation of AKT due to PTEN loss mediates the tumorigenesis process by tumor growth, survival and proliferation [41,42]. Cytokines, angiogenic and growth factors bind to the insulin receptor and activate AKT pathway. Unfortunately, overactivation of AKT signaling pathways is correlated with poor outcome for breast, prostate, endometrial, pancreatic, brain, gastric and melanoma cancers [36]. Moreover, this signaling pathway has been found to be, also, overactivated in oral cancer (Figure 1) [41–46].

Roy NK et al. also identified AKT isoforms specific to oral cancer, immunohistochemical analyzes reporting overexpression for AKT1 and AKT 2, but not for AKT3. In the case of head and neck cancers, genetic changes of AKT 1 and 2 are associated with a low survival rate. AKT1 and 2 isoforms are expressed in different regions of the oral cavity such as the tongue, cheek, and gingiva [47–49].

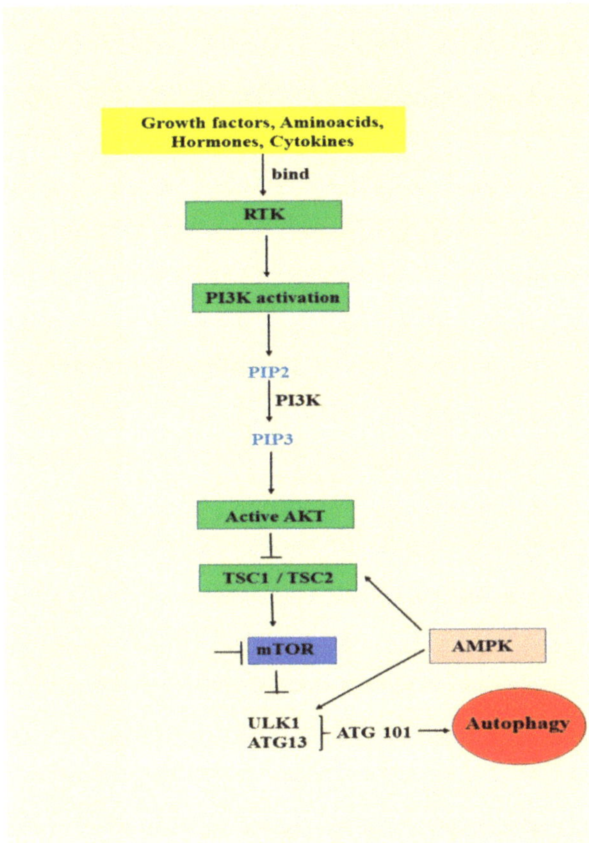

Figure 1. The relationship between the PI3K/AKT/mTOR signaling pathway and autophagy: **RTK** (receptor tyrosine kinase); **PI3K** (phosphatidylinositol 3-kinase); **PTEN** (phosphate and tensin homology); **AKT** (serine/threonine kinase); **TSC** (tuberous sclerosis complex); **mTOR** (mammalian target of rapamycin); **AMPK** (AMP-activated protein kinase); **ULK** (unc-51 like autophagy activating kinase $\frac{1}{2}$); **ATG** (autophagy-related protein 13); **ATG 101** (autophagy-related protein 101).

3.2. *Autophagy—Important Actor on Oral Cancers Scene*

Not surprisingly, the most molecular mechanisms involved in the autophagy regulation are deeply involved within signaling pathways with important roles in cancer control. Autophagy should be regarded as a molecular double-faced Janus God [50–52]. Thus, the tumor suppressors that negatively regulate mTOR (PTEN, AMPK, LKB1, and TSC1/2) will initiate autophagy machinery while, on contrary, oncogenes that activate mTOR (class I PI3K, Ras, and AKT), will inhibit autophagy. The role played by autophagy on the cancer scene depends on the genetic context, microenvironment, tumor type and stage of development [53].

Recent studies have illustrated in oral squamous cell carcinoma (OSCC) tissues or cell lines the existence of aberrant specific ATG protein expression profiles, such asATG9A, ATG5, ATG16L1, LC3 and BECN [53–56]. Experimental results have highlighted interesting correlations between various autophagy genes/proteins expression and OSCC prognosis, opening a challenging way to new biomarkers [51,54,57–61]. ATG9A is a transmembrane protein that regulates membrane delivery during autophagy pathway's initial steps [62]. ATG9A overexpression has been showed to have a significant negative correlation with overall survival in OSCC patients. Consequently, ATG9A presence in the tumor cells cytoplasm should be regarded as a new candidate biomarker for the OSCC recurrence and survival [54,61].

ATG16L1 is also an essential actor in autophagosome formation. Experimental results highlighted its correlation with the unfavorable prognosis of patients with OSCCs. Elevated levels of ATG16L1 expression were detected in keratinizing-type OSCCs and 27 of 90 OSCC tissues [59,62]. Nomura et al. have suggested that ATG16L1 abundant stromal expression was associated with lympho-vascular invasive tumor cells development and positive lymph node status [60].

ATG5 is covalently bound to ATG12. Along with ATG16L1, ATG5 is mainly involved in the phagophore elongation [62]. Dual expression of ATG5 and BECN1 should be regarded as a bad prognostic indicator for OSCC diagnosed patients [55,63].

SQSTM1 is a receptor protein mainly involved in the coordination of selective autophagy and ubiquitination [64]. Liu et al. research revealed that the increased LC3-II expression enhanced SQSTM1 cytoplasmic level. Liu et al. also have shown that excessive SQSTM1 was associated with aggressive clinicopathological features and bad prognosis [59,65]. Moreover, it seemed that excessive SQSTM1 could contribute to glutathione induction, triggering resistance to cytotoxic radiation [66].

BECN1 represents an essential modulator of phagophore nucleation and, also, an important player in tumor suppression molecular mechanism [67]. Specific allelic deletions of BECN1 gene have been found in most human breast, ovarian and prostate cancers [68]. BECN1 and LC3 are two critical players in autophagy induction. Wang Y et al. results revealed reduced levels of BECN1 and LC3 in tongue squamous cell carcinoma tissues and squamous cell carcinoma lines [50,69]. Kapoor et al. also observed low expression of BECN1 mRNA and reduced BECN1 protein levels in other OSCC tissues [56,70]. Wang's group have also shown that reduced BECN1 results in decreased ATG4, ATG5 and LC3-II levels, as well as intensified proliferation, migration and invasion of tongue SCC cells [50,69]. On the contrary, overexpression of BECN1 exerts converse effects [50,69,71]. Autophagy can be considered an important actor in both the pathogenesis and treatment response in oral cancer. Jiang et al. revealed that autophagy could have a significant impact on tumorigenesis and tumor progression in primary salivary gland adenoid cystic carcinoma (ACC) [72]. Liang et al. experimental results revealed a significant correlation between BECN1 and unfavorable prognosis in ACC [73].

3.2.1. Oncogenes and Tumor Suppressors that Control the Autophagy Pathway

It is very important to notice that many autophagy-inducing proteins are either tumor suppressor proteins or oncoproteins (Table 1) [74]. More specifically, it has been highlighted that tumor suppressors that negatively regulate mTOR, (PTEN, AMPK, LKB1, and TSC1/2) initiate autophagy while mTOR activators (such as AKT, class I PI3K, Ras, inhibit autophagy, suggesting that autophagy may have a crucial role in tumor evolution [74].

(1) **mTOR protein kinase** represents the major negative regulator of autophagy [74]. This kinase is involved in many signaling pathways controlling cell growth, mainly downstream of growth factor receptors with tyrosine kinase activity. Constitutive activation of these receptors, activating mutations of Ras, PI3K, AKT and the inactivating mutations of negative regulators, such as PTEN, are all frequently met during cancer development, suggesting that inhibition of autophagy likely contributes to the onset of tumor progression [74]. Martins et al. immunohistochemical investigation in oral epithelial dysplasia revealed a greater expression of AKT and mTOR activated forms, compared to OSCC and non-dysplastic oral tissues [75]. Moreover, mTOR immunohistochemical analysis in both HPV (-) and HPV-associated HNSCC lesions have highlighted its important role as a molecular target in oral cancer [76]. Harsha et al. study revealed a higher expression of AKT and mTOR in human ameloblastoma tissues compared to normal oral mucosa [76]. Matsuo et al. hypothesized that the

downstream protein of AKT/mTOR pathway, GSK3, represents one of the first steps in cervical lymph node metastasis in the OSCC context [77]. Harsha et al. have outlined AKT/mTOR signaling pathway's role in the initiation, development and progression of oral verrucous carcinoma [76]. Ferreira et al. have studied the level of expression of several regulatory proteins in OSCC cells. Their study revealed high levels of AKT and mTOR active forms in OSCC tissues from alveolar ridge, gingiva and hard palate and, leading to the conclusion that AKT/mTOR pathway's activation should be associated with OSCC development [78].

All these studies outlined the AKT/mTOR pathway significance in the molecular landscape of oral cancer initiation and progression. The upregulation or overexpression of this pathway trigger tumor growth and cause poor prognosis, especially by the way they influence autophagy [79]. It has been pointed out that the neutrophil gelatinase-associated lipocalin (NGAL) knockdown induced mTOR activation and, consequently, suppressed autophagy, thereby sustaining oral cancer progression. This study has also revealed the involvement of the AKT/mTOR pathway in NGAL-mediated control of autophagy in oral cancer cells [80].

Table 1. List of oncogene products and tumor suppressors that control the autophagy pathway.

Oncogene Product or Tumor Suppressor	Effect on Autophagy Pathway	Reference
(1) mTOR	Negative regulator	[74–80]
(2) PTEN	Inducer	[46,81]
(3) Beclin-1	Inducer	[74]
(4) DAPK	Inducer	[82]
(5) BCL-2; BCL-XL	Negative regulator	[74]
(6) c-FLIP	Negative regulator	[83,84]
(7) P53	Negative regulator/Inducer	[85–88]

(2) **PTEN (phosphatase and tensin homolog deleted on chromosome 10)** is regarded as the "new guardian of the genome". On the one hand, PTEN plays a significant role in the molecular landscape of cell survival and proliferation. On the other hand, it is deeply involved in the differentiation and apoptosis pathways. PTEN ranks in second place regarding mutations frequency in cancer, after P53 [81].

Several studies have examined the relationship between PTEN and autophagy in many different model systems. For instance, De Amicis et al. reported that in breast cancer cells, progesterone triggered, via its receptor - PR-B, PTEN activation. Moreover, activated PTEN has been shown to induce the downregulation of the PI3K/AKT pathway, consequently stimulating autophagy, which, in turn, led to reduced cell survival [81]. De Amicis et al. have also demonstrated that 5-methoxypsoralen treatment of breast cancer cell lines has induced autophagy by positively regulating Beclin-1, PI3K-III, UVRAG expression and by LC3-I to LC3-II conversion [81]. In conclusion, De Amicis et al. study highlighted PTEN's concrete involvement in the autophagy induction [81].

Downregulation of PTEN has been also reported in the oral cancer context, possibly being caused by epigenetic modifications, mostly hypermethylation. Kurasawa et al. represent one of the research groups that support this molecular mechanism regarding PTEN regulation in OSCC context [46].

The precise molecular mechanisms behind PTEN involvement in the oral cancer molecular landscape are still incompletely elucidated. However, it is very likely that autophagy represents one of PTEN's main targets to regulate, in oral cancer as well.

(3) **Beclin-1**, one of the most important autophagy regulators, functions as a tumor suppressor in mammalian cells. Beclin-1 is included in the class III PI3K complex that promotes autophagy. It is very important to notice that the monoallelic mutations of *Beclin-1* gene have been frequently reported in prostate, ovarian, and breast cancers in humans. These experimental observations outlined the *Beclin-1*

role as a haplo-insufficient tumor suppressor involved in the molecular mechanisms of several human cancers [74].

(4) **The death-associated protein kinase, DAPK**, a kinase that phosphorylates Beclin-1, disrupting the Beclin-1/BCL-2 complex. It has been revealed that DAPK gene, an autophagy inducer, is frequently silenced by methylation in different types of human cancers [82].

(5) **BCL-2 and BCL-XL**, important players in the inhibition of apoptosis, have also been shown to be involved in oncogenesis, as autophagy negative regulators. Although BCL-2 and BCL-XL are not directly involved in mTOR signaling, they can interact with the Beclin-1 BH3 domain and sequester Beclin-1 as an inactive complex in the ER [74].

(6) Recently has been brought to light an important autophagy negative regulator—the **protein c-FLIP (cellular FLICE-like inhibitory protein)** [83]. c-FLIP is, also, an apoptosis-inhibitor of the extrinsic apoptotic pathway by suppressing death receptor-induced caspase 8 activation [83]. Lee et al. study on T lymphocytes revealed that c-FLIP has an important role in both, autophagy and apoptosis regulation. c-FLIP's mission is to prevent Atg3 binding to LC3, consequently, negatively regulating the autophagosome assembly. [84].

(7) A special role is assigned to the **P53 protein**. Considered a true genome guardian, P53 plays a crucial role in the DNA repair mechanism, cell cycle control, cellular differentiation and apoptosis [85]. Sasahira et al. highlighted that P53 somatic mutations have been detected in 10% of oral dysplasia and in 60–80% of OSCC [85]. Furthermore, the GenomeWide Association Study revealed the usual presence of mutated P53 in the cases of human papillomavirus-negative OSCC [85]. It also has been pointed that the overall survival of P53-mutant OSCC patients was much worse, compared with that of OSCC patients with the wild-type P53 [85].

Oikawa et al. study, using a next-generation sequencing in OSCC tissues, revealed that P53, CDKN2A, PIK3CA mutations combined with PIK3CA and AKT1 copy number amplification, triggered distant metastasis and, consequently, a significantly poorer prognosis in the studied group [86].

Recent experimental evidence rigorously sustain that P53 should be regarded either as an inhibitor or an activator of autophagy, depending on its subcellular localization and its downstream signal pathway. This finding gains particular significance as P53 deficiency or mutant variants of P53 that accumulate in the cytoplasm of tumor cells enable activation of autophagy [87].

P53 has the ability to co-regulate autophagy and apoptosis. P53 controls autophagy-related pathways, AMPK/mTOR and Bmf/Beclin-1, and, also, modulates the expression of apoptosis-related genes, Bcl-2 and Apaf1 [88]. Autophagy and apoptosis are strange partners influencing each other. Autophagy and apoptosis cross-talk represents a crucial molecular event to the cell fate. However, their molecular relationship is quite complicated by their contradictory roles under some circumstances.

3.2.2. Autophagy Regarded as a Tumor Suppressor

The first important experimental data sustaining the possible tumor suppressor role of autophagy were obtained in studies targeting Beclin-1. *Beclin-1* gene monoallelic loss on chromosome 17q21, has been reported in 40–75% of human ovary, breast and prostate tumors, suggesting that autophagy may play the role of a tumor suppressor [89]. Furthermore, Beclin-$1^{+/-}$ mice have shown a high incidence of spontaneous tumors, especially lymphoma and hepatocellular carcinoma. Consequently, the experimental evidence presented suggests that beclin-1 functions as a haplo-insufficient tumor suppressor gene. Wei et al. have shown that the EGFR-dependent Beclin-1 phosphorylation on several tyrosine residues, decreased the activity of the Beclin-1/PI3KC3 complex and, consequently, inhibited autophagy in non-small-cell lung carcinoma cells. This effect was reduced in the presence of an EGFR kinase inhibitor. [67,90].

Concerning oral cancer, interesting experimental evidence sustained that activated autophagy was able to induce oral cancer cells survival decline [91–93]. In this context, survivin, a usually expressed protein in head and neck squamous cell carcinoma (HNSCC) patients, should be mentioned. This protein has been associated with poor survival and chemotherapy resistance in HNSC. Zhang et al.

study revealed that survivin overexpression was negatively correlated with the autophagic marker LC3, in human HNSCC cells [91].

Han et al. results suggested that sulfasalazine promoted autophagic cell death via Akt and ERK pathways, having chemotherapeutic potential for the oral cancer treatment [93].

Taken together, the evidence presented above contribute in supporting the hypothesis that autophagy should be regarded as a one of the main actors in tumor suppression, at least in the early stages of oral cancer. However, these evidences also highlight the dual nature of autophagy during tumor development and progression.

The possible molecular strategies that sustain the tumor suppressor role of autophagy:

Autophagic Cell Death

Autophagy represents primarily a mechanism that insures cell survival under stress conditions. However, there is evidence indicating that, under specific conditions, an increase of the autophagic flux may induce cell death, explaining the possible tumor suppressor effects of this Janus God like molecular pathway (Table 2) [94]. Pattingre et al. have shown that the expression of a mutant Beclin-1, unable to interact with BCL-2, induced autophagy to a greater extent compared to the wild-type Beclin-1, triggering cell death (Table 2) [95]. Zhao et al. highlighted that the transcription factor FoxO1 have induced autophagy in a manner independent of its transcriptional activity, triggering autophagic cell death in tumor cells. These results suggest that the cytosolic FoxO1 promoted autophagy acted as a tumor suppressor mechanism (Table 2) [45].

Table 2. Summary of analyzed publications highlighting the tumor suppressor role of autophagy.

Publication Title	Proposed Molecular Mechanism for Sustaining the Tumor Suppressor Role of Autophagy	Reference
Autophagic cell death: the story of a misnomer	**Autophagic cell death**	[94]
Bcl-2 antiapoptotic proteins inhibit Beclin 1-dependent autophagy	**Autophagic cell death**	[95]
Anti-neoplastic activity of the cytosolic FoxO1 results from autophagic cell death	**Autophagic cell death**	[45]
Autophagy mediates the mitotic senescence transition	**Autophagic senescence**	[96]
The dynamic nature of autophagy in cancer	**Autophagic senescence**	[97]
Cancer-related inflammation	**Inflammation downregulation**	[98]
Autophagy promotes tumor cell survival and restricts necrosis, inflammation, and tumorigenesis	**Inflammation downregulation**	[99]
The double-edged sword of autophagy modulation in cancer	**Inflammation downregulation**	[74]
The Roles of Autophagy in Cancer.	**Inflammation downregulation**	[100]
Autophagy in immunity and inflammation	**Inflammation downregulation**	[101]
The Atg5–Atg12 conjugate associates with innate antiviral immune responses.	**Inflammation downregulation**	[102]
Loss of the autophagy protein Atg16L1 enhances endotoxin-induced IL-1β production	**Inflammation downregulation**	[103]

Table 2. *Cont.*

Publication Title	Proposed Molecular Mechanism for Sustaining the Tumor Suppressor Role of Autophagy	Reference
Autophagy in mammalian development and differentiation	**Inflammation downregulation**	[7]
Autophagy in health and disease: A comprehensive review.	**Inflammation downregulation**	[104]
Virus-plus-susceptibility gene interaction determines Crohn's disease gene *Atg16L1* phenotypes in intestine	**Inflammation downregulation**	[105]
Reactive species: a cell damaging rout assisting to chemical carcinogens	**Oxidative stress and genome instability**	[106]
Mitochondrial gateways to cancer	**Oxidative stress and genome instability**	[107]
Autophagy suppresses tumor progression by limiting chromosomal instability	**Oxidative stress and genome instability**	[108]
Oncosuppressive functions of autophagy	**Oxidative stress and genome instability**	[109]
PINK1/Parkin-mediated mitophagy is dependent on VDAC1 and p62/SQSTM1	**Oxidative stress and genome instability**	[110]
A noncanonical mechanism of Nrf2 activation by autophagy deficiency: direct interaction between Keap1 and p62	**Oxidative stress and genome instability**	[111]

Autophagic Senescence

A controversial strategy that may sustain the autophagy's tumor suppressor activity, is its role played in the senescence molecular cascade. Young et al. (Table 2) [96] showed that in fibroblasts autophagy is activated during senescence that has been induced by the oncogene Ras. In this context, autophagy inhibition delayed but did not block the deployment of the oncogene-mediated senescence. These data are important because senescence should be regarded as a major intrinsic barrier against cell malignant transformation, although this barrier protection may be only temporary (Table 2) [97].

Inflammation

The tumor microenvironment is characterized by complex molecular interactions between different cell types coexisting within tumor. The crosstalk between these cells control and regulates tumor progression. In this context, it is important to note that both inflammatory cells and cytokines are main actors because a proinflammatory environment can induce malignant cells survival and proliferation, stimulates angiogenesis, metastasis, and control the response to chemotherapy (Table 2) [98]. Degenhardt et al. have shown that autophagy inhibition in apoptosis-deficient tumor cells promoted local inflammatory reactions and tumor growth (Table 2) [99]. These results led to the hypothesis that autophagy may act as a tumor suppressor by reducing the intensity level of local inflammatory reactions. The anti-inflammatory effect of autophagy has been suggested to be sustained by the removal of cell and corpses (Table 2) [74,100]. Moreover, a complex connection between the immune response and the autophagy has been highlighted, outlining autophagy's role as a subtle but efficient tumor suppressor (Table 2) [101]. For instance, the LC3-conjugation system (LC3, ATG4A–D, ATG7, ATG3), important for isolation membrane elongation and/or complete closure, inhibits type I IFN production [102] and pro-inflammatory cytokine production(Table 2) [103], maintains T cells number [7,104] and is involved in the intestinal immune epithelial cell function (Table 2) [105].

Oxidative Stress and Genomic Instability

One of the most challenging and subtle strategy of autophagy as a tumor suppression is via the regulation of cellular redox homeostasis, by controlling reactive oxygen species (ROS) production. Increased ROS production can induce mutagenesis, upregulating the oncogenes activation and, consequently, initiate carcinogenesis (Table 2) [106]. Mitochondria is regarded as the main source of intracellular ROS. Mitochondrial ROS production increases as these organelles become damaged or age (Table 2) [107]. In this context, autophagy intervenes by selectively degrading the damaged mitochondria, a molecular mechanism known as mitophagy. Consequently, autophagy inhibition will trigger genotoxic effects, genomic instability and oncogenes activation of oncogenes by increased ROS production (Table 2) [108], molecular events reported in autophagy-defective cells [108]. Thus, potentially damaged mitochondria selective removal (mitophagy) reduces excessive ROS production and thereby limits tumor-promoting effects dependent on the production of such species (Table 2) [109].

Moreover, autophagy also responsible for the protein aggregates degradation. Disruptions in the autophagic pathway have been correlated with the accumulation of the autophagy substrate P62 and protein aggregates. These molecular events are considered to induce increased ROS production, ER oxidative stress, and, consequently, activation of the DNA damage response [108]. The selective autophagy substrate, P62, that accumulates when autophagy intensity is reduced, contains: an UBA domain—or binding to polyubiquitinated proteins; a PB1 domain—responsible for protein oligomerization and an LIR domain (LC3-interacting region)—for association with LC3. Due to its structural characteristics, P62 insures selective degradation of both polyubiquitinated proteins and organelles, such as mitochondria (Table 2) [109,110]. Interestingly, Lau et al. reported increased P62 levels in human tumors. Moreover, P62 accumulation induce NRF-2-dependent antioxidant defense upregulation, which, in turn, may contribute to tumor progression (Table 2) [111].

3.2.3. Autophagy Regarded as a Tumor Growth Promoter

The tumor growth promotor face of autophagy is based on the tumoral cells need to adapt to ischemia in a hypoxic and nutrient deprived environment. According to these, autophagy becomes activated in the hypoxic regions of tumors. Degenhardt et al. have reported that autophagy inhibition by monoallelic deletion of *beclin-1* (*Bcn1* $^{+/-}$) induced cell death, specifically in those regions. These findings outlined the autophagy's role as tumor cells' survival promotor, under conditions of metabolic stress (Table 3) [99]. Furthermore, the tumor cells high proliferation rates impose higher biosynthetic, and consequently, bioenergetic needs, compared to non-malignant cells. These elevated requirements can be sustained by inducing autophagy, as a mechanism that will insure both ATP and metabolic intermediates production (Table 3) [100].

Table 3. Summary of analyzed publications highlighting the tumor growth promoter role of autophagy.

Publication Title	Proposed Molecular Mechanism	Reference
Autophagy promotes tumor cell survival and restricts necrosis, inflammation, and tumorigenesis	**Beclin-1 dependent regulation**	[99]
The double-edged sword of autophagy modulation in cancer.	**Stress tolerance increase**	[74]
The Roles of Autophagy in Cancer.	**Inflammation downregulation**	[100]
Targeting GRP75 improves HSP90 inhibitor efficacy by enhancing P53-mediated apoptosis in hepatocellular carcinoma	**Ras activation**	[88]
Autophagy suppresses tumor progression by limiting chromosomal instability	**Ras activation**	[108]
Autophagy opposes P53-mediated tumor barrier to facilitate tumorigenesis in a model of PALB2-associated hereditary breast cancer	**P53 suppression**	[112]

Guo et al. highlighted that in the case of activated Ras oncogene tumor cells, survival is insured by high levels of basal autophagy. These tumor cells become vitally dependent on the autophagy pathway (Table 3) [88]. These findings led to the conclusion that autophagy is able to promote tumor cell survival by increasing the stress tolerance and providing a pathway that insures necessary nutrients in order to support the enhanced energetic requirements of these cells [100].

In the context of activated Ras- driven cancers, Mathew et al. presented autophagy as a mechanism that ensures an adequate mitochondrial metabolism by supplying mitochondrial intermediates, obtained by macromolecules degradation, in both starvation and basal conditions (Table 3) [108]. In conclusion, it can be said that particularly, activated Ras-dependent tumorigenesis seemed to be actually "addicted to autophagy" in order to obtain energetic and metabolic support for rapid tumor growth.

Huo et al. have shown that autophagy partial inhibition by monoallelic loss of Beclin-1 (*Bcn1* $^{+/-}$) also stimulated apoptosis and significantly slowed down tumor growth via P53 activation. Consequently, the authors proposed that autophagy is able to promote tumor growth by P53 suppression when DNA has been damaged (Table 3) [112]. These findings outlined the idea that autophagy can display its face as a tumor progression promoter, also, in a manner independent of activated Ras [112].

Analyzing all that has been presented above, it can be outlined that autophagy should be regarded as a double—faced molecular Janus god. On the one hand, at early stages of tumor development, autophagy is able to act as a tumor suppressor by increasing the damaged proteins and organelles (mostly mitochondria) degradation (Figure 2). In this role, autophagy acts as an efficient regulatory system that controls ROS production, insuring genomic stability. Moreover, autophagy is able to prevent necrotic cell death in apoptosis-defective cells, decreasing in this way the local inflammatory reactions' intensity and, consequently, reducing tumor development. Additionally, sometimes, autophagy may direct the cellular molecular events towards autophagic cell death. On the other hand, especially, at later stages of tumor evolution, under metabolic stress conditions, activated autophagy provides tumor cells nutrients for energy production and metabolic intermediates for biosynthetic pathways, in order to sustain tumoral cells survival and tumor growth. In this context, autophagy, also acts as a promotor of the resistance to cancer therapy (Figure 2).

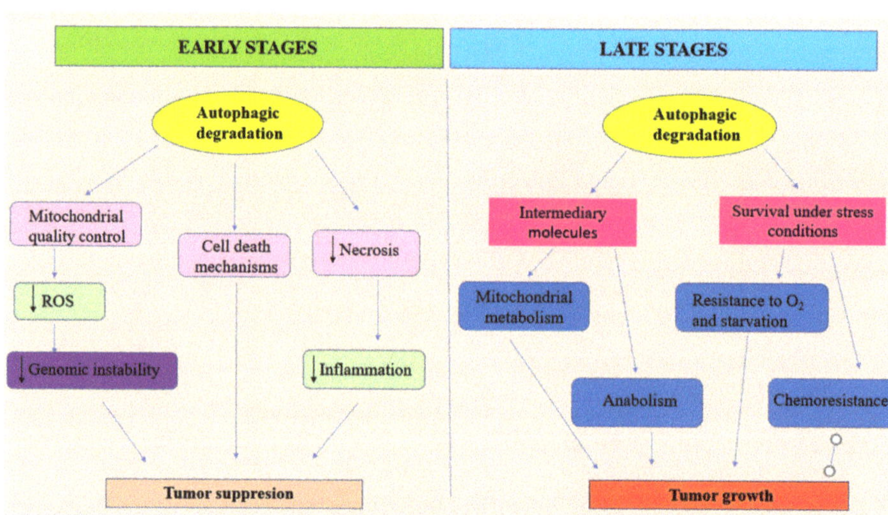

Figure 2. The dual character of autophagy in oral cancer.

- at early stages of tumor development, autophagy plays the role of a tumor suppressor by ensuring damaged proteins and organelles degradation. In this context, autophagy should be regarded as controlling system, able to decreases ROS production and, consequently, maintaining genomic stability. Autophagy also can prevent necrotic cell death in apoptosis-defective cells, ensuring in this way the decrease of local inflammation and tumor growth. In some situations autophagy can lead to apoptotic cell death.

- at later stages of tumor evolution, activated autophagy plays the role of cancer cell survival and tumor growth promoter, by suppling metabolic stressed tumor cells with nutrients, in order to sustain energy generation in mitochondria and biosynthetic pathways. Unfortunately, autophagy represents one of the main actors in developing the resistance to cancer therapy. Adapted from [100].

3.2.4. Autophagy Related Chemoresistance in Oral Cancer

Due to its dual capacity to either induce cell death or promote cell survival, important evidence pointed that autophagy has a two-faced role in response to chemotherapy in cancer. Important experimental evidence has sustained autophagy's potential as a therapeutic target for oral cancer [113–115].

On the one hand, autophagy inhibition can enhance the cisplatin sensitivity in OSCC, hypopharyngeal carcinoma and salivary adenoid cystic carcinoma [114,116–118].

On the other hand, DNA-damaging agents (cisplatin, methotrexate and 5-fluorouracil) are able to induce autophagy with a cytoprotective effect [119,120]. Beclin-1, Atg12-Atg5 and LC3-II enhanced expression together with the autophagosome formation were observed in the methotrexate-resistant SCC-9 cell line compared with the sensitive SCC-9 cell line [121]. Similar results were reported in a laryngeal cancer study, in which exposure to cisplatin induced autophagosomes aggregation in the cytoplasm and enhanced Beclin-1 and LC3II expression [122]. Consequently, the induction of autophagy has attenuated the cisplatin treatment cytotoxicity expression [122]. All these results, taken together, outline the conclusion that, at a certain time, autophagy enhancement may play a key role in the chemoresistance mechanism in head and neck cancers.

Autophagy inhibition should be regarded as a potential target in order to reverse chemoresistance in cancer treatment. However, it must not be forgotten that autophagic cell death could also be induced in oral cancer cells in order to induce tumor cell death. Therapeutic molecules like sulfasalazine thymoquinone and tetrandrine were also shown to have anticancer effects by inducing autophagic cell death. These results highlight the idea that autophagic cell death induction should also be regarded as an alternative approach to destroy tumor cells [76,93,123–127].

In conclusion, understanding how to overcome cytoprotective autophagy and how to take advantage of autophagic cell death is critical in order to enhance the cancer cells' sensitivity to particular therapeutic agents.

4. Conclusions

The important progress made in the molecular landscape of autophagy opened new insights into the pathogenesis of oral cancer.

More and more experimental data emphasizes the duality of autophagy, a tumor suppressor, especially at early stages of tumor development and a tumor promoter, at later stages of tumor evolution. However the exact reason and moment of autophagy's role change, are still unknown.

Also, further studies are imposed in order to better understand the complex molecular interactions between autophagy, immune response, immune response and apoptosis, in the oral cancer context. It can be anticipated that future, more detailed incursions into the autophagy landscape may lead to novel targets' identification, so necessary for elaborating new and efficient therapeutic strategies.

Author Contributions: Conceptualization, T.A., I.M.M. and B.M.; methodology, I.M.M., S.I.I., M.D., S.T. and B.M.; software, R.R., S.T., M.D.; T.A.M., S.I.I.; validation, G.M. and T.A.; formal analysis, S.I.I. and T.A.M.; investigation T.A., M.D., B.M., S.T.; resources R.R., M.D., T.A.M., S.I.I. and S.T.; data curation, T.A., G.M. and I.M.M.; writing—original draft preparation, T.A., D.M., S.I.I.; writing—review and editing, I.M.M., B.M., T.A.M.;

visualization, T.A.M., R.R.; supervision, M.G.; project administration, G.M., T.A., I.M.M. All authors have read and agreed to the published version of the manuscript.

Funding: This research received no external funding.

Conflicts of Interest: The authors declare no conflict of interest.

References

1. Jainkittivong, A.; Aneksuk, V.; Langlais, R.P. Oral Mucosal Conditions in Elderly Dental Patients. *Oral Dis.* **2002**, *8*, 218–223. [CrossRef] [PubMed]
2. Gendreau, L.; Loewy, Z.G. Epidemiology and etiology of den-ture stomatitis. *J. Prosthodont.* **2011**, *20*, 251–260. [CrossRef] [PubMed]
3. Bozdemir, E.; Yilmaz, H.H.; Orhan, H. Oral mucosal lesions and risk factors in elderly dental patients. *J. Dent. Res. Dent. Clin. Dent. Prospect.* **2019**, *13*, 24–30. [CrossRef] [PubMed]
4. Ercalik-Yalcinkaya, S.; Özcan, M. Association between Oral Mucosal Lesions and Hygiene Habits in a Population of Re-movable Prosthesis Wearers. *J. Prosthodont.* **2015**, *24*, 271–278. [CrossRef]
5. Dundar, N.; Ilhan Kal, B. Oral Mucosal Conditions and Risk Factors among Elderly in a Turkish School of Dentistry. *Gerontology* **2007**, *53*, 165–172. [CrossRef]
6. Del Corso, G.; Villa, A.; Tarsitano, A.; Gohel, A. Current trends in oral cancer: A review. *Cell Microenviron.* **2016**, *3*, e1332.
7. Mizushima, N.; Levine, B. Autophagy in mammalian development and differentiation. *Nat. Cell Biol.* **2010**, *12*, 823–830. [CrossRef]
8. Yang, Z.; Klionsky, D.J. Eatenalive: A history of macroautophagy. *Nat. Cell Biol.* **2010**, *12*, 814–822. [CrossRef]
9. Wang, S.; Xia, P.; Rehm, M.; Fan, Z. Autophagy and cell reprogram-ming. *Cell. Mol. Life Sci.* **2015**, *72*, 1699–1713. [CrossRef]
10. Rivera, C. Essentials of oral cancer. *Int. J. Clin. Exp. Pathol.* **2015**, *8*, 11884–11894.
11. Montero, P.H.; Patel, S.G. Cancer of the oral cavity. *Surg. Oncol. Clin. N. Am.* **2015**, *24*, 491–508. [CrossRef] [PubMed]
12. Bray, F.; Ferlay, J.; Soerjomataram, I.; Siegel, R.L.; Torre, L.A.; Jemal, A. Global cancer statistics 2018: Globocan estimates of incidence and mortality worldwide for 36 cancers in 185 countries. *CA Cancer J. Clin.* **2018**, *68*, 394–424. [CrossRef] [PubMed]
13. Kumar, M.; Nanavati, R.; Modi, T.G.; Dobariya, C. Oral cancer: Etiology and risk factors: A review. *J. Cancer Res.* **2016**, *12*, 458–463. [CrossRef] [PubMed]
14. Ong, T.K.; Murphy, C.; Smith, A.B.; Kanatas, A.N.; Mitchell, D.A. Survival after surgery for oral cancer: A 30-year experience. *Br. J. Oral Maxillofac. Surg.* **2017**, *55*, 911–916. [CrossRef] [PubMed]
15. Bais, M.V. Impact of Epigenetic Regulation on Head and Neck Squamous Cell Carcinoma. *J. Dent. Res.* **2019**, *98*, 268–276. [CrossRef] [PubMed]
16. Kaur, J.; Jacobs, R.; Huang, Y.; Salvo, N.; Politis, C. Salivary biomarkers for oral cancer and pre-cancer screening: A review. *Clin. Oral Investig.* **2018**, *22*, 633–640. [CrossRef] [PubMed]
17. Valdez, J.A.; MAS, M.T.B. Impact of oral cancer on quality of life. *Oral Cancer Issue Dent. Clin. N. Am.* **2017**, *62*, 143–154. [CrossRef]
18. D'souza, S.; Addepalli, V. Preventive measures in oral cancer: An overview. *Biomed. Pharm.* **2018**, *107*, 72–80. [CrossRef]
19. Neville, B.W. Oral cancer and precancerous lesions. *Fogorv. Szle.* **2010**, *52*, 195–215. [CrossRef]
20. Lee, Y.C.; Hashibe, M. Tobacco, alcohol, and cancer in low and high income countries. *Ann. Glob. Health* **2014**, *80*, 378–383.
21. Jethwa, A.R.; Khariwala, S.S. Tobacco-related carcinogenesis in head and neck cancer. *Cancer Metastasis Rev.* **2017**, *36*, 411–423. [CrossRef]
22. Gandini, S.E.; Botteri, S.; Iodice, M.; Boniol, A.B.; Lowenfels, P.; Maisonneuve, P.; Boyle, P. Tobacco smoking and cancer: A meta-analysis. *Int. J. Cancer* **2008**, *122*, 155–164. [CrossRef] [PubMed]
23. Choudhari, S.K.; Chaudhary, M.; Gadbail, A.R.; Sharma, A.; Tekade, S. Oxidative and antioxidative mechanisms in oral cancer and precancer: A review. *Oral Oncol.* **2014**, *50*, 10–18. [CrossRef] [PubMed]
24. Boffetta, P.; Hashibe, M. Alcohol and cancer. *Lancet Oncol.* **2006**, *7*, 149–156. [CrossRef]

25. Gaonkar, P.P.; Patankar, S.R.; Tripathi, N.; Sridharan, G. Oral bacterial flora and oral cancer: The possible link? *J. Oral Maxillofac. Pathol.* **2018**, *22*, 234–238. [CrossRef] [PubMed]
26. Candotto, V.D.; Lauritano, M.; Nardone, L.; Baggi, C.; Arcuri, R.; Gatto, R.M.; Gaudio, F.; Spadari, F.; Carinci, F. HPV infection in the oral cavity: Epidemiology, clinical manifestations and relationship with oral cancer. *Oral Implant.* **2017**, *10*, 209–220. [CrossRef] [PubMed]
27. Sritippho, T.P.; Chotjumlong, A.; Iamaroon, A. Roles of Human Papillomaviruses and p16 in Oral Cancer. *Asian Pac. J. Cancer Prev.* **2015**, *16*, 6193–6200. [CrossRef]
28. Lazos, J.P.; Piemonte, E.D.; Lanfranchi, H.E.; Brunotto, M.N. Characterization of Chronic Mechanical Irritation in Oral Cancer. *Int. J. Dent.* **2017**, *2017*, 6784526. [CrossRef]
29. Piemonte, E.J.; Lazos, P.; Belardinelli, D.; Secchi, M.; Brunotto, H.; Lanfranchi-Tizeira, H. Oral cancer associated with chronic mechanical irritation of the oral mucosa. *Med. Oral Patol. Oral Cir. Bucal.* **2018**, *23*, e151–e160. [CrossRef]
30. Yardimci, G.Z.; Kutlubay, B.; Engin, Y.; Tuzun, Y. Precancerous lesions of oral mucosa. *World J. Clin. Cases* **2014**, *2*, 866–872. [CrossRef]
31. Al-Jaber, A.; Al-Nasser, L.; El-Metwally, A. Epidemiology of oral cancer in Arab countries. *Saudi Med. J.* **2016**, *37*, 249–255. [CrossRef] [PubMed]
32. Mohan, M.; Jagannathan, N. Oral field cancerization: An update on current concepts. *Oncol. Rev.* **2014**, *8*, 244. [CrossRef] [PubMed]
33. Paderno, A.R.; Morello, C.; Piazza, C. Tongue carcinoma in young adults: A review of the literature. *Acta Otorhinolaryngol. Ital.* **2018**, *38*, 175–180. [PubMed]
34. Kerawala, C.; Roques, T.; Jeannon, J.P.; Bisase, B. Oral cavity and lip cancer: United Kingdom National Multidisciplinary Guidelines. *J. Laryngol. Otol.* **2016**, *130*, S83–S89. [CrossRef]
35. Tan, Y.Q.; Zhang, J.; Zhou, G. Autophagy and its implication in human oral diseases. *Autophagy* **2017**, *13*, 225–236. [CrossRef]
36. Mauthe, M.; Reggiori, F. ATG proteins: Are we always looking at autophagy? *Autophagy* **2016**, *12*, 2502–2503. [CrossRef]
37. Staal, S.P. Molecular cloning of the akt oncogene and its human homologues AKT1 and AKT2: Amplification of AKT1 in a primary human gastric adenocarcinoma. *Proc. Natl. Acad. Sci. USA* **1987**, *84*, 5034–5037. [CrossRef]
38. Kohn, A.D.; Kovacina, K.S.; Roth, R.A. Insulin stimulates the kinase activity of RAC-PK, a pleckstrin homology domain containing ser/thr kinase. *Embo J.* **1995**, *14*, 4288–4295. [CrossRef]
39. Yang, Z.Z.; Tschopp, O.; Hemmings-Mieszczak, M.; Feng, J.; Brodbeck, D.; Perentes, E.; Hemmings, B.A. Protein kinase B alpha/Akt1 regulates placental development and fetal growth. *J. Biol. Chem.* **2003**, *278*, 32124–321231. [CrossRef]
40. Qin, G.; Li, P.; Xue, Z. Triptolide induces protective autophagy and apoptosis in human cervical cancer cells by downregulating Akt/mTOR activation. *Oncol. Lett.* **2018**, *16*, 3929–3934. [CrossRef]
41. Hers, I.; Vicent, E.E.; Tavare, J.M. AKT signaling in health and disease. *Cell. Signal.* **2011**, *23*, 1515–1527. [CrossRef] [PubMed]
42. Manning, B.D.; Toker, A. AKT/PKB Signaling: Navigating the Network. *Cell* **2017**, *169*, 381–405. [CrossRef] [PubMed]
43. Totan, A.; Miricescu, D.; Stanescu, I.I.; Didilescu, A.; Melescanu-Imre, M.; Tancu, A.M.C.; Totan, C.; Spinu, T.C.; Greabu, M. (PI3K)/AKT signalling pathway–a Pandora's box in oral squamous cell carcinoma. *Rom. J. Med. Pract.* **2019**, *14*, 389–392. [CrossRef]
44. Khalid, A.; Hussain, T.; Manzoor, S.; Saalim, M.; Khaliq, S. PTEN: A potential prognostic marker in virus-induced hepatocellular carcinoma. *Tumour Biol.* **2017**, *39*, 1010428317705754. [CrossRef]
45. Zhao, J.; Chi, J.; Gao, M.; Zhi, J.; Li, Y.; Zheng, X. Loss of PTEN Expression is Associated with High MicroRNA 24 Level and Poor Prognosis in Patients with Tongue Squamous Cell Carcinoma. *J. Oral Maxillofac. Surg.* **2017**, *75*, 1449.e1–1449.e8. [CrossRef]
46. Kurasawa, Y.; Shiiba, M.; Nakamura, M.; Fushimi, K.; Ishigami, T.; Bukawa, H.; Yokoe, H.; Uzawa, K.; Tanzawa, H. PTEN expression and methylation status in oral squamous cell carcinoma. *Oncol. Rep.* **2008**, *19*, 1429–1434.

47. Roy, N.K.; Monisha, J.; Padmavathi, G.; Lalhruaitluanga, H.; Kumar, N.S.; Kumar, A.; Singh, S.D.; Baruah, M.N.; Ahmed, G.N.; Longkumar, I.; et al. Isoform-Specific Role of Akt in Oral Squamous Cell Carcinoma. *Biomolecules* **2019**, *9*, 253. [CrossRef]
48. Xu, D.W.; Zhang, G.Q.; Wang, Z.W.; Xu, X.Y.; Liu, T.X. Autophagy in Tumorigenesis and Cancer Treatment. *Asian Pac. J. Cancer Prev.* **2015**, *16*, 2165–2175. [CrossRef]
49. Wani, W.Y.; Boyer-Guittaut, M.; Dodson, M.; Chatham, J.; Darley-Usmar, V.; Zhang, J. Regulation of autophagy by protein post-translational modification. *Lab. Investig.* **2015**, *95*, 14–25. [CrossRef]
50. Wang, Y.; Wang, C.; Tang, H.; Wang, M.; Weng, J.; Liu, X.; Zhang, R.; Huang, H.; Hou, J. Decrease of autophagy activity promotes malig-nant progression of tongue squamous cell carcinoma. *J. Oral Pathol. Med.* **2013**, *42*, 557–564. [CrossRef]
51. Tang, J.Y.; Hsi, E.; Huang, Y.C.; Hsu, N.C.; Chu, P.Y.; Chai, C.Y. High LC3 expression correlates with poor survival in patients with oral squamous cell carcinoma. *Hum. Pathol.* **2013**, *44*, 2558–2562. [CrossRef] [PubMed]
52. Yu, T.; Zuber, J.; Li, J. Targeting autophagy in skin diseases. *J. Mol. Med.* **2015**, *93*, 31–38. [CrossRef] [PubMed]
53. Sakakura, K.; Takahashi, H.; Kaira, K.; Toyoda, M.; Oyama, T.; Chikamatsu, K. Immunological significance of the accumulation of autophagy components in oral squamous cell carcinoma. *Cancer Sci.* **2015**, *106*, 1–8. [CrossRef] [PubMed]
54. Tang, J.Y.; Fang, Y.Y.; Hsi, E.; Huang, Y.C.; Hsu, N.C.; Yang, W.C.; Chang, H.W.; Chai, C.Y.; Chu, P.Y. Immunopositivity of Beclin-1 and ATG5 as indicators of survival and disease recurrence in oral squamous cell carcinoma. *Anticancer Res.* **2013**, *33*, 5611–5616. [PubMed]
55. Tang, J.Y.; Hsi, E.; Huang, Y.C.; Hsu, N.C.; Yang, W.C.; Chang, H.W.; Chai, C.Y.; Chu, P.Y. Overexpression of autophagy-related 16-like 1 in patients with oral squamous cell carcinoma. *Pathol. Oncol. Res.* **2015**, *21*, 301–305. [CrossRef]
56. Kapoor, V.; Paliwal, D.; Baskar Singh, S.; Mohanti, B.K.; Das, S.N. Deregulation of Beclin 1 in patients with tobacco-related oral squamous cellcarcinoma. *Biochem. Biophys. Res. Commun.* **2012**, *422*, 764–769. [CrossRef]
57. Adhauliya, N.; Kalappanavar, A.N.; Ali, I.M.; Annigeri, R.G. Autophagy: A boon or bane in oral cancer. *Oral. Oncol.* **2016**, *61*, 120–126. [CrossRef]
58. Tang, J.Y.; Hsi, E.; Huang, Y.C.; Hsu, N.C.; Chen, Y.K.; Chu, P.Y.; Chai, C.Y. ATG9A overexpression is associated with disease recurrence and poor survival in patients with oral squamous cell carcinoma. *Virchows Arch.* **2013**, *463*, 737–742. [CrossRef]
59. Liu, J.L.; Chen, F.F.; Lung, J.; Lo, C.H.; Lee, F.H.; Lu, Y.C.; Hung, C.H. Prog- nostic significance of p62/SQSTM1 subcellular localization and LC3Binoralsquamouscellcarcinoma. *Br. J. Cancer* **2014**, *111*, 944–954. [CrossRef]
60. Nomura, H.; Uzawa, K.; Yamano, Y.; Fushimi, K.; Ishigami, T.; Kouzu, Y.; Koike, H.; Siiba, M.; Bukawa, H.; Yokoe, H.; et al. Overexpression and altered subcellular localization of autophagy-related 16-like 1 in human oral squamous-cell carcinoma: Correlation with lymphovas-cular invasion and lymph-node metastasis. *Hum. Pathol.* **2009**, *40*, 83–91. [CrossRef]
61. Ha, J.; Kim, J. Novel pharmacological modulators of autophagy: An updated patent review (2012–2015). *Expert Opin. Pat.* **2016**, *26*, 1273–1289. [CrossRef] [PubMed]
62. Mizushima, N.; Yoshimori, T.; Ohsumi, Y. The role of Atg proteinsin autophagosome formation. *Annu. Rev. Cell Dev. Biol.* **2011**, *27*, 107–132. [CrossRef] [PubMed]
63. Wilson, M.; Dooley, H.; Tooze, S. Wipi2b and atg16l1: Setting the stage for autophagosome formation. *Biochem. Soc. Trans.* **2014**, *42*, 1327–1334. [CrossRef] [PubMed]
64. Ishimura, R.; Tanaka, K.; Komatsu, M. Dissection of the role of p62/ Sqstm1 in activation of Nrf2 during xenophagy. *Febs Lett.* **2014**, *588*, 822–828. [CrossRef]
65. Lamark, T.; Svenning, S.; Johansen, T. Regulation of selective autophagy: The p62/SQSTM1 paradigm. *Essays Biochem.* **2017**, *61*, 609–624. [CrossRef]
66. Inui, T.C.T.; Takikita-Suzuki, M.; Nishikawa, M.; Yamamoto, G.; Okabe, H. Association of p62/SQSTM1 excess and oral carcinogenesis. *PLoS ONE* **2013**, *8*, e74398. [CrossRef]
67. Wei, Y.; Zou, Z.; Becker, N.; Anderson, M.; Sumpter, R.; Xiao, G.; Kinch, L.; Koduru, P.; Christudass, C.S.; Veltri, R.W.; et al. EGFR-mediatedBeclin 1 phosphorylation in autophagy suppression, tumor progression, and tumor chemoresistance. *Cell* **2013**, *154*, 1269–1284. [CrossRef]
68. Lozy, F.; Karantza, V. Autophagy and cancer cell metabolism. *Semin. Cell Dev. Biol.* **2012**, *23*, 395–401. [CrossRef]

69. Levy, J.M.M.; Towers, C.G.; Thorburn, A. Targeting autophagy in cancer. *Nat. Rev. Cancer* **2017**, *17*, 528–542. [CrossRef]
70. Deng, S.; Shanmugam, M.K.; Kumar, A.P.; Yap, C.T.; Sethi, G.; Bishayee, A. Targeting autophagy using natural compounds for cancer prevention and therapy. *Cancer* **2019**, *125*, 1228–1246. [CrossRef]
71. Weng, J.; Wang, C.; Wang, Y.; Tang, H.; Liang, J.; Liu, X.; Huang, H.; Hou, J. Beclin1 inhibits proliferation, migration and invasion in tongue squamous cell carcinoma cell lines. *Oral Oncol.* **2014**, *50*, 983–990. [CrossRef] [PubMed]
72. Jiang, L.; Huang, S.; Li, W.; Zhang, D.; Zhang, S.; Zhang, W.; Zheng, P.; Chen, Z. Expression of autophagy and ER stress-related proteins in primary salivary adenoid cystic carcinoma. *Pathol. Res. Pract.* **2012**, *208*, 635–641. [CrossRef] [PubMed]
73. Liang, L.Z.; Ma, B.; Liang, Y.J.; Liu, H.C.; Zheng, G.S.; Zhang, T.H.; Chu, M.; Xu, P.P.; Su, Y.X.; Liao, G.Q. High expression of the autophagy gene Beclin-1 is associated with favorable prognosis for salivary gland adenoid cystic carcinoma. *J. Oral Pathol. Med.* **2012**, *41*, 621–629. [CrossRef] [PubMed]
74. Avalos, Y.; Canales, J.; Bravo-Sagua, R.; Criollo, A.; Lavandero, S.; Quest, A.F. Tumor suppression and promotion by autophagy. *Biomed. Res. Int.* **2014**, *2014*, 603980. [CrossRef]
75. Martins, F.; De Sousa, S.C.; Dos Santos, E.; Woo, S.B.; Gallottini, M. PI3K-AKT-mTOR pathway proteins are differently expressed in oral carcinogenesis. *J. Oral Pathol. Med.* **2016**, *45*, 746–752. [CrossRef]
76. Choudhary, H.; Banik, K.; Ang, H.L.; Girisa, S.; Vikkurthi, R.; Parama, D.; Rana, V.; Shabnam, B.; Khatoon, E.; Kumar, A.P.; et al. Targeting AKT/mTOR in Oral Cancer: Mechanisms and Advances in Clinical Trials. *Int. J. Mol. Sci.* **2020**, *21*, 3285. [CrossRef]
77. Matsuo, F.S.; Andrade, M.F.; Loyola, A.M.; Da Silva, S.J.; Silva, M.J.B.; Cardoso, S.V.; De Faria, P.R. Pathologic significance of AKT, mTOR, and GSK3beta proteins in oral squamous cell carcinoma-a_ected patients. *Virchows Arch. Int. J. Pathol.* **2018**, *472*, 983–997. [CrossRef]
78. Ferreira, D.M.; Neves, T.J.; Lima, L.G.C.A.; Alves, F.A.; Begnami, M.D. Prognostic implications of the phosphatidylinositol 3-kinase/Akt signaling pathway in oral squamous cell carcinoma: Overexpression of p-mTOR indicates an adverse prognosis. *Appl. Cancer Res.* **2017**, *37*, 41. [CrossRef]
79. Lakshminarayana, S.; Augustine, D.; Rao, R.S.; Patil, S.; Awan, K.H.; Venkatesiah, S.S.; Haragannavar, V.C.; Nambiar, S.; Prasad, K. Molecular pathways of oral cancer that predict prognosis and survival: A systematic review. *J. Carcinog.* **2018**, *17*, 7. [CrossRef]
80. Monisha, J.; Roy, N.K.; Padmavathi, G.; Banik, K.; Bordoloi, D.; Khwairakpam, A.D.; Arfuso, F.; Chinnathambi, A.; Alahmadi, T.A.; Alharbi, S.A.; et al. NGAL is Downregulated in Oral Squamous Cell Carcinoma and Leads to Increased Survival, Proliferation, Migration and Chemoresistance. *Cancers* **2018**, *10*, 228. [CrossRef]
81. De Amicis, A.; Sanctis, S.D.; Cristofaro, S.D.; Franchini, V.; Lista, F.; Regalbuto, E.; Giovenale, E.; Gallerano, G.P.; Nenzi, P.; Bei, R.; et al. Biological effects of in vitro THz radiation exposure in human foetal fibroblasts. *Mutat. Res. Genet. Toxicol. Environ. Mutagen.* **2015**, *793*, 150–160. [CrossRef] [PubMed]
82. Huang, Y.; Chen, L.; Guo, L.; Hupp, T.R.; Lin, Y. Evaluating DAPK as a therapeutic target. *Apoptosis* **2014**, *19*, 371–386. [CrossRef] [PubMed]
83. He, M.X.; He, Y.W. A role for c-FLIP(L) in the regulation of apoptosis, autophagy, and necroptosis in T lymphocytes. *Cell Death Differ.* **2013**, *20*, 188–197. [CrossRef] [PubMed]
84. Lee, J.S.; Li, Q.; Lee, J.Y.; Lee, S.H.; Jeong, J.H.; Lee, H.R.; Chang, H.; Zhou, F.C.; Gao, S.J.; Liang, C.; et al. FLIP-mediated autophagy regulation in cell death control. *Nat. Cell Biol.* **2009**, *11*, 1355–1362. [CrossRef] [PubMed]
85. Sasahira, T.; Kirita, T. Hallmarks of Cancer-Related Newly Prognostic Factors of Oral Squamous Cell Carcinoma. *Int. J. Mol. Sci.* **2018**, *19*, 2413. [CrossRef]
86. Oikawa, Y.; Morita, K.I.; Kayamori, K.; Tanimoto, K.; Sakamoto, K.; Katoh, H.; Ishikawa, S.; Inazawa, J.; Harada, H. Receptor tyros-ine kinase amplification is predictive of distant metas-tasis in patients with oral squamous cell carcinoma. *Cancer Sci.* **2017**, *108*, 256–266. [CrossRef]
87. Mrakovcic, M.; Fröhlich, L.F. P53-Mediated Molecular Control of Autophagy in Tumor Cells. *Biomolecules* **2018**, *8*, 14. [CrossRef]
88. Guo, W.; Yan, L.; Yang, L.; Liu, X.; Qiukai, E.; Gao, P.; Ye, X.; Liu, W.; Zuo, J. Targeting GRP75 improves HSP90 inhibitor efficacy by enhancing P53-mediated apoptosis in hepatocellular carcinoma. *PLoS ONE* **2014**, *9*, e85766. [CrossRef]

89. Liang, X.H.; Jackson, S.; Seaman, M.; Brown, K.; Kempkes, B.; Hibshoosh, H.; Levine, B. Induction of autophagy and inhibition of tumorigenesis by beclin 1. *Nature* **1999**, *402*, 672–676. [CrossRef]
90. Menon, M.B.; Dhamija, S. Beclin 1 phosphorylation–at the center of autophagy regulation. *Front. Cell Dev. Biol.* **2018**, *6*, 137. [CrossRef]
91. Zhang, L.; Zhang, W.; Wang, Y.F.; Liu, B.; Zhang, W.F.; Zhao, Y.F.; Kulkarni, A.B.; Sun, Z.J. Dual induction of apoptotic and autophagic cell death by targeting survivin in head neck squamous cell carcinoma. *Cell Death Dis.* **2015**, *6*, e1771. [CrossRef] [PubMed]
92. Hsieh, M.T.; Chen, H.P.; Lu, C.C.; Chiang, J.H.; Wu, T.S.; Kuo, D.H.; Huang, L.J.; Kuo, S.C.; Yang, J.S. The novel pterostilbene derivative ANK-199 induces autophagic cell death through regulating PI3 kinase class III/beclin 1/Atgrelated proteins in cisplatinresistant CAR human oral cancer cells. *Int. J. Oncol.* **2014**, *45*, 782–794. [CrossRef] [PubMed]
93. Han, H.Y.; Kim, H.; Jeong, S.H.; Lim, D.S.; Ryu, M.H. Sulfasalazine induces autophagic cell death in oral cancer cells via Akt and ERK pathways. *Asian Pac. J. Cancer Prev.* **2014**, *15*, 6939–6944. [CrossRef] [PubMed]
94. Kroemer, G.; Levine, B. Autophagic cell death: The story of a misnomer. *Nat. Rev. Mol. Cell Biol.* **2008**, *9*, 1004–1010. [CrossRef]
95. Pattingre, S.; Tassa, A.; Qu, X.; Garuti, R.; Liang, X.H.; Mizushima, N.; Packer, M.; Schneider, M.D.; Levine, B. Bcl-2 antiapoptotic proteins inhibit Beclin 1-dependent autophagy. *Cell* **2005**, *122*, 927–939. [CrossRef]
96. Young, A.R.; Narita, M.; Ferreira, M.; Kirschner, K.; Sadaie, M.; Darot, J.F.; Tavaré, S.; Arakawa, S.; Shimizu, S.; Watt, F.M.; et al. Autophagy mediates the mitotic senescence transition. *Genes Dev.* **2009**, *23*, 798–803. [CrossRef]
97. Kimmelman, A.C. The dynamic nature of autophagy in cancer. *Genes Dev.* **2011**, *25*, 1999–2010. [CrossRef]
98. Candido, J.; Hagemann, T. Cancer-related inflammation. *J. Clin. Immunol.* **2013**, *33* (Suppl. S1), S79–S84. [CrossRef]
99. Degenhardt, K.; Mathew, R.; Beaudoin, B.; Bray, K.; Anderson, D.; Chen, G.; Mukherjee, C.; Shi, Y.; Gélinas, C.; Fan, Y.; et al. Autophagy promotes tumor cell survival and restricts necrosis, inflammation, and tumorigenesis. *Cancer Cell* **2006**, *10*, 51–64. [CrossRef]
100. Yun, C.; Lee, S. The roles of autophagy in cancer. *Int. J. Mol. Sci.* **2018**, *19*, 3466. [CrossRef]
101. Levine, B.; Mizushima, N.; Virgin, H.W. Autophagy in immunity and inflammation. *Nature* **2011**, *469*, 323–335. [CrossRef] [PubMed]
102. Jounai, N.; Takeshita, F.; Kobiyama, K.; Sawano, A.; Miyawaki, A.; Xin, K.; Ishii, K.J.; Kawai, T.; Akira, S.; Suzuki, K.; et al. The Atg5–Atg12 conjugate associates with innate antiviral immune responses. *Proc. Natl. Acad. Sci. USA* **2007**, *104*, 14050–14055. [CrossRef] [PubMed]
103. Saitoh, T.; Fujita, N.; Jang, M.H.; Uematsu, S.; Yang, B.; Satoh, T.; Omori, H.; Noda, T.; Yamamoto, N.; Komatsu, M.; et al. Loss of the autophagy protein Atg16L1 enhances endotoxin-induced IL-1β production. *Nature* **2008**, *456*, 264–268. [CrossRef] [PubMed]
104. Saha, S.; Panigrahi, D.P.; Patil, S.; Bhutia, S.K. Autophagy in health and disease: A comprehensive review. *Biomed. Pharm.* **2018**, *104*, 485–495. [CrossRef]
105. Cadwell, K.; Patel, K.K.; Maloney, N.S.; Liu, T.; Ng, A.C.Y.; Storer, C.E.; Head, R.D.; Xavier, R.; Stappenbeck, T.S.; Virgin, H.W. Virus-plus-susceptibility gene interaction determines Crohn's disease gene *Atg16L1* phenotypes in intestine. *Cell* **2010**, *141*, 1135–1145. [CrossRef]
106. Goetz, M.E.; Luch, A. Reactive species: A cell damaging rout assisting to chemical carcinogens. *Cancer Lett.* **2008**, *266*, 73–83. [CrossRef]
107. Galluzzi, L.; Morselli, E.; Kepp, O.; Vitale, I.; Rigoni, A.; Vacchelli, E.; Michaud, M.; Zischka, H.; Castedo, M.; Kroemer, G. Mitochondrial gateways to cancer. *Mol. Asp. Med.* **2010**, *31*, 1–20. [CrossRef]
108. Mathew, R.; Kongara, S.; Beaudoin, B.; Karp, C.M.; Bray, K.; Degenhardt, K.; Chen, G.; Jin, S.; White, E. Autophagy suppresses tumor progression by limiting chromosomal instability. *Genes Dev.* **2007**, *21*, 1367–1381. [CrossRef]
109. Morselli, E.; Galluzzi, L.; Kepp, O.; Mariño, G.; Michaud, M.; Vitale, I.; Maiuri, M.C.; Kroemer, G. Oncosuppressive functions of autophagy. *Antioxid. Redox Signal.* **2011**, *14*, 2251–2269. [CrossRef]
110. Geisler, S.; Holmström, K.M.; Skujat, D.; Fiesel, F.C.; Rothfuss, O.C.; Kahle, P.J.; Springer, W. PINK1/Parkin-mediated mitophagy is dependent on VDAC1 and p62/SQSTM1. *Nat. Cell Biol.* **2010**, *12*, 119–131. [CrossRef]

111. Lau, A.; Wang, X.J.; Zhao, F.; Villeneuve, N.F.; Wu, T.; Jiang, T.; Sun, Z.; White, E.; Zhang, D.D. A noncanonical mechanism of Nrf2 activation by autophagy deficiency: Direct interaction between Keap1 and p62. *Mol. Cell. Biol.* **2010**, *30*, 3275–3285. [CrossRef] [PubMed]
112. Huo, Y.; Cai, H.; Teplova, I.; Bowman-Colin, C.; Chen, G.; Price, S.; Barnard, N.; Ganesan, S.; Karantza, V.; White, E.; et al. Autophagy opposes p53-mediated tumor barrier to facilitate tumorigenesis in a model of PALB2-associated hereditary breast cancer. *Cancer Discov.* **2013**, *3*, 894–907. [CrossRef] [PubMed]
113. Nam, H.Y.; Han, M.W.; Chang, H.W.; Lee, Y.S.; Lee, M.; Lee, H.J.; Lee, B.W.; Lee, K.E.; Jung, M.K.; Jeon, H.; et al. Radioresistant cancer cells can beconditioned to enter senescence by mTOR inhibition. *Cancer Res.* **2013**, *73*, 4267–4277. [CrossRef] [PubMed]
114. Jiang, L.; Huang, S.; Zhang, D.; Zhang, B.; Li, K.; Li, W.; Zhang, S.; Zhang, W.; Zheng, P. Inhibition of autophagy augments chemotherapy in human salivary adenoid cystic carcinoma. *J. Oral Pathol. Med.* **2014**, *43*, 265–272. [CrossRef] [PubMed]
115. Yen, C.Y.; Chiang, W.F.; Liu, S.Y.; Cheng, P.C.; Lee, S.Y.; Hong, W.Z.; Lin, P.Y.; Lin, M.H.; Liu, Y.C. Long-term stimulation of areca nut components results in increased chemoresistance through elevated autophagic activity. *J. Oral Pathol. Med.* **2014**, *43*, 91–96. [CrossRef]
116. Shi, T.T.; Yu, X.X.; Yan, L.J.; Xiao, H.T. Research progress of hydroxychloroquine and autophagy inhibitors on cancer. *Cancer Chemother. Pharm.* **2017**, *79*, 287–294. [CrossRef] [PubMed]
117. Quan, H.Y.; Quan, H.Y.; Zhou, L.J.; Li, A.D.; Zhang, Z.B. Mechanism of chloroquine in promoting sensitivity of chemotherapeutics inoral squamous cell carcinoma CAL-27 cell line to cisplatin. *Shanghai Kou Qiang Yi Xue* **2015**, *24*, 30–36.
118. Zhao, X.G.; Sun, R.J.; Yang, X.Y.; Liu, D.; Lei, D.; Jin, T.; Pan, X. Chloroquine-enhanced efficacy of cisplatin in the treatment of hypopharyngeal carcinoma in xenograft mice. *PLoS ONE* **2015**, *10*, e0126147. [CrossRef]
119. Carew, J.S.; Kelly, K.R.; Nawrocki, S.T. Autophagy as a target for cancer therapy: New developments. *Cancer Manag. Res.* **2012**, *4*, 357–365.
120. Notte, A.; Leclere, L.; Michiels, C. Autophagy as a mediator of chemotherapy-induced cell death in cancer. *Biochem. Pharm.* **2011**, *82*, 427–434. [CrossRef]
121. Tsai, C.W.; Lai, F.J.; Sheu, H.M.; Lin, Y.S.; Chang, T.H.; Jan, M.S.; Chen, S.M.; Hsu, P.C.; Huang, T.T.; Huang, T.C.; et al. WWOX suppresses autophagy for inducing apoptosis in methotrexate-treated human squamous cell carcinoma. *Cell Death Dis.* **2013**, *4*, e792. [CrossRef] [PubMed]
122. Kang, R.; Wang, Z.H.; Wang, B.Q.; Zhang, C.; Gao, W.; Feng, Y.; Bai, T.; Zhang, H.; Huang-Pu, H.; Wen, S. Inhibition of autophagy-potentiated chemosensitivity to cisplatin in laryngeal cancer Hep-2cells. *Am. J. Otolaryngol.* **2012**, *33*, 678–684. [CrossRef]
123. Han, H.Y.; Park, B.S.; Lee, G.S.; Jeong, S.H.; Kim, H.; Ryu, M.H. Autophagic cell death by Poncirus trifoliata Rafin., a traditional oriental medicine, in human oral cancer HSC-4 cells. *Evid. Based Complement. Altern. Med.* **2015**, 394263. [CrossRef]
124. Chu, S.C.; Hsieh, Y.S.; Yu, C.C.; Lai, Y.Y.; Chen, P.N. Thymoquinone induces cell death in human squamous carcinoma cells via caspase activation-dependent apoptosis and LC3-II activation-dependent autophagy. *PLoS ONE* **2014**, *9*, e101579. [CrossRef] [PubMed]
125. Huang, A.C.; Lien, J.C.; Lin, M.W.; Yang, J.; Wu, P.; Chang, S.; Lai, T. Tetrandrine induces cell death in SAS human oral cancer cells through caspase activation- dependent apoptosis and LC3-I and LC3-II activation-dependent autophagy. *Int. J. Oncol.* **2013**, *43*, 485–494. [CrossRef] [PubMed]
126. Vander Broek, R.; Mohan, S.; Eytan, D.F.; Chen, Z.; Van Waes, C. The PI 3 K/A kt/m TOR axis in head and neck cancer: Functions, aberrations, cross-talk, and therapies. *Oral Dis.* **2015**, *21*, 815–825. [CrossRef] [PubMed]
127. Khan, T.; Relitti, N.; Brindisi, M.; Magnano, S.; Zisterer, D.; Gemma, S.; Butini, S.; Campiani, G. Autophagy modulators for the treatment of oral and esophageal squamous cell carcinomas. *Med. Res. Rev.* **2020**, *40*, 1002–1060. [CrossRef]

Publisher's Note: MDPI stays neutral with regard to jurisdictional claims in published maps and institutional affiliations.

MDPI
St. Alban-Anlage 66
4052 Basel
Switzerland
Tel. +41 61 683 77 34
Fax +41 61 302 89 18
www.mdpi.com

International Journal of Molecular Sciences Editorial Office
E-mail: ijms@mdpi.com
www.mdpi.com/journal/ijms

MDPI

www.ingramcontent.com/pod-product-compliance
Lightning Source LLC
Chambersburg PA
CBHW061512210926
51850CB00008B/1891
9783036573373